T0332517

Sturm–Liouville and Dirac Operators

Mathematics and Its Applications (*Soviet Series*)

Volume 59

Sturm–Liouville and Dirac Operators

by

B.M. Levitan
Department of Function and Functional Analysis,
Faculty of Mathematics and Mechanics,
Moscow University, U.S.S.R.

and

I. S. Sargsjan
Department of Higher Mathematics,
Polygraphy Institute,
Moscow, U.S.S.R.

KLUWER ACADEMIC PUBLISHERS
DORDRECHT / BOSTON / LONDON

Library of Congress Cataloging-in-Publication Data

```
Levitan, Boris Moiseevich.
    [Operatory Shturma-Liuvillîa i Diraka. English]
    Sturm-Liouville and Dirac operators / by B.M. Levitan, I.S.
Sargsjan ; translated from the Russian by Oleg Efimov.
      p.   cm. -- (Mathematics and its applications (Soviet series) ;
  v. 59)
    Translation of Operatory Shturma-Liuvillîa i Diraka.
    Includes bibliographical references and index.
    ISBN 0-7923-0992-8
    1. Sturm-Liouville equation. 2. Dirac equation.   I. Sargsîan,
  I. S. (Ishkhan Saribekovich) II. Title. III. Series: Mathematics
  and its applications (Kluwer Academic Publishers). Soviet series ;
  59.
  QA379.L483  1990
  515'.35--dc20                                        90-48841
```

ISBN 0-7923-0992-8

Published by Kluwer Academic Publishers,
P.O. Box 17, 3300 AA Dordrecht, The Netherlands.

Kluwer Academic Publishers incorporates
the publishing programmes of
D. Reidel, Martinus Nijhoff, Dr W. Junk and MTP Press.

Sold and distributed in the U.S.A. and Canada
by Kluwer Academic Publishers,
101 Philip Drive, Norwell, MA 02061, U.S.A.

In all other countries, sold and distributed
by Kluwer Academic Publishers Group,
P.O. Box 322, 3300 AH Dordrecht, The Netherlands.

Printed on acid-free paper

Printed in the Netherlands

'Et moi, ..., si j'avait su comment en revenir,
je n'y serais point allé.'

Jules Verne

The series is divergent; therefore we may be
able to do something with it.

O. Heaviside

One service mathematics has rendered the
human race. It has put common sense back
where it belongs, on the topmost shelf next
to the dusty canister labelled 'discarded non-
sense'.

Eric T. Bell

Mathematics is a tool for thought. A highly necessary tool in a world where both feedback and non-linearities abound. Similarly, all kinds of parts of mathematics serve as tools for other parts and for other sciences.

Applying a simple rewriting rule to the quote on the right above one finds such statements as: 'One service topology has rendered mathematical physics ...'; 'One service logic has rendered computer science ...'; 'One service category theory has rendered mathematics ...'. All arguably true. And all statements obtainable this way form part of the raison d'être of this series.

This series, *Mathematics and Its Applications*, started in 1977. Now that over one hundred volumes have appeared it seems opportune to reexamine its scope. At the time I wrote

"Growing specialization and diversification have brought a host of monographs and textbooks on increasingly specialized topics. However, the 'tree' of knowledge of mathematics and related fields does not grow only by putting forth new branches. It also happens, quite often in fact, that branches which were thought to be completely disparate are suddenly seen to be related. Further, the kind and level of sophistication of mathematics applied in various sciences has changed drastically in recent years: measure theory is used (non-trivially) in regional and theoretical economics; algebraic geometry interacts with physics; the Minkowsky lemma, coding theory and the structure of water meet one another in packing and covering theory; quantum fields, crystal defects and mathematical programming profit from homotopy theory; Lie algebras are relevant to filtering; and prediction and electrical engineering can use Stein spaces. And in addition to this there are such new emerging subdisciplines as 'experimental mathematics', 'CFD', 'completely integrable systems', 'chaos, synergetics and large-scale order', which are almost impossible to fit into the existing classification schemes. They draw upon widely different sections of mathematics."

By and large, all this still applies today. It is still true that at first sight mathematics seems rather fragmented and that to find, see, and exploit the deeper underlying interrelations more effort is needed and so are books that can help mathematicians and scientists do so. Accordingly MIA will continue to try to make such books available.

If anything, the description I gave in 1977 is now an understatement. To the examples of interaction areas one should add string theory where Riemann surfaces, algebraic geometry, modular functions, knots, quantum field theory, Kac-Moody algebras, monstrous moonshine (and more) all come together. And to the examples of things which can be usefully applied let me add the topic 'finite geometry'; a combination of words which sounds like it might not even exist, let alone be applicable. And yet it is being applied: to statistics via designs, to radar/sonar detection arrays (via finite projective planes), and to bus connections of VLSI chips (via difference sets). There seems to be no part of (so-called pure) mathematics that is not in immediate danger of being applied. And, accordingly, the applied mathematician needs to be aware of much more. Besides analysis and numerics, the traditional workhorses, he may need all kinds of combinatorics, algebra, probability, and so on.

In addition, the applied scientist needs to cope increasingly with the nonlinear world and the extra mathematical sophistication that this requires. For that is where the rewards are. Linear models are honest and a bit sad and depressing: proportional efforts and results. It is in the non-linear world that infinitesimal inputs may result in macroscopic outputs (or vice versa). To appreciate what I am hinting at: if electronics were linear we would have no fun with transistors and computers; we would have no TV; in fact you would not be reading these lines.

There is also no safety in ignoring such outlandish things as nonstandard analysis, superspace and anticommuting integration, p-adic and ultrametric space. All three have applications in both electrical engineering and physics. Once, complex numbers were equally outlandish, but they frequently proved the shortest path between 'real' results. Similarly, the first two topics named have already provided a number of 'wormhole' paths. There is no telling where all this is leading - fortunately.

Thus the original scope of the series, which for various (sound) reasons now comprises five sub-series: white (Japan), yellow (China), red (USSR), blue (Eastern Europe), and green (everything else), still applies. It has been enlarged a bit to include books treating of the tools from one subdiscipline which are used in others. Thus the series still aims at books dealing with:

- a central concept which plays an important role in several different mathematical and/or scientific specialization areas;
- new applications of the results and ideas from one area of scientific endeavour into another;
- influences which the results, problems and concepts of one field of enquiry have, and have had, on the development of another.

There are few differential operators of more pervasive importance than the Sturm-Liouville operator and, to a lesser extent, the Dirac operator. Few also have been studied this much and have such a large literature.

The book can be used as a text on the general spectral theory of self-adjoint operators but primarily it is a complete systematic text on the spectral theory of these two (classes of) operators: asymptotic behaviour of eigenvalues and eigenfunctions, distribution of eigenvalues, expansions, calculation of regularized traces, inverse problems, etc.

For the one-dimensional Dirac operator this is the first time that its complete spectral theory is presented systematically.

It is a pleasure to welcome such an authoritative volume in this series.

The shortest path between two truths in the real domain passes through the complex domain. J. Hadamard	Never lend books, for no one ever returns them; the only books I have in my library are books that other folk have lent me. Anatole France
La physique ne nous donne pas seulement l'occasion de résoudre des problèmes ... elle nous fait pressentir la solution. H. Poincaré	The function of an expert is not to be more right than other people, but to be wrong for more sophisticated reasons. David Butler

Kyoto, August 1990   Michiel Hazewinkel

Preface

The present monograph is devoted to the theory of Sturm-Liouville operators and one-dimensional Dirac-type operators. Basic problems in the spectral theory of linear operators are discussed for these two concrete classes of ordinary differential operators.

The monograph consists of two parts.

Part I (Chapters 1 to 6) is devoted to the spectral theory of the Sturm-Liouville operator, both in the regular and in the singular case. Major results for the classical Sturm-Liouville problem are discussed in Chapter 1, including the asymptotic behaviour of eigenvalues and eigenfunctions, and eigenfunction expansion theorems. The periodic Sturm-Liouville problem is studied there. The Sturm-Liouville problem on the half-line and on the whole line is considered in Chapter 2. The famous limit-circle and limit-point theorem of H. Weyl is proved, the expansion theorem is deduced, and the resolvent is studied. Chapter 3 is devoted to the study of the spectrum of the Sturm-Liouville operator under various assumptions on the potential. The asymptotic mean-square distribution of the eigenvalues for a purely discrete spectrum is discussed in Chapter 4. The regularized traces for the classical Sturm-Liouville operator with boundary conditions of various kinds are calculated in Chapter 5. Major results related to the reconstruction of the Sturm-Liouville operator both from a spectral function and from two spectra are discussed in Chapter 6.

Part II (Chapters 7 to 12) is devoted to the spectral theory of finite-order differential equations (the one-dimensional analogues of a Dirac system). Its content is similar to that of Chapters 1 to 6.

The monograph is primarily written for all those who want to extend their knowledge of the spectral theory of ordinary differential operators. Nevertheless, the book may also prove useful to the specialist, since it contains discussions of a number of problems not included in the literature up to now.

We hope that it may serve as an introduction to the spectral theory of linear self-adjoint operators in Hilbert space.

The authors would like to express their gratitude to M. V. Fedoryuk, who read the manuscript and gave a lot of valuable advice taken into account in preparing this volume.

B. M. Levitan and I. S. Sargsjan

Contents

Part one

Sturm-Liouville operators

Chapter 1

SPECTRAL THEORY IN THE REGULAR CASE

1.1 Basic properties of the operator

1.1.1

Let L be a linear operator defined on a certain set of elements. An element $y \neq 0$ is called an *eigenelement* of L if $Ly = \lambda y$, and λ is called the corresponding *eigenvalue* of L.

One of the simplest operators often used in applications is

$$L = -\frac{d^2}{dx^2} + q(x).$$

We assume that $q(x)$ is a real-valued and, initially, continuous function on an interval $[a, b]$. The above set of elements (or functions) $y(x)$ is determined for L by the obvious differentiability conditions, and also by certain conditions at the endpoints a, b.

The most important boundary-value conditions for the operator are

1. $y(a) \cos \alpha + y'(a) \sin \alpha = 0$, $y(b) \cos \beta + y'(b) \sin \beta = 0$, where α and β are two arbitrary real numbers, and

2. $y(a) = y(b)$, $y'(a) = y'(b)$.

One of the major aims of this monograph is to study the boundary-value problem

$$Ly(x) = -y'' + q(x)y = \lambda y, \tag{1.1}$$

$$\begin{aligned} y(a) \cos \alpha + y'(a) \sin \alpha &= 0 \\ y(b) \cos \beta + y'(b) \sin \beta &= 0 \end{aligned} \tag{1.2}$$

known as a *Sturm-Liouville problem*. A Sturm-Liouville problem is said to be *regular* if the interval $[a, b]$ is finite, and the function $q(x)$ is summable on it. Otherwise, if $[a, b]$ is infinite, or if $q(x)$ is not summable on the interval, or both, the Sturm-Liouville problem is said to be *singular*.

Note that more general second-order equations

$$y'' + p(x)y' + \{l(x) + \lambda r(x)\}y = 0, \tag{1.3}$$

where the function $r(x)$ is positive on $[a, b]$, are reducible to the form (1.1). If we assume that the first derivative of $p(x)$ and the second derivative of $r(x)$ are continuous, then (1.3) can be reduced to the canonical form

$$\frac{d^2\eta}{d\xi^2} + \{\lambda + q(\xi)\}\eta = 0$$

via the substitution

$$\xi = \int_a^x \sqrt{r(t)}\, dt, \qquad \eta(\xi) = \Phi(x)y(x), \qquad \Phi(x) = \sqrt[4]{r(x)}\exp\left(\frac{1}{2}\int_a^x p(t)\, dt\right), \tag{1.4}$$

which maps the interval $[a, b]$ into $[0, \pi]$, whereas the boundary conditions (1.2) do not change their form.

A transformation like (1.4) is called a *Liouville transformation*.

1.1.2

We consider a boundary-value problem (1.1), (1.2). Without loss of generality, we assume that $a = 0$ and $b = \pi$.

In fact, $[a, b]$ is mapped to $[0, \pi]$ by means of the substitution $t = \dfrac{x - a}{b - a}\pi$, which does not alter the form of (1.1), (1.2).

If the boundary-value problem has a nontrivial solution $y(x, \lambda_1) \not\equiv 0$ for certain λ_1, then λ_1 is called an eigenvalue, and $y(x, \lambda_1)$ is called an *eigenfunction* of (1.1), (1.2).

Lemma 1.1.1 *Two eigenfunctions $y(x, \lambda_1)$ and $y(x, \lambda_2)$ corresponding to different eigenvalues are orthogonal, i.e.,*

$$\int_0^\pi y(x, \lambda_1)y(x, \lambda_2)\, dx = 0, \qquad \lambda_1 \neq \lambda_2.$$

Proof. Let $f(x)$ and $g(x)$ be two continuous and twice differentiable functions. Put

$$Lf = f''(x) - q(x)f(x).$$

Integrating by parts twice, we have

$$\int_0^\pi Lf \cdot g(x)\, dx = W_\pi\{f, g\} - W_0\{f, g\} + \int_0^\pi f(x)Lg(x)\, dx, \tag{1.5}$$

where

$$W_x\{f, g\} = \begin{vmatrix} f(x) & g(x) \\ f'(x) & g'(x) \end{vmatrix}.$$

Let $f(x) = y(x, \lambda_1)$, and $g(x) = y(x, \lambda_2)$. It follows from the boundary conditions (1.2) that $W_0\{f, g\} = W_\pi\{f, g\} = 0$. Therefore, we obtain from (1.5) that

$$(\lambda_1 - \lambda_2) \int_0^\pi y(x, \lambda_1) y(x, \lambda_2) \, dx = 0.$$

The lemma is thus proved, since $\lambda_1 \neq \lambda_2$. ∎

Lemma 1.1.2 *The eigenvalues of the boundary-value problem (1.1), (1.2) are real.*

Proof. Let $\lambda_1 = u + iv$ be a complex eigenvalue. Since $q(x)$ is real-valued and α and β are real, $\lambda_2 = \bar{\lambda}_1 = u - iv$ is also an eigenvalue, corresponding to the eigenfunction $\bar{y}(x, \lambda_1)$. Then, by the previous lemma,

$$\int_0^\pi |y(x, \lambda_1)|^2 \, dx = 0;$$

hence, $y(x, \lambda_1) \equiv 0$. ∎

In the sequel, we need the following well-known theorem (for its proof, cf. the monographs of B. M. Levitan and I. S. Sargsjan (1975) or E. Titchmarsh (1962)).

Theorem 1.1.1 *If $q(x)$ is a continuous function on an interval $[a, b]$, then, for any α, there exists a unique solution $\varphi(x, \lambda)$, $a \leq x \leq b$, of the equation (1.1), such that $\varphi(a, \lambda) = \sin \alpha$ and $\varphi'(a, \lambda) = - \cos \alpha$.*

For any fixed $x \in [a, b]$, $\varphi(x, \lambda)$ is an entire function of λ.

1.2 Asymptotic behaviour of the eigenvalues and eigenfunctions

1.2.1

Put $\cot \alpha = -h$ and $\cot \beta = H^1$. The boundary conditions (1.2) can then be written in the form

$$y'(0) - hy(0) = 0, \qquad y'(\pi) + Hy(\pi) = 0. \tag{2.1}$$

Denote the solutions of (1.1), satisfying the initial conditions

$$\varphi(0, \lambda) = 1, \qquad \varphi'_x(0, \lambda) = h \tag{2.2}$$

and

$$\psi(0, \lambda) = 0, \qquad \psi'_x(0, \lambda) = 1, \tag{2.3}$$

by $\varphi(x, \lambda)$ and $\psi(x, \lambda)$, respectively.

---

[1]Assuming for the present that neither h nor H are infinite. These cases will be considered in the sequel.

Lemma 1.2.1 *Let* $\lambda = s^2$. *Then*

$$\varphi(x, \lambda) = \cos sx + \frac{h}{s} \sin sx + \frac{1}{s} \int_0^x \sin\{s(x - \tau)\} q(\tau) \varphi(\tau, \lambda) \, d\tau, \qquad (2.4)$$

$$\psi(x, \lambda) = \frac{\sin sx}{s} + \frac{1}{s} \int_0^x \sin\{s(x - \tau)\} q(\tau) \psi(\tau, \lambda) \, d\tau. \qquad (2.5)$$

Proof. To prove (2.4), we note that, since $\varphi(x, \lambda)$ satisfies (1.1),

$$\int_0^x \sin\{s(x - \tau)\} q(\tau) \varphi(\tau, \lambda) \, d\tau$$
$$= \int_0^x \sin\{s(x - \tau)\} \varphi_\tau''(\tau, \lambda) \, d\tau + s^2 \int_0^x \sin\{s(x - \tau)\} \varphi(\tau, \lambda) \, d\tau.$$

Twice integrating by parts the first integral on the right-hand side, and taking (2.2) into account, we have

$$\int_0^x \sin\{s(x - \tau)\} q(\tau) \varphi(\tau, \lambda) \, d\tau = -h \sin sx + s\varphi(x, \lambda) - s \cos sx,$$

which is nothing but (2.4). (2.5) can be proved similarly.   ∎

Lemma 1.2.2 *Let* $s = \sigma + it$. *Then there exists* $s_0 > 0$ *such that, for* $|s| > s_0$, *the estimates*

$$\varphi(x, \lambda) = O(e^{|t|x}), \qquad \psi(x, \lambda) = O(|s|^{-1} e^{|t|x}) \qquad (2.6)$$

are valid; more precisely,

$$\varphi(x, \lambda) = \cos sx + O\left(\frac{e^{|t|x}}{|s|}\right),$$

$$\psi(x, \lambda) = \frac{\sin sx}{s} + O\left(\frac{e^{|t|x}}{|s|^2}\right). \qquad (2.7)$$

All estimates are uniform with respect to x *for* $0 \le x \le \pi$.

Proof. Put $\varphi(x, \lambda) = e^{|t|x} f(x)$. It follows from (2.4) that

$$f(x) = \left\{\cos sx + \frac{h}{s} \sin sx\right\} e^{-|t|x} + \frac{1}{s} \int_0^x \sin\{s(x - \tau)\} e^{-|t|(x-\tau)} q(\tau) f(\tau) \, d\tau.$$

Let $\mu = \max_{0 \le x \le \pi} |f(x)|$. We then derive from the last equality that

$$\mu \le 1 + \frac{|h|}{|s|} + \frac{\mu}{|s|} \int_0^x |q(\tau)| \, d\tau;$$

hence,

$$\mu \leq \left(1 + \frac{|h|}{|s|}\right)\left(1 - \frac{1}{|s|}\int_0^\pi |q(\tau)|\,d\tau\right)^{-1},$$

provided that the denominator is positive; this is necessarily true for

$$|s| > \int_0^\pi |q(\tau)|\,d\tau.$$

Therefore (2.6) is proved for $\varphi(x,\lambda)$; it can be proved similarly for $\psi(x,\lambda)$, using (2.5). The estimates (2.7) are obtained by substituting (2.6) in the integrals on the right-hand side of (2.4) and (2.5).

It is also clear that general asymptotic expansions of $\varphi(x,\lambda)$ and $\psi(x,\lambda)$ as functions in s can be obtained by repeating the procedure. ∎

1.2.2

We now derive asymptotic formulas for the eigenvalues and eigenfunctions; in particular, this proves the existence of infinitely many eigenvalues.

First, we again assume that $h \neq \infty$ and $H \neq \infty$. The function $\varphi(x,\lambda)$ obviously satisfies the first boundary condition in (2.1) for any λ. We therefore find the eigenvalues by substituting $\varphi(x,\lambda)$ in the second condition of (2.1).

By Lemma 1.2.1, the eigenvalues are real, i.e., $\operatorname{Im} s = t = 0$. Hence the first estimate in (2.7) takes the form

$$\varphi(x,\lambda) = \cos sx + O(s^{-1}). \tag{2.8}$$

Differentiating (2.4) with respect to x and using (2.8), it is not hard to obtain the estimate

$$\varphi'_x(x,\lambda) = -s\sin sx + O(1). \tag{2.9}$$

Now, substituting the values of $\varphi(x,\lambda)$ and $\varphi'_x(x,\lambda)$, determined by (2.7) and (2.9), in the second condition of (2.1), we arrive at the equation

$$-s\sin s\pi + O(1) = 0 \tag{2.10}$$

which obviously has solutions close to integers for large s; hence, the existence of infinitely many eigenvalues has been proved.

We will show that, from a certain sufficiently large integer N on, only one root of (2.10) is near to each integer $n > N$. To this end, we differentiate the left-hand side of (2.10) with respect to s; this is made possible by formula (2.4) (noting that, by Theorem 1.1.1, $O(1)$ on the left-hand side of (2.10) is actually an analytic function of λ). We obtain $-\pi s\cos s\pi + O(1)$. It is not difficult to see that this expression does not vanish if s is close to large integers.

Let s_n be the nth root of (2.10). It follows from the Sturm theorem (cf. Section 3) and from the asymptotic formulas for the eigenfunctions that s_n should be near to n, and not to any other integer. Another proof, not on the basis of Sturm theory, can be obtained as follows.

We have already noted before that the eigenvalues λ satisfy

$$\varphi(\pi, \lambda) + H\varphi'_x(\pi, \lambda) = \omega(\lambda) = 0.$$

Put $\lambda = s^2$. Then $\omega(\lambda) = \omega_1(s)$ which, according to (2.4), is an entire function of s. It also follows from the asymptotic formulas (2.8) and (2.9) that, for $\sin \pi s \neq 0$,

$$\omega_1(s) = -Hs \sin s\pi \left\{ 1 + O(|s|^{-1}) \right\}. \tag{2.11}$$

Take a circle D_R of radius $R = N + 1/2$ in the s-plane, where N is a natural number. By the Rouché theorem and the asymptotic formula (2.11), there are as many zeros of $\omega_1(s)$ inside D_R as of the function $s \sin s\pi$, i.e., $2(N+1)$. The function $\omega_1(s)$ is even, therefore we only need consider its positive roots. Each positive zero is associated with an eigenvalue, and we have $N + 1$ eigenvalues s_k less than $N + 1/2$. It follows that

$$s_n = n + O(1). \tag{2.12}$$

In fact, let $s_n = m_n + o(1)$, $m_n \neq n$. Then, on the one hand, there are $n + 1$ eigenvalues s_k ($k = 0, 1, \ldots, n$) less than s_n. On the other hand, it follows from the above that there must be $2(m_n + 1)$ zeros of $\omega_1(s)$ in a disc of radius $m_n + 1/2$, i.e., there must be $m_n + 1 \neq n + 1$ eigenvalues s_k less than s_n; a contradiction, and the validity of (2.12) is thus proved.

Put $s_n = n + \delta_n$. The equation (2.10) then takes the form

$$(n + \delta_n) \sin \delta_n \pi + O(1) = 0.$$

Therefore $\sin \delta_n \pi = O(n^{-1})$, i.e., $\delta_n = O(n^{-1})$.

Thus, for large n, the roots of (2.10) are

$$s_n = n + O(n^{-1}) \tag{2.13}.$$

The asymptotic formula (2.13) can be much sharpened if we assume that the derivative of $q(x)$ in (1.1) is bounded. In fact, differentiating (2.4) with respect to x and substituting $\varphi(x, \lambda)$ and $\varphi'_x(x, \lambda)$ in the second boundary condition of (2.1), we obtain after simple transformations that

$$(-s + B) \sin s\pi + A \cos s\pi = 0, \tag{2.14}$$

where

$$A = h + H + \int_0^\pi \left\{ \cos s\tau + \frac{H}{s} \sin s\tau \right\} q(\tau)\varphi(\tau, \lambda)\, d\tau,$$

$$B = \frac{hH}{s} + \int_0^\pi \left\{ \sin s\tau + \frac{H}{s} \right\} q(\tau)\varphi(\tau, \lambda)\, d\tau.$$

Because of (2.8),

$$A = h + H + \frac{1}{2} \int_0^\pi q(\tau)\, d\tau + \frac{1}{2} \int_0^\pi q(\tau) \cos 2s\tau \, d\tau + O\left(\frac{1}{s}\right),$$

$$B = \frac{1}{2} \int_0^\pi q(\tau) \sin 2s\tau \, d\tau + O\left(\frac{1}{s}\right).$$

Since $q(x)$ has bounded derivative by assumption, integrating by parts, we see that

$$\int_0^\pi q(\tau) \cos 2s\tau \, d\tau = O\left(\frac{1}{s}\right), \qquad \int_0^\pi q(\tau) \sin 2s\tau \, d\tau = O\left(\frac{1}{s}\right),$$

and

$$A = h + H + h_1 + O\left(\frac{1}{s}\right), \qquad h_1 = \frac{1}{2} \int_0^\pi q(\tau)\, d\tau, \qquad B = O\left(\frac{1}{s}\right).$$

Consequently, equation (2.14) can be written in the form

$$\tan s\pi = (h + H + h_1 + O(1/s))/(s + O(1/s)).$$

Putting $s_n = n + \delta_n$ again, we obtain

$$\tan \pi \delta_n = \frac{h + H + h_1}{n} + O\left(\frac{1}{n^2}\right).$$

Hence,

$$\delta_n = \frac{h + H + h_1}{\pi n} + O\left(\frac{1}{n^2}\right)$$

and

$$s_n = n + \frac{c}{n} + O\left(\frac{1}{n^2}\right), \tag{2.15}$$

where

$$c = \frac{1}{\pi}\left(h + H + \frac{1}{2} \int_0^\pi q(\tau)\, d\tau \right).$$

Assuming that $q(x) \in C^2([0,\pi])$, we can prove that the more exact asymptotic formula

$$s_n = n + \frac{c}{n} + \frac{c_1}{n^3} + O\left(\frac{1}{n^4}\right) \tag{2.15'}$$

holds, where c_1 is a constant.

Using the formula (2.15), we obtain asymptotic formulas for the eigenfunctions $\varphi(x, \lambda_n) \equiv \varphi_n(x)$. Substituting the expression for $\varphi(x, \lambda)$ from (2.8) in formula (2.4), and since $q(x)$ is differentiable, we obtain

$$\varphi(x, \lambda) = \cos sx + \frac{h}{s} \sin sx + \frac{1}{s} \int_0^\pi \sin\{s(x - \tau)\} \cos s\tau \cdot q(\tau)\, d\tau + O\left(\frac{1}{s^2}\right)$$

$$= \cos sx + \frac{h}{s} \sin sx + \frac{\sin sx}{2s} \int_0^\pi q(\tau)\, d\tau + O\left(\frac{1}{s^2}\right).$$

Substituting s_n for s, we get

$$\varphi(x, \lambda_n) \equiv \varphi_n(x) = \cos nx - \frac{cx}{n} \sin nx + \frac{h}{n} \sin nx + \frac{\sin nx}{2n} \int_0^x q(\tau)\,d\tau + O\left(\frac{1}{n^2}\right)$$

$$= \cos nx + \frac{\beta(x)}{n} \sin nx + O\left(\frac{1}{n^2}\right),$$

relying on formula (2.15), where $\beta(x) = -cx + h + \frac{1}{2}\int_0^x q(\tau)\,d\tau$.

To derive the asymptotic expansion of the normalized eigenfunctions, we consider the integral

$$\alpha_n^2 = \int_0^\pi \varphi_n^2(x)\,dx = \int_0^\pi \cos^2 nx\,dx + \frac{1}{n}\int_0^\pi \beta(x) \sin 2nx\,dx + O\left(\frac{1}{n^2}\right).$$

Since the function $\beta(x)$ is differentiable,

$$\int_0^\pi \beta(x) \sin 2nx\,dx = O\left(\frac{1}{n}\right).$$

Therefore, $\alpha_n^2 = \pi/2 + O\left(\frac{1}{n^2}\right)$,

$$\frac{1}{\alpha_n} = \sqrt{\frac{2}{\pi}}\left\{1 + O\left(\frac{1}{n^2}\right)\right\},$$

and the asymptotic formula for the normalized eigenfunctions

$$v_n(x) = \frac{1}{\alpha_n}\varphi_n(x) = \sqrt{\frac{2}{\pi}}\left\{\cos nx + \frac{\beta(x)}{n}\sin nx\right\} + O\left(\frac{1}{n^2}\right)$$

holds.

1.2.3

We now study the case $h = \infty$, $H \neq \infty$ (the case $h \neq \infty$, $H = \infty$ reduces to this case by the substitution $t = \pi - x$). The first boundary condition in (2.1) takes the form $y(0) = 0$.

The function $\psi(x, \lambda)$ from Item 1 of the present section satisfies this condition. Therefore, we shall determine the eigenvalues in our case by substituting $\psi(x, \lambda)$ in the second boundary condition of (2.1). Differentiating the equality (2.5) with respect to x, we obtain

$$\psi_x'(x, \lambda) = \cos sx + \int_0^x \cos\{s(x - \tau)\}q(\tau)\psi(\tau, \lambda)\,d\tau.$$

It follows from the second boundary condition of (2.1) that

$$\cos s\pi + \int_0^\pi \cos\{s(\pi - \tau)\}q(\tau)\psi(\tau, \lambda)\, d\tau$$

$$+ H\left\{\frac{\sin s\pi}{s} + \frac{1}{s}\int_0^\pi \sin\{s(\pi - \tau)\}q(\tau)\psi(\tau, \lambda)\, d\tau = 0\right\}.$$

We have

$$\cos s\pi + \frac{1}{s}\int_0^\pi \cos\{s(\pi - \tau)\}q(\tau)\sin s\tau\, d\tau + H\frac{\sin s\pi}{s} + O\left(\frac{1}{s^2}\right) = 0, \qquad (2.16)$$

also relying on the estimates (2.7) for $\psi(x, \lambda)$.

Assuming again that $q(x)$ has bounded derivative, we get

$$\int_0^\pi q(\tau)\cos\{s(\pi - \tau)\}\sin s\tau\, d\tau = \frac{\sin s\pi}{2}\int_0^\pi q(\tau)\, d\tau + O\left(\frac{1}{s}\right).$$

Consequently, equation (2.16) takes the form

$$\cos s\pi + \frac{\sin s\pi}{s}\left\{H + \frac{1}{2}\int_0^\pi q(\tau)\, d\tau\right\} + O\left(\frac{1}{s^2}\right) = \cos s\pi + H_1\frac{\sin s\pi}{s} + O\left(\frac{1}{s^2}\right) = 0.$$
$$(2.17)$$

It can be seen that, for large s, the roots of (2.17) should be close to $n + 1/2$, where n is natural. Moreover, we prove as before that, starting with sufficiently large n, there is only one root of the equation near to each $n + 1/2$.

Put $s_n = n + 1/2 + \delta_n$. It follows from (2.17) that

$$\cot(n + \frac{1}{2} + \delta_n)\pi = -\tan \delta_n\pi = -\frac{H_1}{n + 1/2} + O(n^{-2})$$

and $\delta_n = H_1/\pi(n + 1/2) + O(n^{-2})$; therefore,

$$s_n = n + \frac{1}{2} + \frac{H_1}{\pi(n + \frac{1}{2})} + O(n^{-2}),$$

where $H_1 = H + \frac{1}{2}\int_0^\pi q(\tau)\, d\tau$.

Now, substituting s_n in (2.5), we obtain the asymptotic formula

$$\psi_n(x) = \frac{1}{n + 1/2}\sin(n + 1/2)x + O\left(\frac{1}{n^2}\right)$$

for the eigenfunctions $\psi(x, \lambda_n) \equiv \psi_n(x)$.

The formula for the norming constant α_n^{-1} is

$$\frac{1}{\alpha_n} = \sqrt{\frac{2}{\pi}}(n + 1/2)\left\{1 + O\left(\frac{1}{n}\right)\right\}.$$

Therefore, the normalized eigenfunctions are

$$v_n(x) \equiv \frac{1}{\alpha_n}\psi_n(x) = \sqrt{\frac{2}{\pi}}\sin(n + 1/2)x + O\left(\frac{1}{n}\right).$$

1.2.4

Finally, we study the case $h = \infty$ and $H = \infty$, where the boundary conditions (2.1) take the form $y(0) = y(\pi) = 0$; therefore, the function $\psi(x, \lambda)$ of the previous item should also satisfy the condition $\psi(\pi, \lambda) = 0$. It follows from (2.5) that

$$\sin s\pi + \int_0^\pi \sin\{s(\pi - \tau)\}q(\tau)\psi(\tau, \lambda)\, d\tau = 0,$$

or

$$\sin s\pi \left\{1 + \int_0^\pi \cos s\tau \cdot q(\tau)\psi(\tau, \lambda)\, d\tau\right\} - \cos s\pi \int_0^\pi \sin s\tau \cdot q(\tau)\psi(\tau, \lambda)\, d\tau = 0.$$

It follows from the last equation, because of the estimate (2.7) for $\psi(x, \lambda)$, that

$$\sin s\pi - \frac{1}{2s}\cos s\pi \int_0^\pi q(\tau)\, d\tau + O\left(\frac{1}{s^2}\right) = \sin s\pi - \frac{c}{s}\cos s\pi + O\left(\frac{1}{s^2}\right) = 0, \quad (2.18)$$

assuming that $q(x)$ has bounded derivative.

This is an equation of the same form as (2.10). Proceeding as in Item 2, we establish accordingly that the roots of (2.18) are close to integers, and that, beginning with a sufficiently large integer n, there is only one root near to each n. Therefore, the roots s_n of (2.18) are

$$s_n = n + \frac{c}{n} + O\left(\frac{1}{n^2}\right), \quad c = \frac{1}{2\pi}\int_0^\pi q(\tau)\, d\tau. \quad (2.19)$$

Substituting s_n in (2.5), we obtain the asymptotic formula

$$\psi_n(x) = (\sin nx)/n + O(1/n^2)$$

for the eigenfunctions $\psi(x, \lambda_n) \equiv \psi_n(x)$, and the formula

$$v_n(x) = \sqrt{2/\pi}\sin nx + O(1/n)$$

for the normalized eigenfunctions.

1.3   Sturm theory on the zeros of solutions

The deeper study of the eigenfunctions' zero distribution enabled Sturm to prove the existence of infinitely many eigenvalues for the boundary-value problem (1.1), (1.2) in another way.

In order to orient ourselves in the present section, we consider the simplest boundary-value problem

$$y'' + \lambda y = 0, \quad y'(0) = y'(\pi) = 0$$

with the eigenfunctions

$$\varphi_0(x) = 1, \varphi_1(x) = \cos x, \varphi_2(x) = \cos 2x, \ldots, \varphi_n(x) = \cos nx, \ldots$$

and the corresponding eigenvalues

$$\lambda_0 = 0, \lambda_1 = 1^2, \lambda_2 = 2^2, \ldots, \lambda_n = n^2, \ldots.$$

The eigenfunctions are ordered by increasing eigenvalues, beginning with zero. Their zeros are immediately seen to possess the following two properties :

(1) The nth eigenfunction in the interval $[0, \pi]$ has precisely n zeros, and

(2) the zeros of the nth and $(n+1)^{\text{th}}$ eigenfunctions interlace, i.e., there is a zero of the $(n+1)^{\text{th}}$ eigenfunction between any two consecutive zeros of the nth.

These properties also turn out to be valid in the general case.

The following fundamental *Sturm theorem* is the most important part :

Theorem 1.3.1 *Given*

$$u'' + g(x)u = 0 \tag{3.1}$$

$$v'' + h(x)v = 0 \tag{3.2}$$

in the whole interval $[a, b]$, if $g(x) < h(x)$, there is at least one zero of each solution of (3.2) between any two zeros of any nontrivial solution of (3.1).

Proof. Multiplying (3.1) throughout by v, (3.2) by u, and subtracting, we obtain

$$u''v - v''u = \frac{d}{dx}\{u'v - v'u\} = \{h(x) - g(x)\}uv. \tag{3.3}$$

Denote the consecutive zeros of u by x_1 and x_2. Then, integrating the identity (3.3) from x_1 to x_2, we obtain

$$u'(x_2)v(x_2) - u(x_1)v'(x_1) = \int_{x_1}^{x_2} \{h(x) - g(x)\}u(x)v(x)\,dx.$$

Assume that v does not vanish in the interval (x_1, x_2). Without loss of generality, we can assume that $u > 0$ and $v > 0$ in (x_1, x_2). Therefore, the right-hand side of the last equation is positive. Since $u(x) > 0$ by assumption, the function is increasing at the point x_1. Hence, $u'(x_1) > 0$ (it cannot vanish, because then it would follow from the uniqueness theorem for the solutions of (3.1) that $u(x) \equiv 0$, which is impossible). Similarly, $u'(x_2) < 0$. Therefore,

$$u'(x_2)v(x_2) - u'(x_1)v(x_1) \leq 0;$$

a contradiction.   ∎

Corollary 1.3.1 *Any solution of the equation*

$$y'' + g(x)y = 0, \quad -\infty \leq a \leq x \leq b \leq +\infty, \tag{3.4}$$

for $g(x) < -m^2 < 0$ cannot have more than one zero.

Proof. In fact, the equation $y'' - m^2 y = 0$ has the solution e^{mx} which does not vanish anywhere. Therefore, by the theorem, any solution of (3.4) cannot have more than one zero in any finite interval.   ∎

Theorem 1.3.2 *(Comparison Theorem.) Let $u(x)$ be the solution of equation (3.1), satisfying the initial conditions*

$$u(a) = \sin \alpha, \qquad u'(a) = -\cos \alpha, \tag{3.5}$$

and let $v(x)$ be the solution of equation (3.2) with the same initial conditions. Furthermore, suppose that $g(x) < h(x)$ in the whole of $[a, b]$. Then if $u(x)$ has m zeros in $a < x \leq b$, $v(x)$ has not fewer than m zeros in the same interval and the kth zero of $v(x)$ is less than the kth zero of $u(x)$.

Proof. Let x_1 be the zero of the function $u(x)$, nearest[2] to a (but different from it). By the previous theorem, it suffices to show that $v(x)$ has at least one zero in $[a, x_1]$. Assume the contrary. Without loss of generality, we can put $v(x) > 0$ and $u(x) > 0$ in $[a, x_1]$. Since $u(x_1) = 0$, the function $u(x)$ is decreasing in the vicinity of the point x_1. Therefore, $u'(x_1) \leq 0$. Integrating the identity (3.3) from a to x_1, we obtain

$$u'(x_1)v(x_1) = \int_a^{x_1} \{h(x) - g(x)\}u(x)v(x)\, dx.$$

Since, by assumption, $v(x) > 0$ and $u(x) > 0$ in $[a, x_1]$, and $h(x) > g(x)$, the right-hand side of the last equality is positive. However, the expression on the left is less than or equal to zero; a contradiction.   ∎

Let $\varphi(x, \lambda)$ be the function introduced in Section 2. Consider the equation $\varphi(x, \lambda) = 0$, $a \leq x \leq b$. It is obvious that the roots of this equation are functions of λ. We prove that they are continuous.

Lemma 1.3.1 *If x_0, $a < x_0 < b$, is a root of the function $\varphi(x, \lambda_0)$, then, for a sufficiently small number $\epsilon > 0$ there exists a value $\delta > 0$ such that, for $|\lambda - \lambda_0| < \delta$, the function $\varphi(x, \lambda)$ has precisely one zero in the interval $|x - x_0| < \epsilon$.*

---

[2]This least zero exists because a nontrivial solution $y(x)$ of a linear differential equation has only isolated zeros. Indeed, assume that $x_0 \neq \infty$ is the limit point of the zeros $x_1, x_2, \ldots, x_n, \ldots$ of a solution of the differential equation. Then $y(x_0) = 0$ by the continuity of $y(x)$. Furthermore, $(y(x_0) - y(x_n))/(x_0 - x_n) = 0$. Passing to the limit for $n \to \infty$, we obtain $y'(x_0) = 0$, i.e., $y(x) \equiv 0$.

Proof. The zero x_0 of the solution $\varphi(x, \lambda_0)$ of the equation (1.1) is simple, since, if we had $\varphi'_x(x_0, \lambda_0) = 0$, it would follow from the uniqueness theorem for the solution of the problem (1.1), (3.5) that $\varphi(x, \lambda_0) \equiv 0$. Therefore, $\varphi'_x(x_0, \lambda_0) \neq 0$. For definiteness, we put $\varphi'_x(x_0, \lambda_0) > 0$. Let $\epsilon > 0$ be so small that $\varphi'_x(x, \lambda_0) > 0$ in the whole interval $|x - x_0| \leq \epsilon$. Then $\varphi(x_0 - \epsilon, \lambda_0) < 0$ and $\varphi(x_0 + \epsilon, \lambda_0) > 0$.

Furthermore, since $\varphi'_x(x, \lambda)$ is continuous with respect to λ (by Theorem 1.1.1, $\varphi(x, \lambda)$ is an entire function of λ), there is $\delta > 0$ such that, for $|\lambda - \lambda_0| \leq \delta$, $\varphi'_x(x, \lambda)$ also remains positive in the whole interval $|x - x_0| \leq \epsilon$. Therefore, the monotonically increasing function $\varphi(x, \lambda)$ obviously cannot have two zeros in this interval. Furthermore, if we select δ to be so small that, for $|\lambda - \lambda_0| < \delta$, the function $\varphi(x_0 - \epsilon, \lambda)$ remains negative, and $\varphi(x_0 + \epsilon, \lambda)$ positive (which is possible due to the continuity of $\varphi(x, \lambda)$ with respect to λ), then the statement of the lemma follows : The solution $\varphi(x, \lambda)$ for $|\lambda - \lambda_0| < \delta$ has precisely one zero in the interval $[x_0 - \epsilon, x_0 + \epsilon]$. ∎

The following is an important consequence.

Corollary 1.3.2 *As λ varies, $\varphi(x, \lambda)$ can lose or acquire a zero only if the zero gets inside the interval or outside it through one of the endpoints a, b.*

The following proves the existence of infinitely many eigenvalues.

Theorem 1.3.3 *(Sturm Oscillation Theorem.) There is an indefinitely increasing sequence of eigenvalues $\lambda_0, \lambda_1, \ldots, \lambda_n, \ldots$ of the boundary-value problem (1.1), (1.2), and the eigenfunction corresponding to the eigenvalue λ_m has precisely m zeros in the interval $a < x < b$.*

Proof. Let $\varphi(x, \lambda)$ be the solution of equation (1.1), satisfying the initial conditions (3.5). By Theorem 1.3.2, as λ increases, the number of zeros of the function $\varphi(x, \lambda)$ does not decrease. Let $|q(x)| < c$ for $a \leq x \leq b$. Compare (1.1) with the equation $y'' + (\lambda + c)y = 0$. Its solution satisfying the initial conditions (3.5) is the function

$$y = \sin \alpha \cdot \cosh\{(-\lambda - c)^{1/2}(x - a)\} - \cos \alpha \cdot (-\lambda - c)^{1/2} \sinh\{(-\lambda - c)^{1/2}(x - a)\}.$$

For negative values λ which are sufficiently large in absolute value, y obviously does not vanish. Therefore, applying Theorem 1.3.2 again, we see that $\varphi(x, \lambda)$ is not zero for such values of λ.

However, selecting the equation $y'' + (\lambda - c)y = 0$, we see that the number of zeros of $\varphi(x, \lambda)$ in the interval $[a, b]$ increases indefinitely if λ is positive and indefinitely increasing.

Consider the equation $\varphi(x, \lambda) = 0$. It follows from Lemma 1.3.1 that its roots depend continuously on λ. On the other hand, by Theorem 1.3.2, as λ increases, each zero of $\varphi(x, \lambda)$ shifts to the left, but cannot cross the point a, since the number of zeros does not decrease. By the corollary to Lemma 1.3.1, new zeros enter through the point b. Let μ_0 be the first value of the parameter λ for which $\varphi(b, \mu_0) = 0$.

It is evident that such a value exists. Let μ_1 be the second value of λ for which $\varphi(b, \mu_1) = 0$, etc. The sequence of values $\mu_0, \mu_1, \ldots, \mu_m, \ldots$ has the property that the function $\varphi(x, \mu_m)$ has precisely m zeros inside the interval $[a, b]$, and $\varphi(b, \mu_m) = 0$. If $\sin \beta = 0$, then the second boundary condition in (1.2) holds (the first one holds because of (3.5)); therefore μ_m are eigenvalues, and the theorem is proved in this case.

Now, suppose that $\sin \beta \neq 0$, and that $u(x)$, $v(x)$ are the functions considered in Theorem 1.3.2. Then

$$
\begin{aligned}
\frac{d}{dx} \left\{ u^2 \left(\frac{u'}{u} - \frac{v'}{v} \right) \right\} &= 2uu' \left(\frac{u'}{u} - \frac{v'}{v} \right) + u^2 \left(\frac{u''}{u} - \frac{v''}{v} \right) - u^2 \left(\frac{u'^2}{u^2} - \frac{v'^2}{v^2} \right) \\
&= \frac{(u'v - uv')^2}{v^2} + u^2 \{ h(x) - g(x) \} > 0.
\end{aligned}
\tag{3.6}
$$

Therefore, the function $u^2(u'/u - v'/v)$ increases monotonically in any interval where v does not vanish. Suppose that $u(x)$ and $v(x)$ have the same number of zeros inside $[a, b]$.

Let x_ν be the root of $u(x)$, which is nearest to the point b. We show that the function $v(x)$ cannot have any zeros for $x_\nu \leq x \leq b$. In fact, by Theorem 1.3.2, there are at least ν zeros of $v(x)$ between a and x_ν. If $v(x)$ vanished for $x_\nu \leq x \leq b$, it would have more zeros in the whole interval $[a, b]$ than $u(x)$, contrary to the assumption.

Integrating the relation (3.6) from x_ν to b, we obtain

$$
u^2(b) \left\{ \frac{u'(b)}{u(b)} - \frac{v'(b)}{v(b)} \right\} \geq u^2(x_\nu) \left\{ \frac{u'(x_\nu)}{u(x_\nu)} - \frac{v'(x_\nu)}{v(x_\nu)} \right\} = 0.
$$

Therefore,

$$
\frac{u'(b)}{u(b)} > \frac{v'(b)}{v(b)}.
\tag{3.7}
$$

Let $\varphi(x, \lambda') = u(x)$ and $\varphi(x, \lambda'') = v(x)$, where $\mu_m < \lambda' < \lambda'' < \mu_{m+1}$. According to (3.7), the function $\varphi'(b, \lambda)/\varphi(b, \lambda)$ decreases monotonically in the interval (μ_m, μ_{m+1}). Since $\varphi(b, \mu_m) = \varphi(b, \mu_{m+1}) = 0$, the function should decrease from $+\infty$ to $-\infty$. Therefore, there exists a value λ_m inside (μ_m, μ_{m+1}), for which $\varphi'(b, \lambda_m)/\varphi(b, \lambda_m) = -\cot \beta$, and the second condition in (2.1) holds. Hence, λ_m is an eigenvalue, and $\varphi(x, \lambda_m)$ has as many zeros in (a, b) as $\varphi(x, \mu_m)$, i.e., m. ∎

1.4 The periodic and the semi-periodic problem

1.4.1

Consider the equation

$$
y'' + \{ \lambda - q(x) \} y = 0,
\tag{4.1}
$$

where $q(x)$ is a real-valued and periodic function with period a, i.e., $q(x+a) = q(x)$. Because of the periodicity, it is natural to consider boundary-value problems for (4.1) with the boundary conditions

$$y(0) = y(a), \qquad y'(0) = y'(a) \tag{4.2}$$

$$y(0) = -y(a), \qquad y'(0) = -y'(a). \tag{4.3}$$

The problem (4.1), (4.2) is said to be *periodic*, and (4.1), (4.3) *semi-periodic*. It is easy to see (with the use of Green's identity) that both problems are self-adjoint.

Denote by $\varphi(x, \lambda)$ and $\vartheta(x, \lambda)$ the solutions of (4.1) with the initial conditions $\varphi(0, \lambda) = \vartheta'(0, \lambda) = 0$, $\varphi'(0, \lambda) = \vartheta(0, \lambda) = 1$. Let $y(x, \lambda)$ be the eigenfunctions of the problem (4.1), (4.2) or (4.1), (4.3), corresponding to the eigenvalue λ. Since $y(x, \lambda)$ is a solution of (4.1) and $\varphi(x, \lambda)$ and $\vartheta(x, \lambda)$ are two linearly independent solutions of the equation (4.1),

$$y(x, \lambda) = C_1\vartheta(x, \lambda) + C_2\varphi(x, \lambda), \tag{4.4}$$

where C_1 and C_2 are two constants. Substituting the expression (4.4) in the boundary conditions (4.2), we obtain

$$C_1\vartheta(a, \lambda) + C_2\varphi(a, \lambda) = C_1, \qquad C_1\vartheta'(a, \lambda) + C_2\varphi'(a, \lambda) = C_2;$$

hence,

$$\begin{aligned} C_1[\vartheta(a, \lambda) - 1] + C_2\varphi(a, \lambda) = 0, \\ C_1\vartheta'(a, \lambda) + C_2[\varphi'(a, \lambda) - 1] = 0. \end{aligned} \tag{4.5}$$

For the system (4.5) to have a nontrivial solution, it is necessary and sufficient that

$$\begin{vmatrix} \vartheta(a, \lambda) - 1 & \varphi(a, \lambda) \\ \vartheta'(a, \lambda) & \varphi'(a, \lambda) - 1 \end{vmatrix} = 0. \tag{4.6}$$

Since the Wronskian of (4.1) is constant and

$$W\{\varphi(x, \lambda), \vartheta(x, \lambda)\} = \varphi'(x, \lambda)\vartheta(x, \lambda) - \varphi(x, \lambda)\vartheta'(x, \lambda) = -1,$$

it follows from (4.6) that $\varphi'(a, \lambda) + \vartheta(a, \lambda) - 2 = 0$. Therefore, the eigenvalues of the periodic problem (4.1), (4.2) satisfy the equation $F(\lambda) = 2$, where $F(\lambda) = \vartheta(a, \lambda) + \varphi'(a, \lambda)$.

It can be shown similarly that the eigenvalues of the semi-periodic problem (4.1), (4.3) satisfy the equation $F(\lambda) = -2$.

That the eigenvalues of the periodic and semi-periodic problems are real is proved in a similar way as in Section 1.

1.4.2   The multiplicity of the eigenvalues

In contrast to the problems for the equation (4.1) with separated boundary conditions, the eigenvalues of the periodic and semi-periodic problems may turn out to be repeated, but of multiplicity not exceeding two. The following result offers a rather simple test for a repeated eigenvalue.

Lemma 1.4.1 *For an eigenvalue λ^0 of the periodic (semi-periodic) problem to be repeated, it is necessary and sufficient that*

$$\begin{aligned} \vartheta(a,\lambda^0) = \varphi'(a,\lambda^0) = 1, \quad & \vartheta'(a,\lambda^0) = \varphi(a,\lambda^0) = 0. \\ (\vartheta(a,\lambda^0) = \varphi'(a,\lambda^0) = -1, \quad & \vartheta'(a,\lambda^0) = \varphi(a,\lambda^0) = 0.) \end{aligned} \tag{4.7}$$

Proof. (Necessity.) Let λ^0 be a repeated eigenvalue of, e.g., the periodic problem (4.1), (4.2). Then there exist two linearly independent solutions $y_1(x,\lambda^0)$ and $y_2(x,\lambda^0)$ of the equation (4.1), satisfying the boundary conditions (4.2). Any other solution of (4.1) is a linear combination of $y_1(x,\lambda^0)$, $y_2(x,\lambda^0)$, and therefore also satisfies (4.2). In particular, this remains valid for the solutions $\vartheta(x,\lambda^0)$ and $\varphi(x,\lambda^0)$. Therefore,

$$\begin{aligned} \vartheta(a,\lambda^0) = \vartheta(0,\lambda^0) = 1, \quad & \vartheta'(a,\lambda^0) = \vartheta'(0,\lambda^0) = 0, \\ \varphi(a,\lambda^0) = \varphi(0,\lambda^0) = 0, \quad & \varphi'(a,\lambda^0) = \varphi'(0,\lambda^0) = 1, \end{aligned}$$

and the conditions (4.7) hold.
(Sufficiency.) Let (4.7) be fulfilled. Then

$$\begin{aligned} \vartheta(a,\lambda^0) = 1 = \vartheta(0,\lambda^0), \quad & \vartheta'(a,\lambda^0) = 0 = \vartheta'(0,\lambda^0), \\ \varphi(a,\lambda^0) = 0 = \varphi(0,\lambda^0), \quad & \varphi'(a,\lambda^0) = 1 = \varphi'(0,\lambda^0). \end{aligned}$$

Therefore, $\vartheta(x,\lambda^0)$, $\varphi(x,\lambda^0)$ satisfy (4.2) and are eigenfunctions of the periodic problem.

The proof for the semi-periodic problem is similar.   ∎

Remark   To see that an eigenvalue of the periodic or semi-periodic problem is repeated, it suffices to verify the second condition in (4.7), or that $\vartheta'(a,\lambda^0) = \varphi(a,\lambda^0) = 0$.

Indeed, if it were so, we would have $\vartheta(a,\lambda^0)\varphi'(a,\lambda^0) = 1$, because the Wronskian is constant. Moreover, $\varphi'(a,\lambda^0) + \vartheta(a,\lambda^0) = \pm 2$. Therefore, $\varphi'(a,\lambda^0)$, $\vartheta(a,\lambda^0)$ are the roots of the quadratic equation $X^2 \mp 2X + 1 = (X \mp 1)^2 = 0$, and $\vartheta(a,\lambda^0) = \varphi'(a,\lambda^0) = \pm 1$.

To prove that the multiplicity condition for an eigenvalue is related to the zeros of the function $F(\lambda) - 2$ (or $F(\lambda) + 2$) is not so simple.

Theorem 1.4.1 *A number λ^0 is a multiple root of the equation $F(\lambda) \mp 2 = 0$ if and only if*

$$\varphi(a,\lambda^0) = \vartheta'(a,\lambda^0) = 0. \tag{4.8}$$

Proof. [3] If the conditions (4.8) are fulfilled, then, relying on the Remark after Lemma 1.4.1, $\vartheta(a, \lambda^0) = \varphi'(a, \lambda^0) = \pm 1$.

We write the Wronskian identity in the form

$$-\vartheta'\varphi = 1 - \varphi'\vartheta = 1 - \frac{1}{4}\{(\vartheta + \varphi')^2 - (\vartheta - \varphi')^2\}. \tag{4.9}$$

Since $\varphi(a, \lambda)$ and $\vartheta'(a, \lambda)$ have a root for $\lambda = \lambda^0$ (relying on (4.8)), the product $\varphi(a, \lambda)\vartheta(a, \lambda)$ has a multiple root at $\lambda = \lambda_0$. Therefore, the right-hand side of (4.9) also has a multiple root for $\lambda = \lambda^0$. However, the right-hand side is

$$\frac{1}{4}[2 - (\vartheta + \varphi')][2 + (\vartheta + \varphi')] + \frac{1}{4}(\vartheta - \varphi')^2. \tag{4.10}$$

For $\lambda = \lambda^0$, the functions $\vartheta(a, \lambda)$ and $\varphi'(a, \lambda)$ are either 1 or -1; therefore, their difference $\vartheta(a, \lambda) - \varphi'(a, \lambda)$ has a root for $\lambda = \lambda^0$, and $(\vartheta - \varphi')^2$ has a multiple root at the point $\lambda = \lambda^0$. It then follows from (4.10) that the function $[2 - (\vartheta + \varphi')][2 + (\vartheta + \varphi')]$ also has a multiple root for $\lambda = \lambda^0$. But if λ^0 is an eigenvalue of, e.g., the periodic problem, then $2 + (\vartheta + \varphi') = 4$, and therefore $2 - (\vartheta + \varphi') = 2 - F(\lambda)$ has a multiple root for $\lambda = \lambda^0$.

Conversely, let $F'(\lambda^0) = 0$, i.e., let λ^0 be a multiple root of the equation $F(\lambda) = \pm 2$. Differentiating the basic equation (4.1) with respect to λ, we obtain

$$\frac{\partial^3 y}{\partial x^2 \partial \lambda} + \{\lambda - q(x)\}\frac{\partial y}{\partial \lambda} = -y \tag{4.11}$$

for both $\varphi(x, \lambda)$ and $\vartheta(x, \lambda)$; also,

$$\frac{\partial y(0, \lambda)}{\partial \lambda} = 0, \qquad \frac{\partial y'_x(0, \lambda)}{\partial \lambda} = 0, \tag{4.12}$$

since the initial conditions for $\varphi(x, \lambda)$ and $\vartheta(x, \lambda)$ for $x = 0$ do not depend on λ. We now determine $\partial y/\partial \lambda$ as a solution of the nonhomogeneous equation (4.11) with the initial conditions (4.12). The method of variation of the constants leads to the formulas

$$\frac{\partial \varphi(x, \lambda)}{\partial \lambda} = \int_0^x \{\varphi(\xi, \lambda)\vartheta(x, \lambda) - \varphi(x, \lambda)\vartheta(\xi, \lambda)\}\varphi(\xi, \lambda)\, d\xi, \tag{4.13}$$

$$\frac{\partial \vartheta(x, \lambda)}{\partial \lambda} = \int_0^x \{\varphi(\xi, \lambda)\vartheta(x, \lambda) - \varphi(x, \lambda)\vartheta(\xi, \lambda)\}\vartheta(\xi, \lambda)\, d\xi. \tag{4.14}$$

Differentiating (4.13) with respect to x, we have

$$\frac{\partial \varphi'(x, \lambda)}{\partial \lambda} = \int_0^x \{\varphi(\xi, \lambda)\vartheta'(x, \lambda) - \varphi'(x, \lambda)\vartheta(\xi, \lambda)\}\varphi(\xi, \lambda)\, d\xi. \tag{4.15}$$

---

[3] In the sequel, we will often write φ, φ', ϑ, ϑ' instead of $\varphi(a, \lambda)$, $\varphi'(a, \lambda)$, $\vartheta(a, \lambda)$ and $\vartheta'(a, \lambda)$, respectively.

Now, putting $x = a$ in (4.14), (4.15), adding them together, and taking into account that $\vartheta(a, \lambda) + \varphi'(a, \lambda) = F(\lambda)$, we obtain

$$\frac{dF(\lambda)}{d\lambda} = \int_0^a \{\vartheta'\varphi^2(\xi, \lambda) - (\vartheta - \varphi')\vartheta(\xi, \lambda)\varphi(\xi, \lambda) - \varphi\vartheta^2(\xi, \lambda)\} \, d\xi. \qquad (4.16)$$

Let $\sqrt{\lambda} = s = \sigma + it$. By Lemma 1.2.2, we have

$$\vartheta = \cos as + O(|s|^{-1}e^{at}), \qquad \vartheta' = -s \sin as + O(e^{at}),$$
$$\varphi = s^{-1}\sin as + O(|s|^{-2}e^{at}), \qquad \varphi' = \cos as + O(|s|^{-1}e^{at}).$$

Therefore, for large $|\lambda|$,

$$F(\lambda) = 2\cos as + O(|s|^{-1}e^{at}).$$

As Real $\lambda \to +\infty$, the function $F(\lambda)$ oscillates between 2 and -2. As Real $\lambda \to -\infty$, the number $s = i\sqrt{|\lambda|}$ is pure imaginary, and $F(\lambda) \sim 2\cosh a\sqrt{|\lambda|} \to +\infty$. It follows that there exists at least one zero λ_0 of the function $F(\lambda) - 2$; however, it is not quite clear whether $F(\lambda)$ takes the values ± 2 to the right of the point λ_0. This follows from further analysis.

Let λ be such that $-2 < F(\lambda) < 2$. Then

$$\vartheta^2 + \varphi'^2 + 2\vartheta\varphi' < 4 = 4(\vartheta\varphi' - \varphi\vartheta').$$
$$(\vartheta - \varphi')^2 = (\vartheta + \varphi')^2 - 4\vartheta\varphi' < 4 - 4\vartheta\varphi' = 4(1 - \vartheta\varphi') = -4\varphi\vartheta'.$$

Hence, $\varphi \neq 0$, $\vartheta' \neq 0$ and φ, ϑ' have opposite signs.

Transform the formula (4.16) as

$$\frac{dF}{d\lambda} = -\varphi \int_0^a \left\{\vartheta(\xi, \lambda) - \frac{\vartheta - \varphi'}{2\varphi}\varphi(\xi, \lambda)\right\}^2 d\xi - \frac{4 - (\vartheta + \varphi')^2}{4\varphi} \int_0^a \varphi^2(\xi, \lambda) \, d\xi, \quad (4.17)$$

where the right-hand side is not zero, with sign opposite to that of $\varphi(a, \lambda)$. Therefore, $F(\lambda)$ can have neither a maximum nor a minimum at such a point, and if $F'(\lambda_0) \neq 0$, then $F(\lambda)$ monotonically decreases from 2 for $\lambda = \lambda_0$ until -2 for a certain μ_0. In the general case, the curve $y = F(\lambda)$ meets the straight line $y = -2$ at the point μ_0; in exceptional cases, a contact may occur.

Thus, in addition to μ_0, generally speaking, another zero μ_1 of the function $F(\lambda) + 2$ appears, a zero λ_1 of the function $F(\lambda) - 2$, etc., and we write

$$\lambda_0 < \mu_0 < \mu_1 < \lambda_1 < \lambda_2 < \mu_3 < \ldots.$$

It also follows from the above that $\varphi > 0$ for $\lambda_0 < \lambda < \mu_0$ and $\varphi < 0$ for $\mu_1 < \lambda < \lambda_1$, etc.

We now assume that λ^0 is a zero of $F(\lambda) - 2$ of order higher than one. Then $\vartheta + \varphi' = 2$, and if $\varphi \neq 0$, we obtain

$$\frac{dF}{d\lambda} = -\varphi \int_0^a \left\{ \vartheta(\xi, \lambda) - \frac{\vartheta - \varphi'}{2\varphi} \varphi(\xi, \lambda) \right\}^2 d\xi \neq 0$$

from the formula (4.17), for the integrand cannot be identically zero. Hence, $dF/d\lambda = 0$ implies $\varphi(a, \lambda^0) = 0$. Similarly, if $\vartheta' \neq 0$, then

$$\frac{dF}{d\lambda} = \vartheta' \int_0^a \left\{ \varphi(\xi, \lambda) + \frac{\vartheta - \varphi'}{2\vartheta'} \vartheta(\xi, \lambda) \right\}^2 d\xi,$$

relying on the formula (4.16). Therefore, if $dF/d\lambda = 0$, then $\vartheta'(a, \lambda^0) = 0$.

Thus, if λ^0 is a zero of the function $F(\lambda) - 2$ of order higher than one, then $\varphi(a, \lambda^0) = \vartheta'(a, \lambda^0) = 0$; hence, λ^0 is a repeated eigenvalue, and the theorem is thus proved.

In conclusion, we show that the functions $F(\lambda) \pm 2$ cannot have zeros of order higher than two.

Differentiating the equation (4.11) and the initial conditions (4.12) with respect to λ again, we obtain the problem

$$\frac{\partial^4 y}{\partial x^2 \partial \lambda^2} + \{\lambda - q(x)\} \frac{\partial^2 y}{\partial \lambda^2} = -2 \frac{\partial y}{\partial \lambda},$$

$$\frac{\partial^2 y(0, \lambda)}{\partial \lambda^2} = 0, \qquad \frac{\partial^2 y'(0, \lambda)}{\partial \lambda^2} = 0$$

for the functions $\vartheta(x, \lambda)$, $\varphi(x, \lambda)$, which, when solved using the method of variation of the constants, yields the expressions

$$\frac{\partial^2 \varphi(x, \lambda)}{\partial \lambda^2} = 2 \int_0^x \{\vartheta(x, \lambda)\varphi(\xi, \lambda) - \varphi(x, \lambda)\vartheta(\xi, \lambda)\} \frac{\partial \varphi(\xi, \lambda)}{\partial \lambda} d\xi, \qquad (4.18)$$

$$\frac{\partial^2 \vartheta(x, \lambda)}{\partial \lambda^2} = 2 \int_0^x \{\vartheta(x, \lambda)\varphi(\xi, \lambda) - \varphi(x, \lambda)\vartheta(\xi, \lambda)\} \frac{\partial \vartheta(\xi, \lambda)}{\partial \lambda} d\xi. \qquad (4.19)$$

Now, differentiating the formula (4.18) with respect to x, adding it to (4.19) and putting $x = a$, we find

$$\frac{d^2 F}{d\lambda^2} = \pm 2 \int_0^a \left\{ \varphi(\xi, \lambda) \frac{\partial \vartheta(\xi, \lambda)}{\partial \lambda} - \vartheta(\xi, \lambda) \frac{\partial \varphi(\xi, \lambda)}{\partial \lambda} \right\} d\xi$$

$$= \mp 2 \int_0^a d\xi \int_0^\xi \{\varphi(\xi, \lambda)\vartheta(t, \lambda) - \vartheta(\xi, \lambda)\varphi(t, \lambda)\}^2 dt,$$

also using the condition (4.7) and the formulas (4.13), (4.14) in the last transformation.

Since the right-hand side of the last equation is not zero, the functions $F(\lambda) \pm 2$ cannot have zeros of order higher than two. Also, $F(\lambda)$ has a maximum (or minimum) at a zero of order two, so that, e.g., the point λ^0 cannot be a zero of order two. ∎

1.4.3   The zeros of eigenfunctions

Consider two auxiliary problems

$$y(0) = y(a) = 0 \tag{4.20}$$

$$y'(0) = y'(a) = 0 \tag{4.21}$$

for the equation (4.1).

Let $\nu_1 < \nu_2 < \nu_3 < \ldots$ be the eigenvalues of the problem (4.1), (4.20), and let $\tau_0 < \tau_1 < \tau_2 < \ldots$ be those of (4.1), (4.21). It is obvious that ν_n are the zeros of the function $\varphi(a, \lambda)$, and that τ_n are the zeros of $\vartheta'(a, \lambda)$.

Theorem 1.4.2 *In each of the intervals* $[\mu_0, \mu_1], (\lambda_1, \lambda_2), [\mu_2, \mu_3], (\lambda_3, \lambda_4), \ldots$, *which are called* gaps, *there lie exactly one* ν_k *and one* τ_k, $k = 1, 2, \ldots$, *and the value* $\tau_0 \in (-\infty, \lambda_0]$.

Proof. It follows from the identity (4.9) that

$$-\varphi(a, \lambda)\vartheta'(a, \lambda) = 1 - \frac{1}{4}\{[\vartheta(a, \lambda) + \varphi'(a, \lambda)]^2 - [\vartheta(a, \lambda) - \varphi'(a, \lambda)]^2\}$$

i.e.,

$$[\vartheta(a, \nu_k) + \varphi'(a, \nu_k)]^2 = 4 + [\vartheta(a, \nu_k) - \varphi'(a, \nu_k)]^2 \geq 4,$$

for $\lambda = \nu_k$, and a similar inequality holds for $\lambda = \tau_k$, $k = 0, 1, 2, \ldots$. Therefore, ν_k and τ_k can get only in the intervals specified in the theorem.

We show that not more than one ν_k (τ_k) can get into each gap, referring to Sturm Theorem 1.3.1, which states that the function $\varphi(x, \nu_n)$ has precisely $n - 1$ zeros in the interval $(0, a)$. Since $\varphi'(0, \nu_n) = 1$, $\varphi(a, \nu_n) > 0$ for even n and $\varphi'(a, \nu_n) < 0$ for odd n. By the Wronskian identity, $\vartheta(a, \nu_n)\varphi'(a, \nu_n) = 1$; consequently,

$$F(\nu_n) = \vartheta(a, \nu_n) + \varphi'(a, \nu_n) = \vartheta(a, \nu_n) + 1/\vartheta(a, \nu_n) \geq 2$$

if n is even, and $F(\nu_n) \leq -2$ if n is odd. It follows that ν_n and ν_{n+1} cannot get into the same gap. A similar conclusion also holds for τ_n.

It remains to show that, in fact, one ν_n and one τ_n get into each gap, except for the trivial gap $(-\infty, \lambda_0]$ which only contains τ_0.

Consider two consecutive roots λ' and λ'' of the function $F^2(\lambda) - 4$, with $F(\lambda) > 2$ or $F(\lambda) < -2$ in between them. By the formula (4.17), $\varphi(a, \lambda)F'(\lambda) \leq 0$ both for $\lambda = \lambda'$ and $\lambda = \lambda''$. Since $F'(\lambda'), F'(\lambda'')$ have opposite signs, $\varphi(a, \lambda')$ and $\varphi(a, \lambda'')$ are also of opposite signs; therefore, $\varphi(a, \lambda)$ has at least one zero in the interval (λ', λ''). However, as we have seen above, there cannot be more than one zero.

Furthermore, ν_1 cannot belong to the interval $(-\infty, \lambda_0]$. In fact, $\varphi(a, \lambda)$ is greater than zero for large negative λ. Hence, if $\varphi(a, \lambda)$ vanished on $(-\infty, \lambda_0]$, then this function might be positive on the interval (λ_0, μ_0). For the same reason, τ_0 is in the interval $(-\infty, \lambda_0]$. Indeed, for large negative λ, $\vartheta'(a, \lambda) > 0$. Therefore, if the first zero of the function $\vartheta'(a, \lambda)$ were in the interval $[\mu_0, \mu_1]$, then $\vartheta'(a, \lambda)$ could not be negative on (λ_0, μ_0). (Recall that φ and ϑ' are of opposite signs, $\varphi(a, \lambda)$ being positive on (λ_0, μ_0).) ∎

Theorem 1.4.3 *(Eigenfunction Zero Theorem.) Let $\eta_0(x), \eta_1(x), \eta_2(x), \ldots$ be the eigenfunctions of the periodic problem, and let $\xi_1(x), \xi_2(x), \ldots$ be those of the semiperiodic problem. Then*

1. *$\eta_0(x)$ cannot have zeros on the interval $[0, a]$,*

2. *$\eta_{2m+1}(x)$ and $\eta_{2m+2}(x)$ have precisely $2m + 2$ zeros on $[0, a)$, $m = 0, 1, 2, \ldots,$*

3. *$\xi_{2m+1}(x)$ and $\xi_{2m+2}(x)$ have precisely $2m + 1$ zeros on $[0, a)$, $m = 0, 1, 2, \ldots.$*

Remark Items 2 and 3 deal with the interval $[0, a)$, and not $[0, a]$, for, if $x = 0$ is a zero of the periodic (semi-periodic) solution, then $x = a$ is also a zero, and we should take only one zero into account.

Proof. Since $\varphi(x, \nu_1)$ has no zeros on $(0, a)$ and $\lambda_0 < \nu_1$, $\eta_0(x)$ cannot have more than one zero on $[0, a]$. Because of the periodicity, $\eta_0(x)$ should have an even number of zeros on $[0, a]$. Therefore, $\eta_0(x)$ has no zeros in $[0, a)$ and, relying on the periodicity, no zeros in $[0, a]$.

We now consider $\eta_{2m+1}(x)$. We have $\nu_{2m+1} < \lambda_{2m+1} \leq \nu_{2m+2}$. Furthermore, the function $\varphi(x, \nu_{2m+1})$ has $2m + 2$ zeros, and $\varphi(x, \nu_{2m+2})$ has $2m + 1$ zeros in $[0, a)$. Therefore, by the Sturm comparison theorem, $\eta_{2m+1}(x)$ has no fewer than $2m + 1$ and no more than $2m + 2$ zeros in $[0, a)$. Relying on the periodicity of the boundary condition, $\eta_{2m+1}(x)$ must have an even number of zeros, i.e., $2m + 2$, on $[0, a)$. The proof is similar both for $\eta_{2m+2}(x)$ and for $\xi_n(x)$. ∎

Replace $q(x)$ by $q(x + t)$ in the equation (4.1), where t is any real number. The function $q(x + t)$ is also periodic with the same period as $q(x)$. It is easy to see that $\eta(x + t)$ (resp. $\xi(x + t)$) are also eigenfunctions, as well as $\eta(x)$ (resp. $\xi(x)$), with the same eigenvalues. If $f(x)$ is a periodic or semi-periodic function with period a, then the function $f(x + t)$ also has this property for any real t. Therefore, $\eta_n(x + t)$, $n = 0, 1, 2, \ldots$, form a complete set of periodic eigenfunctions with eigenvalues $\lambda_0, \lambda_1, \lambda_2, \ldots$, and $\xi_n(x + t)$ a complete set of semi-periodic eigenfunctions with eigenvalues $\mu_1, \mu_2, \mu_3, \ldots.$

Thus, if we replace $q(x)$ by $q(x + t)$, the eigenvalues of the periodic and semi-periodic problems and, therefore, the function $F(\lambda)$ remain unaltered. However, the eigenvalues of the auxiliary problems (4.1), (4.20) and (4.1), (4.21), $\nu_k = \nu_k(t)$ and $\tau_k = \tau_k(t)$, depend on t, remaining inside the corresponding gaps.

Theorem 1.4.4 *As t varies (it being sufficient that t changes from 0 to a), $\nu_k(t)$ and $\tau_k(t)$, $k = 1, 2, 3, \ldots$, sweep out the whole gap.*

Proof. We have already noted above that $\nu_k(t)$ and $\tau_k(t)$ cannot lie outside the gaps. Since $\nu_k(t)$ is a continuous function, it suffices to show that the gap endpoints are reached.

Let x_0 be a zero of the function $\xi_{2k+1}(x)$, and $t = x_0$. Then $0 = \xi_{2k+1}(x_0) = -\xi_{2k+1}(x_0 + a) = 0$, i.e., $\xi_{2k+1}(x_0 + x)$ is an eigenfunction. Furthermore, $\xi_{2k+1}(x)$ has $2k$ zeros in the interval $(x_0, x_0 + a)$. Therefore, $\xi_{2k+1}(x)$ is an eigenfunction of the problem (4.1), (4.20) with the corresponding eigenvalue $\nu_{2k+1}(x_0)$. Hence, $\nu_{2k+1}(x_0) = \mu_{2k+1}$. Similarly, considering the function $\xi_{2k+2}(x)$, we prove that $\nu_{2k+1}(t)$ attains μ_{2k+2}. For $\nu_{2k+2}(t)$, the reasoning is similar. For $\tau_k(t)$, the proof is also similar, and the corresponding equalities hold if t is a zero of $\eta_k'(x)$.   ∎

Remark We can show that $\tau_0(t)$ have a finite lower bound, i.e., they do not sweep out the whole trivial gap $(-\infty, a)$.

Corollary 1.4.1 *If $q(x)$ is a smooth and periodic function, then the eigenvalues of the semi-periodic problem (μ_{2k}, μ_{2k+1}), $k = 1, 2, 3, \ldots$, have the same asymptotic expansion as ν_{2k+1}, whereas those of the periodic problem $(\lambda_{2k-1}, \lambda_{2k})$, $k = 1, 2, 3, \ldots$, have the same asymptotic behaviour as ν_{2k}.*

Proof. In fact, if $q(x)$ is differentiable sufficiently many times, we can prove the asymptotic expansion
$$\sqrt{\nu_n} = n + \frac{a_1}{n^3} + \frac{a_2}{n^5} + \cdots$$
for ν_n (cf. Chapter 5), where the numbers a_i are expressed as integrals of polynomials in $q(x)$ and its derivatives, taken over $(0, a)$. If $q(x)$ is a smooth and periodic function, then, for $q(x + t)$, a_i do not depend on t. The corollary then follows by Theorem 1.4.4.   ∎

1.5   Proof of the expansion theorem by the method of integral equations

1.5.1

In Sections 2 and 3, we have proved the existence of infinitely many eigenvalues by two different methods, one due to Liouville, the other to Sturm. The boundary-value problem

$$y'' + \{\lambda - q(x)\}y = 0 \tag{5.1}$$

$$y(0)\cos\alpha + y'(0)\sin\alpha = 0 \tag{5.2}$$

$$y(\pi)\cos\beta + y'(\pi)\sin\beta = 0 \tag{5.3}$$

is accordingly said to be *Sturm-Liouville*. However, neither Sturm nor Liouville could prove the completeness of the eigenfunctions. This was done by V. A. Steklov.

At present, there are several methods to prove that a set of eigenfunctions is complete, and the most important are the *method of integral equations* (or the *method of*

Green's function), the *method of contour integration* and the *finite-difference method.* Here, we discuss the method of integral equations; in Section 7, the method of contour integration. To the reader who wants to familiarize himself with the finite-difference method, we recommend the monograph by B. M. Levitan and I. S. Sargsjan (1975).

Returning to the problem (5.1–3), suppose that λ is a fixed complex number. Let $u(x, \lambda)$ be a solution of the equation (5.1), satisfying the initial conditions

$$u(0, \lambda) = \sin \alpha, \qquad u'(0, \lambda) = -\cos \alpha,$$

and let $v(x, \lambda)$ be another solution of (5.1), satisfying

$$v(\pi, \lambda) = \sin \beta, \qquad v'(\pi, \lambda) = -\cos \beta.$$

If $u(x, \lambda)$ and $v(x, \lambda)$ are linearly independent, i.e., if $u(x, \lambda)$ is not an eigenfunction of the problem (5.1–3) (if $u = cv$, u satisfies the boundary conditions (5.2) and (5.3), and is, therefore, an eigenfunction), then the Wronskian $W\{u, v\} \neq 0$. Conversely, if, for certain λ, the Wronskian is zero, then $u = cv$; hence, u is an eigenfunction. Thus, the eigenvalues of (5.1–3) coincide with the zeros of the Wronskian. Since the coefficient of the first derivative in the equation (5.1) is zero in this case, relying on the familiar Liouville formula, $W\{u, v\}$ does not depend on x, i.e., $W\{u, v\} = \omega(\lambda)$.

We introduce the function

$$G(x, t; \lambda) = \begin{cases} \dfrac{1}{\omega(\lambda)} u(x, \lambda) v(t, \lambda), & x \leq t \\[2mm] \dfrac{1}{\omega(\lambda)} u(t, \lambda) v(x, \lambda), & x \geq t \end{cases}$$

called *Green's function* for the boundary-value problem (5.1–3). It is symmetric with respect to x and t, and real-valued for real λ.

We show that the function

$$y(x, \lambda) = \int_0^\pi G(x, t; \lambda) f(t)\, dt \tag{5.4}$$

called a *resolvent* is a solution of the equation

$$y'' + \{\lambda - q(x)\} y = f(x) \tag{5.5}$$

(where $f(x) \not\equiv 0$ is a continuous function), satisfying the boundary conditions (5.2), (5.3). In fact, because of the definition of $G(x, t; \lambda)$, (5.4) can be rewritten in the form

$$y(x, \lambda) = \frac{1}{\omega(\lambda)} \left\{ v(x, \lambda) \int_0^x u(t, \lambda) f(t)\, dt + u(x, \lambda) \int_x^\pi v(t, \lambda) f(t)\, dt \right\}. \tag{5.4'}$$

Therefore,

$$
\begin{aligned}
y''(x,\lambda) &= \frac{1}{\omega(\lambda)}\left\{v''(x,\lambda)\int_0^x u(t,\lambda)f(t)\,dt + u''(x,\lambda)\int_x^\pi v(t,\lambda)f(t)\,dt\right\} \\
&\quad + \frac{1}{\omega(\lambda)}\{v'(x,\lambda)u(x,\lambda)f(x) - u'(x,\lambda)v(x,\lambda)f(x)\} \\
&= \frac{q(x)-\lambda}{\omega(\lambda)}\left\{v(x,\lambda)\int_0^x u(t,\lambda)f(t)\,dt + u(x,\lambda)\int_x^\pi v(t,\lambda)f(t)\,dt\right\} + f(x) \\
&= \{q(x)-\lambda\}y(x,\lambda) + f(x),
\end{aligned}
$$

i.e., $y'' + \{\lambda - q(x)\}y = f(x)$.

It can be directly checked that the function $y(x,\lambda)$ satisfies the boundary conditions (5.2), (5.3).

Thus, *if λ is not an eigenvalue of the homogeneous problem (5.1–3), then the nonhomogeneous problem (5.2), (5.3), (5.5) is solvable for any function $f(x)$ and the solution is given by the formula (5.4). Conversely, if λ is an eigenvalue of the homogeneous problem, then, generally speaking, the nonhomogeneous problem is unsolvable.*

If λ is not an eigenvalue of the homogeneous problem, then the nonhomogeneous problem (5.20), (5.5) has a unique solution. Indeed, it is obvious that the difference of two solutions of the nonhomogeneous problem is an eigenfunction of the homogeneous problem, and, by assumption, it must be identically zero.

We can assume that $\lambda = 0$ is not an eigenvalue. Otherwise, we take a fixed number η, and consider the boundary-value problem

$$
\begin{aligned}
y'' + \{(\lambda + \eta) - q(x)\}y &= 0 \\
y(0)\cos\alpha + y'(0)\sin\alpha &= 0 \\
y(\pi)\cos\beta + y'(\pi)\sin\beta &= 0
\end{aligned}
$$

with the same eigenfunctions as for the problem (5.1–3). All the eigenvalues are shifted through η to the right. It is evident that η can be selected so that 0 is not an eigenvalue of the new problem.

Put $G(x,t;0) = G(x,t)$. Then the function

$$
y(x) = \int_0^\pi G(x,t)f(t)\,dt
$$

is a solution of the equation $y'' - q(x)y = f(x)$, and satisfies the initial conditions (5.2), (5.3). We rewrite (5.5) in the form

$$
y'' - q(x) = f(x) - \lambda y.
$$

We can assert, relying on the above, that the problem (5.2), (5.5) is equivalent to the integral equation

$$
y(x) + \lambda \int_0^\pi G(x,t)y(t)\,dt = \int_0^\pi G(x,t)f(t)\,dt.
$$

In particular, the homogeneous problem ($f(x) \equiv 0$) is equivalent to the integral equation

$$y(x) + \lambda \int_0^\pi G(x,t)y(t)\, dt = 0. \tag{5.6}$$

1.5.2

Denote by $\lambda_0, \lambda_1, \lambda_2, \ldots, \lambda_n, \ldots$ the collection of all the eigenvalues of the problem (5.1–3), and by $v_0(x), v_1(x), v_2(x), \ldots, v_n(x), \ldots$ the corresponding normalized eigenfunctions. Consider the kernel

$$H(x,\xi) = \sum_{n=0}^\infty \frac{v_n(x)v_n(\xi)}{\lambda_n}.$$

By the asymptotic formulas for the eigenvalues, obtained in Section 2, the series for $H(x,\xi)$ converges absolutely and uniformly; therefore, $H(x,\xi)$ is continuous.

Consider the kernel

$$Q(x,\xi) = G(x,\xi) + H(x,\xi) = G(x,\xi) + \sum_{n=0}^\infty \frac{v_n(x)v_n(\xi)}{\lambda_n}$$

which is obviously continuous and symmetric.

By the familiar theorem in the theory of integral equations, any symmetric kernel $Q(x,\xi)$ which is not identically zero has at least one eigenfunction[4], i.e., there is a number λ_0 and a function $u(x) \not\equiv 0$ satisfying the equation

$$u(x) + \lambda_0 \int_0^\pi Q(x,\xi)u(\xi)\, d\xi = 0. \tag{5.7}$$

Thus, if we show that the kernel has no eigenfunctions, we obtain that $Q(x,\xi) \equiv 0$, i.e.,

$$G(x,\xi) = -\sum_{n=1}^\infty \frac{v_n(x)v_n(\xi)}{\lambda_n}. \tag{5.8}$$

Hence, to obtain the completeness of the eigenfunctions is now easy. It follows from the equation (5.6) that

$$\int_0^\pi G(x,\xi)v_n(\xi)\, d\xi = -\frac{1}{\lambda_n}v_n(x);$$

therefore,

$$\int_0^\pi Q(x,\xi)v_n(\xi)\, d\xi = 0,$$

i.e., the kernel $Q(x,\xi)$ is orthogonal to all eigenfunctions of the boundary-value problem (5.1–3).

---

[4]See I. G. Petrovsky (1948, p. 68).

Let $u(x)$ be a solution of the integral equation (5.7). We show that $u(x)$ is orthogonal to all $v_n(x)$. In fact, it follows from (5.7) that

$$
\begin{aligned}
0 &= \int_0^\pi u(x)v_n(x)\,dx + \lambda_0 \int_0^\pi v_n(x)\left\{\int_0^\pi Q(x,\xi)u(\xi)\,d\xi\right\}dx \\
&= \int_0^\pi u(x)v_n(x)\,dx + \lambda_0 \int_0^\pi u(\xi)\left\{\int_0^\pi Q(x,\xi)v_n(x)\,dx\right\}d\xi \\
&= \int_0^\pi u(x)v_n(x)\,dx;
\end{aligned}
$$

hence,

$$
0 = u(x) + \lambda_0 \int_0^\pi Q(x,\xi)u(\xi)\,d\xi = u(x) + \lambda_0 \int_0^\pi G(x,\xi)u(\xi)\,d\xi,
$$

i.e., $u(x)$ is an eigenfunction of the boundary-value problem (5.1–3). Since it is orthogonal to all $v_n(x)$, it is also orthogonal to itself, with the consequence that $u(x) \equiv 0$ and $Q(x,\xi) = 0$. The formula (5.8) is thus proved.

Theorem 1.5.1 *(Expansion Theorem.) If $f(x)$ has a continuous second derivative and satisfies the boundary conditions (5.2), (5.3), then $f(x)$ can be expanded into an absolutely and uniformly convergent Fourier series of eigenfunctions of the boundary-value problem (5.1–3), viz.*

$$
f(x) = \sum_{n=0}^\infty a_n v_n(x), \qquad a_n = \int_0^\pi f(x)v_n(x)\,dx. \tag{5.9}
$$

Proof. Put $f''(x) - q(x)f(x) = h(x)$. Then, relying on (5.4) and (5.8), we have

$$
f(x) = \int_0^\pi G(x,\xi)h(\xi)\,d\xi = -\sum_{n=0}^\infty v_n(x)\frac{1}{\lambda_n}\int_0^\pi v_n(\xi)h(\xi)\,d\xi \equiv \sum_{n=0}^\infty a_n v_n(x).
$$

It follows from the orthogonality and normalization of the functions $v_n(x)$ that

$$
a_n = \int_0^\pi f(x)v_n(x)\,dx.
$$

∎

Theorem 1.5.2 *For any square-integrable function $f(x)$ in the interval $[0,\pi]$, the Parseval equality*

$$
\int_0^\pi f^2(x)\,dx = \sum_{n=0}^\infty a_n^2 \tag{5.10}
$$

holds.

Proof. If $f(x)$ satisfies the conditions of Theorem 1.5.1, then (5.10) follows immediately from the uniform convergence of the series (5.9). Indeed,

$$\int_0^\pi f^2(x)\,dx = \sum_{n=0}^\infty a_n \int_0^\pi f(x)v_n(x)\,dx = \sum_{n=0}^\infty a_n^2. \tag{5.11}$$

The extending the Parseval equality to arbitrary square-integrable functions is done by the following method.

Let $f(x)$ be an arbitrary square-integrable function in the interval $[0,\pi]$. There exists a sequence of twice (even infinitely) differentiable functions $f_k(x)$ converging in mean-square to $f(x)$. We can assume that the $f_k(x)$ are identically zero in the neighborhoods of the points $x = 0$ and $x = \pi$. By (5.11), we have

$$\int_0^\pi \{f_k(x) - f_l(x)\}^2\,dx = \sum_{n=0}^\infty \{a_n^{(k)} - a_n^{(l)}\}^2, \tag{5.12}$$

where

$$a_n^{(k)} = \int_0^\pi f_k(x)v_n(x)\,dx.$$

As $k, l \to \infty$, the left-hand side of (5.12) tends to zero. Therefore, the right-hand side also tends to zero. It follows from the Cauchy-Schwarz inequality that

$$|a_n - a_n^{(k)}| \le \left\{ \int_0^\pi [f(x) - f_k(x)]^2\,dx \right\}^{1/2}.$$

Hence, it follows from the convergence in the mean of $f_k(x)$ to $f(x)$ that

$$\lim_{k \to \infty} a_n^{(k)} = a_n, \quad n = 0, 1, 2, \dots.$$

Let $N > 0$ be a fixed positive integer. It follows from (5.12) that

$$\sum_{n=0}^N \{a_n^{(k)} - a_n^{(l)}\}^2 \le \int_0^\pi \{f_k(x) - f_l(x)\}^2\,dx.$$

Letting $k \to \infty$, we obtain

$$\sum_{n=0}^N \{a_n - a_n^{(l)}\}^2 \le \int_0^\pi \{f(x) - f_l(x)\}^2\,dx,$$

whence, if $N \to \infty$,

$$\sum_{n=0}^\infty \{a_n - a_n^{(l)}\}^2 \le \int_0^\pi \{f(x) - f_l(x)\}^2\,dx.$$

In particular, it follows (by the Minkowski inequality) that the series $\sum_{n=0}^{\infty} a_n^2$ converges. Since

$$\left| \sum_{n=0}^{\infty} a_n^2 - \sum_{n=0}^{\infty} \{a_n^{(l)}\}^2 \right| = \left| \sum_{n=0}^{\infty} \{a_n - a_n^{(l)}\}\{a_n + a_n^{(l)}\} \right|$$

$$\leq \left(\sum_{n=0}^{\infty} |a_n - a_n^{(l)}|^2 \right)^{1/2} \left(\sum_{n=0}^{\infty} |a_n + a_n^{(l)}|^2 \right)^{1/2},$$

we infer from the above that $\sum_{n=0}^{\infty} \{a_n^{(l)}\}^2 \to \sum_{n=0}^{\infty} a_n^2$ as $l \to \infty$. On the other hand, we derive from the convergence in the mean of $f_l(x)$ to $f(x)$ that

$$\int_0^{\pi} f_l^2(x)\, dx \to \int_0^{\pi} f^2(x)\, dx.$$

Therefore, passing to the limit (as $l \to \infty$) in the equality

$$\int_0^{\pi} f_l^2(x)\, dx = \sum_{n=0}^{\infty} \{a_n^{(l)}\}^2,$$

we obtain (5.10).   ∎

1.5.3

We now return to the formula

$$y(x, \lambda) = \int_0^{\pi} G(x, t; \lambda) f(t)\, dt \tag{5.13}$$

whose right-hand side has been called the resolvent. The resolvent is known to exist for all λ which are not eigenvalues. Now, we show how to obtain an expansion into a Fourier series of the resolvent, given the expansion of $f(x)$.

Since the function $y(x, \lambda)$ determined by the formula (5.13) satisfies the boundary conditions (5.2), (5.3), integrating by parts, we find

$$\int_0^{\pi} \{y''(x, \lambda) - q(x)y(x, \lambda)\}v_n(x)\, dx = \int_0^{\pi} \{v_n''(x) - q(x)v_n(x)\}y(x, \lambda)\, dx$$

$$= -\lambda_n \int_0^{\pi} y(x, \lambda)v_n(x)\, dx \equiv -\lambda_n d_n(\lambda). \tag{5.14}$$

Let

$$y(x, \lambda) = \sum_{n=0}^{\infty} d_n(\lambda)v_n(x), \qquad a_n = \int_0^{\pi} f(x)v_n(x)\, dx.$$

Since $y(x, \lambda)$ satisfies the equation $y'' + \{\lambda - q(x)\}y = f(x)$, we have

$$a_n = \int_0^{\pi} \{y'' + [\lambda - q(x)]y\}v_n(x)\, dx = -\lambda_n d_n(\lambda) + \lambda d_n(\lambda)$$

by 5.14); hence, $d_n(\lambda) = a_n/(\lambda - \lambda_n)$ and the expansion of the resolvent is

$$y(x, \lambda) = \int_0^\pi G(x, t; \lambda) f(t)\, dt = \sum_{n=0}^\infty \frac{a_n}{\lambda - \lambda_n} v_n(x). \tag{5.15}$$

An important formula can now be derived from the above. Substituting

$$a_n = \int_0^\pi f(t) v_n(t)\, dt$$

on the right-hand side, we see that

$$\int_0^\pi G(x, t; \lambda) f(t)\, dt = \sum_{n=0}^\infty \frac{v_n(x)}{\lambda - \lambda_n} \int_0^\pi f(t) v_n(t)\, dt.$$

Since $f(t)$ is arbitrary,

$$G(x, t; z) = \sum_{n=0}^\infty \frac{v_n(x) v_n(t)}{z - \lambda_n}.$$

Putting $t = x$, integrating with respect to x from 0 to π, and taking into account that the eigenfunctions $v_n(x)$ are normalized, we obtain

$$\int_0^\pi G(x, x; z)\, dx = \sum_{n=0}^\infty \frac{1}{z - \lambda_n}. \tag{5.16}$$

Put $N(\lambda) = \sum_{0 \le \lambda_n < \lambda} 1$, $N(\lambda)$ is the number of eigenvalues λ_n less than λ. Then (5.16) takes the form

$$\int_0^\pi G(x, x; z)\, dx = \int_0^\infty \frac{dN(\lambda)}{z - \lambda} \tag{5.17}$$

which is called the *Carleman equation.*

1.6   Proof of the expansion theorem in the periodic case

To prove that the functions satisfying the periodic conditions can be expanded into a series of eigenfunctions of the boundary-value problem

$$y'' + \{\lambda - q(x)\} y = 0 \tag{6.1}$$

$$y(0) = y(\pi), \quad y'(0) = y'(\pi) \tag{6.2}$$

we could proceed, as in Section 5, on the basis of the asymptotic formulas for the eigenvalues and eigenfunctions, a priori obtaining more exact approximations. However, this method would be very impractical. Instead, we use the applicability of the argument in Section 5 if only the the uniform and absolute convergence of

$$\sum_{n=0}^\infty \frac{v_n(x) v_n(t)}{\lambda_n} \tag{6.3}$$

is proved.

By Theorem 1.4.2 and the asymptotic formulas (2.15), (2.19) we get some information about the eigenvalues λ_n of the problem (6.1), (6.2), viz. $\lambda_n/n^2 \to 1$. Therefore, to prove that (6.3) converges uniformly and absolutely, it suffices to prove that the eigenfunctions $v_n(x)$ are bounded.

For simplicity, we exclude the case of a double eigenvalue. Put $\lambda = s^2$, and rewrite the equation (6.1) as $y'' + s^2 y = q(x)y$. Applying the method of variation of the constants, we obtain

$$y(x, \lambda) = c_1 \sin sx + c_2 \cos sx + \frac{1}{s} \int_0^x y(t, \lambda) q(t) \sin s(x - t)\, dt,$$

where the coefficients c_1 and c_2 are determined, up to a constant multiplier, by the relation

$$c_2 = c_1 \sin s\pi + c_2 \cos s\pi + \frac{1}{s} \int_0^\pi y(t, \lambda) q(t) \sin s(\pi - t)\, dt.$$

We select them so that the function $y(x, \lambda)$ may be normalized, viz.

$$\int_0^\pi y^2(x, \lambda)\, dx = 1,$$

and then, for $s_n^2 = \lambda_n$, a normalized eigenfunction. Indeed, since double roots are excluded, λ_n is never equal to any eigenvalue μ_n, for all n (see Section 4); therefore, no solution of the equation $y'' + \{\lambda - q(x)\}y = 0$ (except the trivial solution $y \equiv 0$) can be zero at the endpoints of the interval. We conclude that, for any solution y other than $v_n(x)$, for which $y(0) = y(\pi)$, we can select c such that $v_n(0) - cy(0) = v_n(\pi) - cy(\pi) = 0$, and then $v_n(x) - cy(x) \equiv 0$. Hence, for $s_n^2 = \lambda_n$, the normalized function $y(x, \lambda)$ is identical with the normalized eigenfunction.

Applying the Cauchy-Schwarz inequality, we now obtain

$$\left(\int_0^x y(t, \lambda) q(t) \sin s(x - t)\, dt \right)^2 \leq \int_0^\pi y^2(t, \lambda)\, dt \cdot \int_0^\pi q^2(t)\, dt \leq C.$$

Therefore,

$$y(x, \lambda) = c_1(s) \sin sx + c_2(s) \cos sx + O(s^{-1}).$$

Transposing $O(s^{-1})$ on the left-hand side, squaring and integrating, we obtain

$$\int_0^\pi \left\{ y(x, \lambda) - O\left(\frac{1}{s}\right) \right\}^2 dx = 1 - 2 \int_0^\pi y(x, \lambda) O\left(\frac{1}{s}\right) dx + \int_0^\pi O^2\left(\frac{1}{s}\right) dx$$

$$= c_1^2 \left[\frac{\pi}{2} + O\left(\frac{1}{s}\right) \right] + c_2^2 \left[\frac{\pi}{2} + O\left(\frac{1}{s}\right) \right] + 2c_1 c_2 O\left(\frac{1}{s}\right). \quad (6.4)$$

Furthermore,

$$\left(\int_0^\pi y(x, \lambda) O\left(\frac{1}{s}\right) dx \right)^2 \leq \int_0^\pi y^2(x, \lambda)\, dx \cdot \int_0^\pi O^2\left(\frac{1}{s}\right) dx = O\left(\frac{1}{s^2}\right).$$

Since the numbers c_1 and c_2 are symmetrically involved in the equality (6.4), we can assume, e.g., that $|c_1| \geq |c_2|$. Putting $\eta = (c_2/c_1)^2 \leq 1$, we obtain

$$c_1^2 \left(\frac{\pi}{2} + \eta \frac{\pi}{2} + O\left(\frac{1}{s}\right) \right) \leq 1 + O\left(\frac{1}{s}\right),$$

which shows that c_1 (and also c_2) is bounded. This exactly proves that the eigenfunctions are bounded. Thus, we have shown that the series (6.3) converges uniformly and absolutely. All conclusions of the preceding section now remain valid; in particular, the formula

$$G(x,t) = - \sum_{n=0}^{\infty} \frac{v_n(x)v_n(t)}{\lambda_n}$$

holds.

Theorem 1.6.1 *Each continuous function $f(x)$ with continuous second derivative can be expanded into a series of the eigenfunctions of the problem (6.1), (6.2) if it satisfies the boundary conditions (6.2).*

The proof coincides verbatim with that of Theorem 1.5.1.

1.7   Proof of the expansion theorem by the method of contour integration

First, some formal reasoning.

Assume that the eigenfunction expansion of the form (5.9) has already been proved. As is shown in Section 5, the resolvent $y(x, \lambda)$ determined by (5.4) then admits the eigenfunction expansion (5.15), and satisfies (5.5). Solving this, we can determine the terms in the expansion of the function $f(x)$ if we find the residues of the poles of $y(x, \lambda)$, as follows from (5.15). It has been shown in Section 5, Item 1, that the zeros of the function $\omega(\lambda)$ coincide with the eigenvalues of the problem (5.1–3), and are, therefore, simple. Let $v(x, \lambda_n) = k_n u(x, \lambda_n)$. It follows from the boundary conditions that the constant k_n is finite, and different from zero. Therefore, $y(x, \lambda)$ has the residue

$$\frac{k_n}{\omega'(\lambda_n)} u(x, \lambda_n) \int_0^\pi u(t, \lambda_n) f(t)\, dt$$

for $\lambda = \lambda_n$.

The above formal reasoning shows that the expansion

$$f(x) = \sum_{n=0}^{\infty} \frac{k_n}{\omega'(\lambda_n)} u(x, \lambda_n) \int_0^\pi u(t, \lambda_n) f(t)\, dt$$

should hold.

The general method to be used in the sequel is as follows : The function $y(x, \lambda)$ is determined by the formula (5.4); integrating over expanding paths in the complex λ-plane, we obtain $f(x)$ in the limit; however, contracting the integration path onto the real axis with the singularities of $y(x, \lambda)$, we obtain the eigenfunction expansion of $f(x)$.

Theorem 1.7.1 *Let a function* $f(x) \in \mathcal{L}(0, \pi)$. *Then, for* $0 < x < \pi$, *the eigenfunction expansion for the boundary-value problem (5.1–3) has the same type of convergence as the usual trigonometric Fourier series; in particular, this expansion converges to* $\frac{1}{2}\{f(x + 0) + f(x - 0)\}$ *if* $f(x)$ *has bounded variation in the neighborhood of the point* x.

Proof. Consider the integral

$$\frac{1}{2\pi i} \int_C y(x, \lambda) \, d\lambda \qquad (7.1)$$

($y(x, \lambda)$ being determined by the formula (5.4′)) taken over a closed contour C in the λ-plane, described as follows. Let $\lambda = s^2$. The upper half of the contour is associated with a quarter of the boundary of the square in the s-plane, formed by the straight line segments ($s = \sigma + it$)

$$\sigma = n + \frac{1}{2}, \quad 0 \le t \le n + \frac{1}{2},$$

$$t = n + \frac{1}{2}, \quad 0 \le \sigma \le n + \frac{1}{2}.$$

The lower half is constructed symmetrically about the real axis. It is easy to see that (7.1) is equal to the finite sum of the series of eigenfunctions. In fact, there are only finitely many eigenvalues inside the integration path, i.e., finitely many poles of the function $y(x, \lambda)$. By the familiar theorem in the theory of residues, (7.1) then equals the sum of the residues at the poles of $y(x, \lambda)$, the latter being terms in the expansion of $f(x)$.

Consider the case $\sin \alpha \ne 0$, $\sin \beta \ne 0$. By Lemma 1.2.2,

$$
\begin{aligned}
u(y, \lambda) &= \cos sy \cdot \sin \alpha + O(|s|^{-1} e^{ty}) \\
v(x, \lambda) &= \cos s(\pi - x) \cdot \sin \beta + O(|s|^{-1} e^{t(\pi - x)}) \\
u'(y, \lambda) &= -s \sin sy \cdot \sin \alpha + O(e^{ty}) \\
v'(x, \lambda) &= s \sin s(\pi - x) \cdot \sin \beta + O(e^{t(\pi - x)})
\end{aligned}
$$

on the boundary segments. Then

$$\omega(\lambda) = s \sin s\pi \cdot \sin \alpha \cdot \sin \beta + O(e^{t\pi}).$$

On the same segments, $|\sin s\pi| > A e^{t\pi}$; therefore,

$$\frac{1}{\omega(\lambda)} = \frac{1}{s \sin s\pi \cdot \sin \alpha \cdot \sin \beta} \left\{1 + O\left(\frac{1}{|s|}\right)\right\},$$

and

$$\frac{v(x,\lambda)u(y,\lambda)}{\omega(\lambda)} = \frac{\cos\{s(\pi-x)\}\cos sy}{s\sin s\pi} + O\left(\frac{e^{t(y-x)}}{|s|^2}\right)$$

$$\frac{v(x,y)}{\omega(\lambda)}\int_0^x u(y,\lambda)f(\lambda)\,dy = \int_0^x \frac{\cos\{s(\pi-x)\}\cos sy}{s\sin s\pi}f(y)\,dy$$

$$+O\left\{\frac{1}{|s|^2}\int_0^x e^{t(y-x)}|f(y)|\,dy\right\}.$$

Let $0 < \delta < x$. Then the last term in this formula is reduced to

$$O\left(\frac{e^{-\delta t}}{|s|^2}\right) + O\left(\frac{1}{|s|^2}\int_{x-\delta}^x |f(y)|\,dy\right).$$

Since $d\lambda/ds = 2s$, the latter term makes the addend

$$\int_C \left\{e^{-\delta t} + \int_{x-\delta}^x |f(y)|\,dy\right\}O\left(\frac{1}{|s|}\right)|ds| = O\left\{\int_C e^{-\delta t}\left|\frac{ds}{s}\right|\right\} + O\left\{\int_{x-\delta}^x |f(y)|\,dy\right\}$$

appear in (7.1). The second term on the right can be made as small as we please on account of δ, while the first term tends to zero for fixed δ as $n \to \infty$, being of the form

$$O\left\{\frac{1}{n}\int_0^{n+1/2} e^{-\delta t}\,dt\right\} + O\left\{\frac{1}{n}\int_0^{n+1/2} e^{-\delta(n+1/2)}\,d\sigma\right\} = O\left(\frac{1}{n\delta}\right) + O(e^{-\delta(n+1/2)}).$$

A similar study can be made for the other addend in $y(x,\lambda)$, where $x \le y \le \pi$. Finally, we obtain that (7.1) equals

$$\frac{1}{2\pi i}\int_C \left\{\int_0^x \frac{\cos\{s(\pi-x)\}\cos sy}{s\sin s\pi}f(y)\,dy + \int_x^\pi \frac{\cos\{s(\pi-y)\}\cos sx}{s\sin s\pi}f(y)\,dy\right\}d\lambda + o(1).$$

$$(7.2)$$

The first term in this formula precisely coincides with the expression which would be obtained for the corresponding problem in expanding $f(x)$ into its Fourier cosine series, and, therefore, equals a partial sum of this series. Certainly, this can be verified directly by residue-theoretic means. The theorem on the relation of the eigenfunction expansion to the expansion into a Fourier series then follows. If $\sin\alpha = 0$ or $\sin\beta = 0$, Theorem 1.7.1 is proved similarly.

In the case where $f(x)$ has bounded variation in the neighborhood of the point x, the above result can be obtained directly without relying on the theory of Fourier series. First of all, we have

$$\frac{\cos\{s(\pi-x)\}\cos sy}{\sin s\pi} = O\left\{\frac{e^{t(\pi-x)+ty}}{e^{t\pi}}\right\} = O\{e^{-t(x-y)}\}.$$

Therefore, the part of the expression (7.2) with $0 \leq y \leq x - \delta$ tends to zero as before. For $x - \delta \leq y \leq x$, we have

$$\frac{\cos\{s(\pi - x)\} \cos sy}{\sin s\pi} = \frac{e^{-is(\pi-x)}\{1 + O(e^{-2t(\pi-x)})\} e^{-isy}\{1 + O(e^{-2ty})\}}{2ie^{-is\pi}\{1 + O(e^{-2t\pi})\}}$$

$$= -\frac{1}{2}ie^{is(x-y)}\{1 + O(e^{-2\delta t})\}$$

if $\delta < \pi - x$. The term involving $O(e^{-2\delta t})$ as before leads to an addend which tends to zero. The principal term in (7.2) is given by the integral

$$-\frac{i}{2s}\int_{x-\delta}^{x} e^{is(x-y)} f(y)\, dy$$

$$= -\frac{i}{2s}f(x - 0)\int_{x-\delta}^{x} e^{is(x-y)}\, dy + \frac{i}{2s}\int_{x-\delta}^{x} e^{is(x-y)}\{f(x - 0) - f(y)\}\, dy$$

$$= f(x - 0)\frac{1 - e^{is\delta}}{2s^2} + \frac{i}{2s}\int_{x-\delta}^{x} e^{is(x-y)}\{f(x - 0) - f(y)\}\, dy.$$

The first addend, $(1/2\lambda)f(x - 0)$, gives $\frac{1}{2}f(x - 0)$ as required. The integral of the second term yields $O(e^{-\delta t}|\lambda|^{-1})$ and, therefore, tends to zero. Furthermore, since $f(x)$ has bounded variation, we can put $f(x - 0) - f(y) = g(y) - h(y)$, where $g(y)$ and $h(y)$ are two positive and monotonically decreasing functions tending to zero as $y \to x$. By the second mean-value theorem,

$$\int_{x-\delta}^{x} \text{Real}\{e^{is(x-y)}\}g(y)\, dy = g(x - \delta)\int_{x-\delta}^{\xi} \text{Real}\{e^{is(x-y)}\}\, dy$$

$$= g(x - \delta)\, \text{Real}\int_{x-\delta}^{\xi} e^{is(x-y)}\, dy = O\left\{\frac{g(x - \delta)}{|s|}\right\}.$$

This term provides for the addend

$$\int O\left(\frac{g(x - \delta)}{|\lambda|}\right) d|\lambda| = O\{g(x - \delta)\}$$

in (7.1), which can be made as small as we please on account of δ. We study the terms involving $h(y)$, and those with $y > x$, in the same way.   ∎

Notes

Sections 1 and 2 The transformation of the general second-order equation to the canonical form (1.1) and the asymptotic formulas for the eigenvalues and eigenfunctions are classical and mostly due to Liouville.

Section 3 The results are classical and were obtained by Sturm.

Section 4 The essential results in the theory were established by G. Hamel (1913), O. Haupt (1919) and E. Titchmarsh (1950); see also G. Hoheisel (1951).

Section 5 The results are classical. Green's function was first systematically applied to the study of boundary-value problems by D. Hilbert (1904). Note that the completeness and expansion theorems for the classical Sturm-Liouville problem were first proved by V. A. Steklov (1896).

Section 6 See, e.g., G. Hoheisel (1951).

Section 7 The idea to apply contour integration to the proof of the expansion theorem is due to Cauchy. This section was written on the basis of Section 1.9 of the monograph of E. Titchmarsh (1962). See the monograph of B. M. Levitan and I. S. Sargsjan (1975) for other methods of proof of the expansion theorem.

Chapter 2

SPECTRAL THEORY IN THE SINGULAR CASE

2.1 The Parseval equation on the half-line

2.1.1

As was noted in Section 1 of Chapter 1, a Sturm-Liouville problem is said to be *singular* if either the interval $[a, b]$ is infinite or if the function $q(x)$ on $[a, b]$ is not summable (or both). Here, we obtain the expansion theorem for the singular problem, considering it as the limit of regular ones.

We start by considering the case where the interval is the semi-axis $[0, \infty)$ and $q(x)$ is continuous in each finite interval $0 \le x \le b$. Thus, we consider the equation

$$y'' + \{\lambda - q(x)\}y = 0, \quad 0 \le x < \infty. \tag{1.1}$$

Let α be an arbitrary real number, and let $y(x, \lambda)$ be a solution of (1.1), satisfying the initial conditions

$$y(0, \lambda) = \sin \alpha, \qquad y'(0, \lambda) = -\cos \alpha. \tag{1.2}$$

It is obvious that $y(x, \lambda)$ satisfies the boundary condition

$$y(0, \lambda) \cos \alpha + y'(0, \lambda) \sin \alpha = 0. \tag{1.3}$$

Let b be an arbitrary positive number (increasing indefinitely in the sequel), and β an arbitrary real number. Subjoin the boundary condition

$$y(b, \lambda) \cos \beta + y'(b, \lambda) \sin \beta = 0 \tag{1.4}$$

to the problem (1.1), (1.3). Then (1.1), (1.3), (1.4) is a regular Sturm-Liouville problem. Let $\lambda_{n,b}$ be its eigenvalues, and let $y_{n,b}(x) = y(x, \lambda_{n,b})$ be the corresponding eigenfunctions satisfying the initial conditions (1.2). If $f(x) \in \mathcal{L}^2(0, b)$ and

$$\alpha_{n,b}^2 = \int_0^b y_{n,b}^2(x)\, dx,$$

38

then, by the Parseval equation (5.10) in Chapter 1 (the eigenfunctions

$$v_n(x) = \frac{1}{\alpha_{n,b}} y_{n,b}(x)$$

being normalized), we have

$$\int_0^b f^2(x)\,dx = \sum_{n=0}^{\infty} \frac{1}{\alpha_{n,b}^2} \left(\int_0^b f(x) y_{n,b}(x)\,dx \right)^2.$$ (1.5)

If we introduce the monotonically increasing step function

$$\rho_b(\lambda) = \begin{cases} - \displaystyle\sum_{\lambda < \lambda_{n,b} \le 0} \frac{1}{\alpha_{n,b}^2}, & \lambda \le 0 \\[3ex] \displaystyle\sum_{0 < \lambda_{n,b} \le \lambda} \frac{1}{\alpha_{n,b}^2}, & \lambda > 0 \end{cases}$$

the equality (1.5) can be written in the form

$$\int_0^b f^2(x)\,dx = \int_{-\infty}^{\infty} F^2(\lambda)\,d\rho_b(\lambda),$$ (1.6)

where

$$F(\lambda) = \int_0^b f(x) y(x,\lambda)\,dx.$$

We show that the Parseval equation for the singular problem (1.1), (1.3) is obtained from (1.6) as $b \to \infty$. To this end, we prove the following.

Lemma 2.1.1 *For any positive number N, there exists a constant positive number $A = A(N)$ not depending on b, so that*

$$\bigvee_{-N}^{N} \{\rho_b(\lambda)\} = \sum_{-N < \lambda_{n,b} < N} \frac{1}{\alpha_{n,b}^2} = \rho_b(N) - \rho_b(-N) < A,$$ (1.7)

i.e., in each finite interval in the domain of λ, the variation of the functions $\rho_b(\lambda)$ is uniformly bounded (with respect to b).

Proof. We first consider the case $\sin \alpha \ne 0$. Due to the continuity of the function $y(x,\lambda)$ in the domain $-N \le \lambda \le N$, $0 \le x \le a$ (a being an arbitrary fixed positive number) and the condition $y(0,\lambda) = \sin \alpha$, there exists a small positive number h such that, for $|\lambda| < N$,

$$\left(\frac{1}{h} \int_0^h y(x,\lambda)\,dx \right)^2 > \frac{1}{2} \sin^2 \alpha.$$ (1.8)

If we apply the equality (1.6) to the function

$$f_h(x) = \begin{cases} 1/h, & 0 \le x < h \\ \\ 0, & x \ge h \end{cases}$$

then, using (1.8), we obtain

$$\int_0^h f_h^2(x)\, dx = \frac{1}{h} = \int_{-\infty}^{\infty} \left\{ \frac{1}{h} \int_0^h y(x,\lambda)\, dx \right\}^2 d\rho_b(\lambda) \ge \int_{-N}^{N} \left\{ \frac{1}{h} \int_0^h y(x,\lambda)\, dx \right\}^2 d\rho_b(\lambda)$$

$$> \frac{1}{2} \sin^2 \alpha \cdot \int_{-N}^{N} d\rho_b(\lambda) = \frac{1}{2} \sin^2 \alpha \cdot \{\rho_b(N) - \rho_b(-N)\},$$

and the inequality (1.7) follows.

When $\sin \alpha = 0$, we obtain the estimate (1.7), applying the Parseval equation (1.6) to the function

$$f_h(x) = \begin{cases} 1/h^2, & 0 \le x < h \\ \\ 0, & x \ge h \end{cases}$$

and the lemma is thus proved.   ∎

We now show that, using Lemma 2.1.1 and the familiar theorems on passage to the limit under the Stieltjes integral sign, we can derive the Parseval equation for the problem (1.1), (1.3).

First, let the function $f_n(x)$ vanish outside the interval $0 \le x \le n$, $n < b$, let it have a continuous second derivative, and satisfy the boundary condition (1.3). Applying the Parseval equation (1.6), we obtain

$$\int_0^n f_n^2(x)\, dx = \int_{-\infty}^{\infty} F_n^2(\lambda)\, d\rho_b(\lambda), \tag{1.9}$$

where

$$F_n(\lambda) = \int_0^n f_n(x) y(x,\lambda)\, dx.$$

Since both $f_n(x)$ and $y(x,\lambda)$ satisfy the boundary condition (1.3) and $f_n(x)$ is identically zero in the neighborhood of the point b, it follows from Green's identity that

$$F_n(\lambda) = -\frac{1}{\lambda} \int_0^b f_n(x)\{y''(x,\lambda) - q(x)y(x,\lambda)\}\, dx$$

$$= -\frac{1}{\lambda} \int_0^b y(x,\lambda)\{f_n''(x) - q(x)f_n(x)\}\, dx.$$

Therefore, for arbitrary finite $N > 0$,

$$\int_{|\lambda|>N} F_n^2(\lambda)\, d\rho_b(\lambda) \le \frac{1}{N^2} \int_{|\lambda|>N} \left\{ \int_a^b y(x,\lambda)[f_n''(x) - q(x)f_n(x)]\, dx \right\}^2 d\rho_b(\lambda)$$

$$< \frac{1}{N^2} \int_{-\infty}^{\infty} \left\{ \int_0^b y(x,\lambda)[f_n''(x) - q(x)f_n(x)]\, dx \right\}^2 d\rho_b(\lambda)$$

$$= \frac{1}{N^2} \int_0^n \{f_n''(x) - q(x)f_n(x)\}^2\, dx.$$

We obtain from the last estimate and from the equality (1.9) that

$$\left| \int_0^n f_n^2(x)\, dx - \int_{-N}^{N} F_n^2(\lambda)\, d\rho_b(\lambda) \right| < \frac{1}{N^2} \int_0^n \{f_n''(x) - q(x)f_n(x)\}^2\, dx. \qquad (1.10)$$

By Lemma 2.1.1, the set of monotone functions $\{\rho_b(\lambda)\}$ $(-N \le \lambda \le N)$ is bounded. Hence, we can select a sequence $\{b_k\}$ for which the functions $\rho_b(\lambda)$ converge to a monotone function $\rho(\lambda)$.

Passing to the limit in the inequality (1.10) with respect to b_k, we obtain

$$\left| \int_0^n f_n^2(x)\, dx - \int_{-N}^{N} F_n^2(\lambda)\, d\rho(\lambda) \right| \le \frac{1}{N^2} \int_0^n \{f_n''(x) - q(x)f_n(x)\}^2\, dx.$$

Finally, letting $N \to \infty$, we get

$$\int_0^n f_n^2(x)\, dx = \int_{-\infty}^{\infty} F_n^2(\lambda)\, d\rho(\lambda),$$

which is the Parseval equation for functions satisfying the above conditions.

Now, let $f(x)$ be an arbitrary square-integrable function $f(x) \in \mathcal{L}^2(0, \infty)$. It is known that there exists a sequence of functions $\{f_n(x)\}$ satisfying the preceding conditions (i.e., they vanish outside the interval $[0, n]$, satisfy the boundary condition (1.3), and have a continuous second derivative) and such that

$$\lim_{n \to \infty} \int_0^{\infty} \{f(x) - f_n(x)\}^2\, dx = 0.$$

Since

$$\int_0^{\infty} \{f_n(x) - f_m(x)\}^2\, dx \to 0, \quad n, m \to \infty,$$

we have

$$\int_{-\infty}^{\infty} \{F_n(\lambda) - F_m(\lambda)\}^2\, d\rho(\lambda) = \int_0^{\infty} \{f_n(x) - f_m(x)\}^2\, dx \to 0.$$

The space of square-integrable functions (with respect to the monotone function $\rho(\lambda)$) being complete, it follows from the last equality that a limit function $F(\lambda)$ exists and that it satisfies the Parseval equation

$$\int_0^{\infty} f^2(x)\, dx = \int_{-\infty}^{\infty} F^2(\lambda)\, d\rho(\lambda).$$

It remains to show that the functions

$$F_n(\lambda) = \int_0^n f(x)y(x,\lambda)\,dx$$

converge in the mean to $F(\lambda)$ as $n \to \infty$.

Let $g(x)$ be another square-integrable function, and let $G(x)$ be constructed from $g(x)$ in the way $F(\lambda)$ is from f. It is obvious that

$$\int_0^\infty \{f(x) - g(x)\}^2\,dx = \int_{-\infty}^\infty \{F(\lambda) - G(\lambda)\}^2\,d\rho(\lambda).$$

Now, putting $g(x) = f(x)$ for $0 \le x \le n$, and $g(x) = 0$ for $x > n$, we obtain

$$\int_{-\infty}^\infty \{F(\lambda) - F_n(\lambda)\}^2\,d\rho(\lambda) = \int_n^\infty f^2(x)\,dx \to 0, \quad n \to \infty,$$

which proves the mean convergence of $F_n(\lambda)$ to $F(\lambda)$.

We have thus proved the following.

Theorem 2.1.1 *Let $f(x) \in \mathcal{L}^2(0, \infty)$. There exist a monotonically increasing function $\rho(\lambda)$ not depending on the function $f(x)$ and a function $F(\lambda)$ (the generalized Fourier transform of $f(x)$) such that the equality*

$$\int_0^\infty f^2(x)\,dx = \int_{-\infty}^\infty F^2(\lambda)\,d\rho(\lambda)$$

holds.

The function $\rho(\lambda)$ is often said to be *spectral.* We shall see in the sequel that it need not be unique.

The function $F(\lambda)$ is the mean-square limit of the sequence of continuous functions

$$F_n(\lambda) = \int_0^n f(x)y(x,\lambda)\,dx,$$

i.e.,

$$\lim_{n\to\infty} \int_{-\infty}^\infty \{F(\lambda) - F_n(\lambda)\}^2\,d\rho(\lambda) = 0$$

which will in the sequel be shortened to

$$F(\lambda) = \underset{n\to\infty}{\text{l.i.m.}} \int_0^n f(x)y(x,\lambda)\,dx.$$

2.1.2

Let two functions $f(x), g(x) \in \mathcal{L}^2(0, \infty)$, and let $F(\lambda)$, $G(\lambda)$ be their Fourier transforms. It is obvious that $f(x) \pm g(x)$ have the functions $F(\lambda) \pm G(\lambda)$ as their Fourier transforms. Therefore,

$$\int_0^\infty \{f(x) + g(x)\}^2 \, dx = \int_{-\infty}^\infty \{F(\lambda) + G(\lambda)\}^2 \, d\rho(\lambda),$$

$$\int_0^\infty \{f(x) - g(x)\}^2 \, dx = \int_{-\infty}^\infty \{F(\lambda) - G(\lambda)\}^2 \, d\rho(\lambda).$$

Subtracting the lower equality from the upper, we obtain

$$\int_0^\infty f(x)g(x) \, dx = \int_{-\infty}^\infty F(\lambda)G(\lambda) \, d\rho(\lambda) \tag{1.11}$$

which is called the *generalized Parseval equation*.

Theorem 2.1.2 *(Expansion Theorem.) Let $f(x) \in \mathcal{L}^2(0, \infty)$ be a continuous function $(0 \le x < \infty)$, and let the integral*

$$\int_{-\infty}^\infty F(\lambda)y(x, \lambda) \, d\rho(\lambda)$$

converge absolutely and uniformly in x in each finite interval. Then

$$f(x) = \int_{-\infty}^\infty F(\lambda)y(x, \lambda) \, d\rho(\lambda). \tag{1.12}$$

Proof. Let $g(x)$ be a continuous function vanishing outside a finite interval $[0, n]$. Then the equality (1.11) is written as

$$\int_0^\infty f(x)g(x) \, dx = \int_{-\infty}^\infty F(\lambda) \left\{ \int_0^n g(x)y(x, \lambda) \, dx \right\} \, d\rho(\lambda).$$

Relying on the absolute convergence, we can change the order of integration in the last integral. We obtain

$$\int_0^n f(x)g(x) \, dx = \int_0^n g(x) \left\{ \int_{-\infty}^\infty F(\lambda)y(x, \lambda) \, d\rho(\lambda) \right\} \, dx.$$

Since the continuous function $g(x)$ is arbitrary, and $f(x)$ and

$$\int_{-\infty}^\infty F(\lambda)y(x, \lambda) \, d\rho(\lambda)$$

are continuous (the continuity of the latter follows from the assumed uniform convergence of the integral), we have

$$f(x) = \int_{-\infty}^\infty F(\lambda)y(x, \lambda) d\rho(\lambda)$$

as required.   ∎

2.2   The limit-circle and limit-point cases

2.2.1

In this section, we still consider the interval $[0, \infty)$ and a function $q(x)$ which is continuous in each finite interval.

Let $F(x)$ satisfy the equation

$$y'' + \{\lambda - q(x)\}y = 0, \tag{2.1}$$

and let $G(x)$ satisfy the same equation, but with λ replaced by λ'. If b is a fixed number, then, by the identity (1.5), Chapter 1,

$$
\begin{aligned}
(\lambda' - \lambda) \int_0^b F(x)G(x)\,dx &= \int_0^b F(x)[q(x)G(x) - G''(x)]\,dx \\
&\quad - \int_0^b G(x)[q(x)F(x) - F''(x)]\,dx \\
&= -\int_0^b \{F(x)G''(x) - G(x)F''(x)\}\,dx \\
&= W_0\{F, G\} - W_b\{F, G\}. \tag{2.2}
\end{aligned}
$$

In particular, if $\lambda = u + iv$, $\lambda' = \bar{\lambda} = u - iv$, then, $q(x)$ being real, $G(x) = \bar{F}(x)$; hence,

$$2v \int_0^b |F(x)|^2\,dx = iW_0\{F, \bar{F}\} - iW_b\{F, \bar{F}\} \tag{2.3}$$

because of (2.2).

Denote by $\varphi(x) = \varphi(x, \lambda)$ and $\vartheta(x) = \vartheta(x, \lambda)$ the solutions of (2.1) with the initial conditions

$$
\begin{aligned}
\varphi(0) &= \sin \alpha, \quad \varphi'(0) = -\cos \alpha \\
\vartheta(0) &= \cos \alpha, \quad \vartheta'(0) = \sin \alpha
\end{aligned}
$$

where α is a real number.

Since the term with the first derivative in (2.1) is zero, by the familiar Liouville formula, the Wronskian is constant; therefore, $W_x\{\varphi, \vartheta\} = W_0\{\varphi, \vartheta\} = \sin \alpha^2 + \cos \alpha^2 = 1$ and the general solution of (2.1) can be represented as $\vartheta(x) + l\varphi(x)$, up to a constant multiplier.

Consider the solutions satisfying the condition

$$\{\vartheta(b) + l\varphi(b)\}\cos \beta + \{\vartheta'(b) + l\varphi'(b)\}\sin \beta = 0 \tag{2.4}$$

at the point $x = b$ with real β, and determine

$$l = -\frac{\vartheta(b)\cot \beta + \vartheta'(b)}{\varphi(b)\cot \beta + \varphi'(b)}. \tag{2.4'}$$

If b is fixed and $\cot\beta$ takes all values from $-\infty$ to $+\infty$, then l describes a circle C_b in the complex plane. Replace $\cot\beta$ by a complex variable z, and put

$$l = l(\lambda, z) = -\frac{\vartheta(b)z + \vartheta'(b)}{\varphi(b)z + \varphi'(b)}. \tag{2.5}$$

The point $l = \infty$ is associated with $z = -\varphi'(b)/\varphi(b)$. Therefore, the center of C_b is associated with the point $\bar{z} = -\bar{\varphi}'(b)/\bar{\varphi}(b)$. Substituting $\bar{z}$ in the formula for l, we see that the center of C_b is at the point $-W_b(\vartheta, \bar{\varphi})/W_b(\varphi, \bar{\varphi})$. Furthermore,

$$\mathrm{Im}\left\{-\frac{\varphi'(b)}{\varphi(b)}\right\} = \frac{1}{2}i\left\{\frac{\varphi'(b)}{\varphi(b)} - \frac{\bar{\varphi}'(b)}{\bar{\varphi}(b)}\right\} = -\frac{1}{2}i\frac{W_b\{\varphi, \bar{\varphi}\}}{|\varphi(b)|^2}.$$

Since $W_0\{\varphi, \bar{\varphi}\} = 0$ due to (2.3), the last expression has the same sign as v. We have thus proved the following.

Lemma 2.2.1 *If $v > 0$, then the upper z-half-plane is associated with the exterior of the circle C_b.*

Since the point $-\vartheta'(b)/\varphi'(b)$ is on C_b (for $z = 0$), the radius is

$$r_b = \left|\frac{\vartheta'(b)}{\varphi'(b)} - \frac{W_b\{\vartheta, \bar{\varphi}\}}{W_b\{\varphi, \bar{\varphi}\}}\right| = \left|\frac{W_b\{\vartheta, \varphi\}}{W_b\{\varphi, \bar{\varphi}\}}\right| = \left(2v\int_0^b |\varphi(x)|^2\, dx\right)^{-1}. \tag{2.6}$$

Furthermore, by Lemma 2.2.1, l is inside C_b if $\mathrm{Im}\, z < 0$, or if $i(z - \bar{z}) > 0$. Solving (2.5) for z and substituting in $i(z - \bar{z}) > 0$, we obtain

$$i\left\{-\frac{l\varphi'(b) + \vartheta'(b)}{l\varphi(b) + \vartheta(b)} - \frac{\bar{l}\bar{\varphi}'(b) + \bar{\vartheta}'(b)}{\bar{l}\bar{\varphi}(b) + \bar{\vartheta}(b)}\right\} > 0$$

and then

$$i[|l|^2 W_b\{\varphi, \bar{\varphi}\} + lW_b\{\varphi, \bar{\vartheta}\} + \bar{l}W_b\{\vartheta, \bar{\varphi}\} + W_b\{\vartheta, \bar{\vartheta}\}] = iW_b\{\vartheta + l\varphi, \bar{\vartheta} + \bar{l}\bar{\varphi}\} > 0.$$

Therefore, putting $F = \vartheta + l\varphi$ in the identity (2.3), we obtain

$$2v\int_0^b |\vartheta(x) + l\varphi(x)|^2\, dx < iW_0\{\vartheta + l\varphi, \bar{\vartheta} + \bar{l}\bar{\varphi}\}.$$

Since $W_0\{\vartheta, \bar{\vartheta}\} = W_0\{\varphi, \bar{\varphi}\} = 0$ and $W_0\{\vartheta, \bar{\varphi}\} = W_0\{\varphi, \bar{\vartheta}\} = 1$, we have

$$W_0\{\vartheta + l\varphi, \bar{\vartheta} + \bar{l}\bar{\varphi}\} = l - \bar{l} = 2i\,\mathrm{Im}\, l.$$

Thus, l is inside C_b if $v > 0$ and

$$\int_0^b |\vartheta(x) + l\varphi(x)|^2\, dx < -\frac{\mathrm{Im}\, l}{v}.$$

The same is obtained if $v < 0$ (using the inequality $\operatorname{Im} z > 0$). In both cases, the sign of $\operatorname{Im} l$ is opposite to that of v. Therefore, if l is inside C_b and $0 < b' < b$, then

$$\int_0^{b'} |\vartheta(x) + l\varphi(x)|^2\, dx < \int_0^b |\vartheta(x) + l\varphi(x)|^2\, dx < -\frac{\operatorname{Im} l}{v}.$$

Therefore, l is also inside $C_{b'}$ and, if $b' < b$, then $C_{b'}$ contains C_b. Hence, as $b \to \infty$, the circles C_b converge either to a limit circle or to a limit point. Let $m = m(\lambda)$ be the limit point or any point on the limit circle. Then, for any b,

$$\int_0^b |\vartheta(x) + m\varphi(x)|^2\, dx \leq -\frac{\operatorname{Im}(m)}{v}.$$

Therefore,

$$\int_0^\infty |\vartheta(x) + m\varphi(x)|^2\, dx \leq -\frac{\operatorname{Im}(m)}{v}. \tag{2.7}$$

Thus, the following holds.

Theorem 2.2.1 *For all nonreal λ, there exists a square-integrable solution*

$$\psi(x, \lambda) = \vartheta(x, \lambda) + m(\lambda)\varphi(x, \lambda)$$

of the equation (2.1) on the interval $[0, \infty)$.

The function $\psi(x, \lambda)$ is called a *Weyl solution.*

In the limit-circle case, r_b tends to a finite limit as $b \to \infty$. Therefore, it follows from (2.6) that $\varphi(x, \lambda) \in \mathcal{L}^2(0, \infty)$, i.e., each solution of (2.1) belongs to the space $\mathcal{L}^2(0, \infty)$.

The question naturally arises : Do the limit-circle and limit-point classifications depend on a particular value of λ_0, or do they only depend on the operator $-d^2/dx^2 + q(x)$? The answer is supplied by the following.

Theorem 2.2.2 *If each solution of (2.1) for some complex λ^0 is in the class $\mathcal{L}^2(0, \infty)$, then, for any complex λ, each solution of (2.1) also belongs to $\mathcal{L}^2(0, \infty)$. In other words, if we are in the limit-circle case for some λ^0, then we are in this case for all λ.*

Proof. Let two linearly independent solutions φ and ψ of the equation $-y'' + q(x)y = \lambda_0 y$ be in the class $\mathcal{L}^2(0, \infty)$, where $\operatorname{Im} \lambda_0 \neq 0$, and let χ be an arbitrary solution of $-y'' + q(x)y = \lambda y$, which we write in the form

$$-y'' + q(x)y = \lambda_0 y + (\lambda - \lambda_0)y.$$

Multiplying φ by a constant if necessary, so that $W\{\varphi, \bar{\psi}\} = 1$, and applying the variation of constants, we obtain

$$\chi(x) = c_1\varphi(x) + c_2\psi(x) + (\lambda - \lambda_0)\int_c^x \{\varphi(x)\psi(t) - \varphi(t)\psi(x)\}\chi(t)\, dt, \tag{2.8}$$

where c, c_1 and c_2 are three constants. If we use the notation

$$\|\chi(x)\|_c = \left\{ \int_c^x |\chi(t)|^2 \, dt \right\}^{1/2}$$

and if M is such that $\|\varphi(x)\|_c \leq M$, $\|\psi(x)\|_c \leq M$ for all $x \leq c$, then the Cauchy-Schwarz inequality yields

$$\int_c^x |\{\varphi(x)\psi(t) - \varphi(t)\psi(x)\}\chi(t) \, dt \leq M\{|\varphi(x)| + |\psi(x)|\}\|\chi(x)\|_c,$$

which, applied to the formula (2.8), gives, by the Minkowski inequality,

$$\|\chi(x)\|_c \leq \{|c_1| + |c_2|\}M + 2|\lambda - \lambda_0|M^2 \|\chi(x)\|_c.$$

If c is so large that $|\lambda - \lambda_0|M^2 < \frac{1}{4}$, then $\|\chi(x)\|_c \leq 2\{|c_1| + |c_2|\}M$. Since the right-hand side does not depend on x, $\chi(x) \in \mathcal{L}^2(0, \infty)$ and the proof is complete. ∎

A useful sufficient condition for the operator $-d^2/dx^2 + q(x)$ to be in the limit-point case is given by the following.

Theorem 2.2.3 *If $q(x) \geq -kx^2$, where k is a positive constant, then we are in the limit-point case for the operator $-d^2/dx^2 + q(x)$.*

Proof. We show that the equation $Ly \equiv -y'' + q(x)y = 0$ does not have two linearly independent solutions in the class $\mathcal{L}^2(0, \infty)$. Assume that $\varphi(x)$ is a real solution of $Ly = 0$, and that $\varphi(x) \in \mathcal{L}^2(0, \infty)$. We obtain

$$\int_c^x \frac{\varphi''(t)\varphi(t)}{t^2} \, dt = \int_c^x \frac{q(t)}{t^2}\varphi^2(t) \, dt \geq -k \int_c^x \varphi^2(t) \, dt$$

from $\varphi''(x) = q(x)\varphi(x)$, $(c > 0)$. Integrating by parts and using the fact that $\varphi(x) \in \mathcal{L}^2(0, \infty)$, we prove that there is a constant k_1 such that

$$-\frac{\varphi'(x)\varphi(x)}{x^2} + \int_c^x \frac{\{\varphi'(t)\}^2}{t^2} \, dt - 2 \int_c^x \frac{\varphi'(t)\varphi(t)}{t^3} \, dt < k_1. \tag{2.9}$$

Let

$$H(x) = \int_c^x \frac{\{\varphi'(t)\}^2}{t^2} \, dt.$$

Applying the Cauchy-Schwarz inequality, we obtain

$$\left| 2 \int_c^x \frac{\varphi'(t)\varphi(t)}{t^3} \, dt \right|^2 \leq k_2 \left(\int_c^x \frac{|\varphi'(t)| \cdot |\varphi(t)|}{t} \, dt \right)^2 \leq k_2 H(x) \int_c^x \varphi^2(t) \, dt. \tag{2.10}$$

Thus, it follows from (2.9), because of (2.10), that there is a constant k_3 such that

$$-\frac{\varphi'(x)\varphi(x)}{x^2} + H(x) - k_3 H^{1/2}(x) < k_1.$$

If $H(x) \to \infty$ as $x \to \infty$, it follows that

$$\frac{\varphi'(x)\varphi(x)}{x^2} > \frac{1}{2}H(x)$$

for all sufficiently large x, which means that the functions $\varphi(x)$ and $\varphi'(x)$ have the same sign for large x, contrary to $\varphi(x) \in \mathcal{L}^2(0, \infty)$. Thus, $H(x)$ remains finite, so that

$$\int_c^\infty \frac{\{\varphi'(t)\}^2}{t^2}\, dt < +\infty. \tag{2.11}$$

Assume that $\varphi(x)$ and $\psi(x)$ are two linearly independent solutions of $Ly = 0$ in the class $\mathcal{L}^2(0, \infty)$, i.e., that we are in the limit-circle case for L. We can also assume that these solutions are real and that $W_x\{\varphi, \psi\} = 1$. It then follows from the definition of $W_x\{\varphi, \psi\}$ that

$$\varphi(x)\frac{\psi'(x)}{x} - \psi(x)\frac{\varphi'(x)}{x} = \frac{1}{x}.$$

Owing to (2.11) and the Cauchy-Schwarz inequality, the left-hand side is integrable on the interval (c, ∞) while the right-hand side is obviously nonintegrable. Therefore, we cannot be in the limit-circle case. ∎

2.2.2

For given β, $l = l(\lambda)$ is an analytic function of λ. It is even meromorphic with poles on the real axis. In fact, in view of (2.4), the poles are zeros of the function $\varphi(b, \lambda)\cos\beta + \varphi'(b, \lambda)\sin\beta$, the eigenvalues of the operator $-d^2/dx^2 + q(x)$ in the finite interval $[a, b]$, and consequently simple.

It follows from the argument of the preceding item that, for given λ, the l-plane domain covered with the circle C_b decreases as b increases. Therefore, the family of analytic functions $l(\lambda) = l(\lambda, b, \beta)$ is uniformly bounded in each bounded domain wholly in the upper (resp. lower) λ-half-plane. By the familiar theorem in the theory of analytic functions, this family is normal[1] (i.e., a uniformly convergent subsequence can be selected from each infinite sequence) in each of the above domains.

If, for certain nonreal λ, we are in the limit-point case (by Theorem 2.2.2, the same then holds for all other nonreal λ), then, as is easily seen, the family of $l(\lambda)$ has a unique limit $m(\lambda)$, which is an analytic function by the Weierstrass theorem.

In the limit-point case, the above limit is not unique. It follows from the normality of the family that we can pick out an indefinitely increasing sequence of numbers b_k and a sequence of numbers β_k, so that $\lim_{k\to\infty} l(\lambda, b_k, \beta_k) = m(\lambda)$ exists in each of the λ-half-planes, and is an analytic function. It is obvious that the numbers $m(\lambda)$ are on the limit circles.

Put $\psi_b(x, \lambda) = \vartheta(x) + l\varphi(x)$, $l = l(\lambda, b)$. Then the following holds.

---

[1]See A. I. Markushevich (1950, p. 294).

Lemma 2.2.2 *For each nonreal λ,*

$$\psi_b(x, \lambda) \to \psi(x, \lambda)$$

$$\int_0^b |\psi_b(x, \lambda)|^2 \, dx \to \int_0^\infty |\psi(x, \lambda)|^2 \, dx, \quad b \to \infty.$$

Proof. It is obvious that

$$\psi_b(x, \lambda) = \psi(x, \lambda) + [l(\lambda) - m(\lambda)]\varphi(x, \lambda),$$

where $\psi(x, \lambda) \in \mathcal{L}^2(0, \infty)$.

In the limit-circle case, $l(\lambda) \to m(\lambda)$; therefore, $\psi_b(x, \lambda) \to \psi(x, \lambda)$, and, since $\varphi(x, \lambda) \in \mathcal{L}^2(0, \infty)$,

$$\int_0^b |\psi_b(x, \lambda)|^2 \, dx \to \int_0^b |\psi(x, \lambda)|^2 \, dx.$$

In the limit-point case, according to (2.6),

$$|l(\lambda) - m(\lambda)| \le r_b = \left(2v \int_0^b |\varphi(x, \lambda)|^2 \, dx \right)^{-1}, \quad (v \ne 0).$$

As $r_b \to 0$, $\psi_b(x, \lambda) \to \psi(x, \lambda)$. Moreover,

$$\int_0^b |\{l(\lambda) - m(\lambda)\}\varphi(x, \lambda)|^2 \, dx = |l(\lambda) - m(\lambda)|^2 \int_0^b |\varphi(x, \lambda)|^2 \, dx$$

$$\le \left(4v^2 \int_0^b |\varphi(x, \lambda)|^2 \, dx \right)^{-1}.$$

Therefore,

$$\int_0^b |\psi_b(x, \lambda)|^2 \, dx \to \int_0^b |\psi(x, \lambda)|^2 \, dx$$

in this case, too. ∎

2.3 Integral representation of the resolvent

Let $f(x) \in \mathcal{L}^2(0, \infty)$. We put

$$\Phi(x, \lambda) = \psi(x, \lambda) \int_0^x \varphi(y, \lambda)f(y) \, dy + \varphi(x, \lambda) \int_x^\infty \psi(y, \lambda)f(y) \, dy,$$

where λ is nonreal and $\varphi(x, \lambda), \psi(x, \lambda)$ are determined in Section 2.

If $f(x)$ is a continuous function, then

$$\Phi_x'(x, \lambda) = \psi_x'(x, \lambda) \int_0^x \varphi(y, \lambda)f(y) \, dy + \varphi_x'(x, \lambda) \int_x^\infty \psi(y, \lambda)f(y) \, dy,$$

$$\begin{aligned}
\Phi_x''(x, \lambda) &= \psi''(x, \lambda) \int_0^x \varphi(y, \lambda)f(y) \, dy + \varphi''(x, \lambda) \int_x^\infty \psi(y, \lambda)f(y) \, dy \\
&\quad + \{\psi'(x, \lambda)\varphi(x, \lambda) - \varphi'(x, \lambda)\psi(x, \lambda)\}f(x) \\
&= \{q(x) - \lambda\}\Phi(x, \lambda) + f(x),
\end{aligned}$$

since

$$\varphi(x,\lambda)\psi'(x,\lambda) - \varphi'(x,\lambda)\psi(x,\lambda) = W_x\{\varphi,\psi\} = W_0\{\varphi,\psi\}$$
$$= \begin{vmatrix} \sin\alpha & \cos\alpha + m(\lambda)\sin\alpha \\ -\cos\alpha & \sin\alpha - m(\lambda)\cos\alpha \end{vmatrix} = 1.$$

Thus, $\Phi''(x,\lambda) + \{\lambda - q(x)\}\Phi(x,\lambda) = f(x)$; furthermore,

$$\Phi(0,\lambda) = \varphi(0,\lambda)\int_0^\infty \psi(y,\lambda)f(y)\,dy$$
$$\Phi'_x(0,\lambda) = \varphi'_x(0,\lambda)\int_0^\infty \psi(y,\lambda)f(y)\,dy$$

so that $\Phi(x,\lambda)$ satisfies the boundary condition

$$\Phi(0,\lambda)\cos\alpha + \Phi'_x(0,\lambda)\sin\alpha = 0.$$

Let $\lambda_{n,b}$ and $\varphi_{n,b}(x)$ be the eigenvalues and eigenfunctions of the boundary-value problem (1.1), (1.3), (1.4), and let $l(\lambda)$, $\psi_b(x,\lambda)$ be as in Section 2. Putting

$$G_b(x,y;z) = \begin{cases} \psi_b(x,z)\varphi(y,z), & y \le x \\[2mm] \varphi(x,z)\psi_b(y,z), & y > x \end{cases}$$

$$R_{z,b}f = \int_0^b G_b(x,y;z)f(y)\,dy, \quad z = u + iv$$

and

$$\alpha_n^2 = \int_0^b \varphi_{n,b}^2(x)\,dx, \tag{3.1}$$

we get (see Chapter 1, Section 5)

$$R_{z,b}f = \sum_{n=1}^\infty \frac{\varphi_{n,b}(x)\int_0^b f(y)\varphi_{n,b}(y)\,dy}{\alpha_n^2(z - \lambda_{n,b})} = \int_{-\infty}^\infty \frac{\varphi(x,\lambda)}{z - \lambda}\left\{\int_0^b f(y)\varphi(y,\lambda)\,dy\right\}d\rho_b(\lambda), \tag{3.2}$$

where the function $\rho_b(\lambda)$ is determined in Section 1.

Lemma 2.3.1 *For each nonreal z and fixed x,*

$$\int_{-\infty}^\infty \left|\frac{\varphi(x,\lambda)}{z - \lambda}\right|^2 d\rho_b(\lambda) < K.$$

Proof. Putting $f(y) = \varphi_{n,b}(y)/\alpha_n$ in the first part of (3.2) and taking into account the definition of α_n in (3.1), since the eigenfunctions $\varphi_{n,b}(x)$ are orthogonal, we obtain

$$\frac{1}{\alpha_n}\int_0^b G_b(x,y;z)\varphi_{n,b}(y)\,dy = \frac{\varphi_{n,b}(x)}{\alpha_n(z - \lambda_{n,b})}. \tag{3.4}$$

Applying the Parseval equation to $G_b(x, y; z)$ (as a function of y) and taking into account (3.4), we now find that

$$\int_0^b |G_b(x, y; z)|^2 \, dy = \sum_{n=1}^\infty \frac{|\varphi_{n,b}(x)|^2}{\alpha_n^2 |z - \lambda_{n,b}|^2} = \int_{-\infty}^\infty \left| \frac{\varphi(x, \lambda)}{z - \lambda} \right|^2 d\rho_b(\lambda).$$

The statement of the lemma follows, since the integral on the left converges by Lemma 2.2.2. ∎

Corollary 2.3.1 *Let the function $\rho(\lambda)$ assume the same values as in Section 1. Then*

$$\int_{-\infty}^\infty \left| \frac{\varphi(x, \lambda)}{z - \lambda} \right| d\rho(\lambda) \le K, \qquad (3.5)$$

with the same constant K as in the inequality (3.3).

Proof. In fact, for arbitrary $a > 0$, it follows from (3.3) that

$$\int_{-a}^a \left| \frac{\varphi(x, \lambda)}{z - \lambda} \right|^2 d\rho_b(\lambda) < K.$$

Letting $b \to \infty$ and then $a \to \infty$, we obtain (3.5). ∎

Corollary 2.3.2 *The inequalities*

$$\int_{-\infty}^{-a} \frac{d\rho(\lambda)}{|z - \lambda|^2} < \infty, \qquad \int_a^\infty \frac{d\rho(\lambda)}{|z - \lambda|^2} < \infty \qquad (3.6)$$

hold for $a > 0$.

Proof. In fact, if $\sin \alpha \ne 0$, then, putting $x = 0$ in (3.5), we obtain

$$\int_{-\infty}^\infty \frac{d\rho(\lambda)}{|z - \lambda|^2} < \infty;$$

hence, (3.6) holds for any $a > 0$.

If $\sin \alpha = 0$, then, differentiating (3.4) with respect to x, we obtain

$$\frac{1}{\alpha_n} \int_0^b \frac{\partial}{\partial x} G_b(x, y; z) \varphi_{n,b}(y) \, dy = \frac{\varphi_{n,b}'(x)}{\alpha_n(z - \lambda_{n,b})}.$$

Therefore, by the Parseval equation,

$$\int_0^b \left| \frac{\partial}{\partial x} G_b(x, y; z) \right|^2 dy = \int_{-\infty}^\infty \left| \frac{\varphi'(x, \lambda)}{z - \lambda} \right|^2 d\rho_b(\lambda).$$

Proceeding similarly, we can easily derive (3.6) in this case, too. ∎

Lemma 2.3.2 *Let* $f(x) \in \mathcal{L}^2(0, \infty)$, *and let*

$$R_z f = \int_0^\infty G(x, y; z) f(y) \, dy,$$

where

$$G(x, y; z) = \begin{cases} \psi(x, z)\varphi(y, z), & y \le x \\ \\ \varphi(x, z)\psi(y, z), & y > x \end{cases}$$

with $\varphi(x, z)$ *and* $\psi(x, z)$ *having the same values as in Section 2. Then*

$$\int_0^\infty |R_z f(x)|^2 \, dx \le \frac{1}{v^2} \int_0^\infty f^2(x) \, dx, \quad z = u + iv.$$

Proof. For each $b > 0$, it follows from (3.2) and the Parseval equation that

$$\int_0^b |R_{z,b} f(x)|^2 \, dx = \sum_{n=1}^\infty \frac{1}{\alpha_n^2 |z - \lambda_{n,b}|^2} \left\{ \int_0^b f(y)\varphi_{n,b}(y) \, dy \right\}^2$$

$$\le \frac{1}{v^2} \sum_{n=1}^\infty \frac{1}{\alpha_n^2} \left\{ \int_0^b f(y)\varphi_{n,b}(y) \, dy \right\}^2 = \frac{1}{v^2} \int_0^b f^2(y) \, dy.$$

Let $a > 0$ be fixed. If $a < b$, then

$$\int_0^a |R_{z,b} f(x)|^2 \, dx \le \int_0^b |R_{z,b} f(x)|^2 \, dx \le \frac{1}{v^2} \int_0^b f^2(y) \, dy.$$

Letting $b \to \infty$, we obtain

$$\int_0^a |R_z f|^2 \, dx \le \frac{1}{v^2} \int_0^\infty f^2(y) \, dy.$$

Since a is arbitrary, the proof is complete. ∎

Theorem 2.3.1 *(Integral Representation of the Resolvent.) For each function* $f(x) \in \mathcal{L}^2(0, \infty)$ *and for each nonreal* z,

$$R_z f = \int_{-\infty}^\infty \frac{\varphi(x, \lambda) F(\lambda)}{z - \lambda} \, d\rho(\lambda), \tag{3.7}$$

where

$$F(\lambda) = \underset{n \to \infty}{\text{l.i.m.}} \int_0^n f(x)\varphi(x, \lambda) \, dx.$$

Proof. We first assume that $f(x) = f_n(x)$ vanishes outside the finite interval $[0, n]$, that it satisfies the boundary condition (1.3) and that it has a continuous second derivative. Let $b > n$, and let a be an arbitrary positive number. We put

$$F_n(\lambda) = \int_0^n f_n(x)\varphi(x, \lambda) \, dx.$$

The right-hand side of the equality (3.2) can then be rewritten in the form

$$R_{z,b}f_n = \int_{-\infty}^{\infty} \frac{\varphi(x,\lambda)F_n(\lambda)}{z-\lambda}\, d\rho_b(\lambda)$$

$$= \left\{ \int_{-\infty}^{-a} + \int_{-a}^{a} + \int_{a}^{\infty} \right\} \frac{\varphi(x,\lambda)F_n(\lambda)}{z-\lambda}\, d\rho_b(\lambda) \equiv I_1 + I_2 + I_3. \quad (3.8)$$

We now estimate I_1. By (3.2), we have

$$
\begin{aligned}
I_1 &= \int_{-\infty}^{-a} \frac{\varphi(x,\lambda)F_n(\lambda)}{z-\lambda}\, d\rho_b(\lambda) \\
&= \sum_{\lambda_{k,b}<-a} \frac{\varphi_{k,b}(x)}{\alpha_k^2(z-\lambda_{k,b})} \int_0^n f_n(y)\varphi_{k,b}(y)\, dy \\
&\le \left(\sum_{\lambda_{k,b}<-a} \frac{\varphi_{k,b}^2(x)}{\alpha_k^2|z-\lambda_{k,b}|^2} \right)^{1/2} \left(\sum_{\lambda_{k,b}<-a} \frac{1}{\alpha_k^2}\left[\int_0^n f_n(x)\varphi_{k,b}(x)\, dx\right]^2 \right)^{1/2}. \quad (3.9)
\end{aligned}
$$

Integrating twice by parts, we obtain

$$
\begin{aligned}
\int_0^n f_n(x)\varphi_{k,b}(x)\, dx &= -\frac{1}{\lambda_{k,b}} \int_0^n f_n(x)\{\varphi_{k,b}''(x) - q(x)\varphi_{k,b}(x)\}\, dx \\
&= -\frac{1}{\lambda_{k,b}} \int_0^n \{f_n''(x) - q(x)f(x)\}\varphi_{k,b}(x)\, dx. \quad (3.10)
\end{aligned}
$$

Since the first sum on the right-hand side of (3.9) converges by Lemma 2.3.1, taking into account (3.10), we deduce from (3.9) that

$$I_1 \le K^{1/2}\frac{1}{a}\left(\sum_{\lambda_{k,b}<-a} \frac{1}{\alpha_k^2}\left\{\int_0^n [f_n''(x) - q(x)f(x)]\varphi_{k,b}(x)\, dx\right\}^2 \right)^{1/2}$$

Therefore, by the Bessel inequality,

$$I_1 \le \frac{K^{1/2}}{a}\left(\int_0^n \{f_n''(x) - q(x)f_n(x)\}^2\, dx \right)^{1/2} = \frac{C}{a}.$$

It is proved similarly that $I_3 \le C/a$. It follows that both I_1 and I_3 tend to zero as $a \to \infty$ uniformly with respect to b. Therefore, we can use the generalization of the Helly selection theorem, and obtain from the equality (3.8)[2]

$$R_z f = \int_{-\infty}^{\infty} \frac{\varphi(x,\lambda)F_n(\lambda)}{z-\lambda}\, d\rho(\lambda) \quad (3.11)$$

---

[2]The convergence of $R_{z,b}f_n$ to $R_z f_n$ follows from the first part of Lemma 2.2.2. Note that, from now onwards, we consider those $\rho(\lambda)$ which are associated with the sequences of b_k and β_k such that $\lim_{k\to\infty} l(\lambda, b_k, \beta_k) = m(\lambda)$ (see Section 2).

(cf. the proof of Theorem 2.1.1).

If $f(x)$ is an arbitrary square-integrable function on $[0, \infty)$, then, as is well known, we can select a sequence of functions $\{f_n(x)\}_1^\infty$ fulfilling the preceding conditions and converging in mean square to $f(x)$ as $n \to \infty$. By the Parseval equation, the sequence of Fourier transforms $F_n(\lambda)$ of $f_n(x)$ also converges in mean square (for the measure $\rho(\lambda)$) to the Fourier transform of the function $f(x)$, which we denote by $F(\lambda)$. Therefore, by Corollary 1 to Lemma 2.3.1 and by Lemma 2.3.2, we can pass to the limit in the equality (3.11) as $n \to \infty$, and the theorem is thus proved. ∎

Remark By a similar argument, we can obtain the formula

$$\int_0^\infty R_z(f)g(x)\,dx = \int_{-\infty}^\infty \frac{F(\lambda)G(\lambda)}{z - \lambda}\,d\rho(\lambda), \qquad (3.7')$$

where

$$F(\lambda) = \underset{n\to\infty}{\text{l.i.m.}} \int_0^n f(x)\varphi(x, \lambda)\,dx$$

$$G(\lambda) = \underset{n\to\infty}{\text{l.i.m.}} \int_0^n g(x)\varphi(x, \lambda)\,dx.$$

Theorem 2.3.2 *Let $f(x)$ fulfil the conditions*

1. $f(x), \{f''(x) - q(x)f(x)\} \in \mathcal{L}^2(0, \infty)$,
2. $f(0)\cos\alpha + f'(0)\sin\alpha = 0$,
3. $\lim_{x\to\infty} W\{f, E_\lambda\} = 0$, *where*

$$E_\lambda(x) = \int_{+0}^\lambda \varphi(x, \lambda)\,d\rho(\lambda), \quad (\lambda \neq 0).$$

We put

$$g(\lambda) = \int_0^\infty f(y)E_\lambda(y)\,dy,$$

where the integral

$$f(x) = \int_{-\infty}^\infty \varphi(x, \lambda)\,dg(\lambda)$$

converges absolutely.

Proof. If $f(x)$ vanishes outside a finite interval, then, by changing the order of integration, we obtain

$$g(\lambda) = \int_0^\lambda F(\lambda)\,d\rho(\lambda), \qquad F(\lambda) = \int_0^\infty f(x)\varphi(x, \lambda)\,dx.$$

This formula for $g(\lambda)$ also remains valid in the general case, relying on the mean convergence of the Fourier transform. Therefore, the formula (3.7) can be written as

$$R_z f = \int_{-\infty}^\infty \frac{\varphi(x, \lambda)\,dg(\lambda)}{z - \lambda}.$$

The integral converges absolutely by Lemma 2.3.1. Let $h(\lambda)$ be determined from $Lf = f''(x) - q(x)f(x)$ as $g(\lambda)$ is from $f(x)$. It follows from Condition 3 and from Green's identity that

$$h(\lambda) = \int_0^\infty \{f''(x) - q(x)f(x)\} E_\lambda(x)\, dx = -\int_0^\infty f(x) \left\{ \int_{+0}^\lambda \mu\varphi(x,\mu)\, d\rho(\mu) \right\}\, dx.$$

If $f(x)$ vanishes outside a finite interval, then

$$h(\lambda) = -\int_0^\lambda \mu F(\mu)\, d\rho(\mu) = -\int_{+0}^\lambda \mu\, dg(\mu). \tag{3.12}$$

The formula remains valid in the general case, too, for

$$\int_{+0}^\lambda \mu\varphi(x,\mu)\, d\rho(\mu) \in \mathcal{L}^2(0,\infty).$$

Indeed,

$$\int_0^b \left\{ \int_{+0}^\lambda \mu\varphi(x,\mu)\, d\rho_b(\mu) \right\}^2 dx = \int_0^b \left\{ \sum_{0 \le \lambda_{n,b} \le \lambda} \frac{\lambda_{n,b}}{\alpha_n^2} \varphi_{n,b}(x) \right\}^2 dx$$

$$= \sum_{0 \le \lambda_{n,b} \le \lambda} \frac{\lambda_{n,b}^2}{\alpha_n^2} = \int_{+0}^\lambda \mu^2\, d\rho_b(\mu).$$

Hence,

$$\int_0^a \left\{ \int_{+0}^\lambda \mu\varphi(x,\mu)\, d\rho_b(\mu) \right\}^2 dx \le \int_{+0}^\lambda \mu^2\, d\rho_b(\mu),$$

where $a < b$. Letting $b \to \infty$ and then $a \to \infty$, we obtain the above statement. Now, differentiating (3.12), we get

$$dh(\lambda) = -\lambda\, dg(\lambda), \qquad dg(\lambda) = -dh(\lambda)/\lambda. \tag{3.13}$$

Since the integral $\int_{-\infty}^\infty \varphi(x,\lambda)\, dh(\lambda)/\lambda$ converges absolutely because of Condition 1, $\int_{-\infty}^\infty \varphi(x,\lambda) dg(\lambda)$ also converges absolutely thanks to (3.13). Therefore, Theorem 2.3.2 immediately follows from Theorem 2.1.2. ∎

2.4   The Weyl-Titchmarsh function

With the help of the integral representation of the resolvent, obtained in Section 3 (Theorem 2.3.1), we can derive useful formulas for the functions $\rho(\lambda)$ and $m(z)$. The former, introduced in Section 1, was called the *spectral function* of the boundary-value problem. Its formula can be taken as the basis for determining the spectrum of

the Sturm-Liouville operator (see Chapter 3). $m(z)$ introduced in Section 2 is called
the *Weyl-Titchmarsh function*. $\rho(\lambda)$ and the Weyl-Titchmarsh function are closely
related, and, as is shown in the sequel, each can be expressed in terms of the other.

We first prove some auxiliary lemmas.

Let $\psi(x, \lambda)$ be the function considered in Section 2 (see Theorem 2.2.1).

Lemma 2.4.1 *For any fixed nonreal λ and λ', the equality*

$$\lim_{x \to \infty} W\{\psi(x, \lambda), \psi(x, \lambda')\} = 0$$

holds.

Proof. The function $\vartheta(x, \lambda) + l(\lambda)\varphi(x, \lambda)$ satisfies the boundary value condition (2.4).
Therefore,

$$W_b\{\vartheta(x, \lambda) + l(\lambda)\varphi(x, \lambda), \vartheta(x, \lambda') + l(\lambda')\varphi(x, \lambda')\} = 0.$$

It follows that

$$W_b\{\psi(x, \lambda) + [l(\lambda) - m(\lambda)]\varphi(x, \lambda), \psi(x, \lambda') + [l(\lambda') - m(\lambda')]\varphi(x, \lambda')\} = 0,$$

or, in greater detail,

$$\begin{aligned}
W_b\{\psi(x, \lambda), \psi(x, \lambda')\} &+ [l(\lambda) - m(\lambda)]W_b\{\varphi(x, \lambda), \psi(x, \lambda')\} \\
&+ [l(\lambda') - m(\lambda')]W_b\{\psi(x, \lambda), \varphi(x, \lambda')\} \\
&+ [l(\lambda) - m(\lambda)][l(\lambda') - m(\lambda')]W_b\{\varphi(x, \lambda), \varphi(x, \lambda')\} = 0. \quad (4.1)
\end{aligned}$$

It follows from the identity (2.2) that

$$W_b\{\varphi(x, \lambda), \psi(x, \lambda')\} = (\lambda' - \lambda)\int_0^b \varphi(x, \lambda)\psi(x, \lambda')\, dx + W_0\{\varphi(x, \lambda), \psi(x, \lambda')\}.$$

Therefore, by Theorem 2.2.1, as $b \to \infty$,

$$W_b\{\varphi(x, \lambda), \psi(x, \lambda')\} = O\left\{\int_a^b |\varphi(x, \lambda)|^2\, dx\right\} + O(1).$$

In the limit-point case,

$$|l(\lambda) - m(\lambda)| \le r_b = \left\{2v\int_0^b |\varphi(x, \lambda)|^2\, dx\right\}^{-1},$$

so that

$$\lim_{b \to \infty}[l(\lambda) - m(\lambda)]W_b\{\varphi(x, \lambda), \psi(x, \lambda')\} = 0.$$

This also occurs in the limit-circle case if $l(\lambda) \to m(\lambda)$, since the integral

$$\int_0^b |\varphi(x, \lambda)|^2\, dx$$

remains bounded. The other addends in (4.1) are estimated similarly, and the proof
is thus complete.   ∎

Lemma 2.4.2 *With the notation of Lemma 2.4.1, we have*

$$\int_0^\infty \psi(x,\lambda)\psi(x,\lambda')\,dx = \frac{m(\lambda) - m(\lambda')}{\lambda' - \lambda}. \tag{4.2}$$

Proof. By the identity (2.2),

$$(\lambda' - \lambda)\int_0^b \psi(x,\lambda)\psi(x,\lambda')\,dx = W_0\{\psi(x,\lambda),\psi(x,\lambda')\} - W_b\{\psi(x,\lambda),\psi(x,\lambda')\}. \tag{4.3}$$

Using the conditions (2.3), we see that the first term on the right is

$$[\cos\alpha + m(\lambda)\sin\alpha][\sin\alpha - m(\lambda')\cos\alpha] - [\cos\alpha + m(\lambda')\sin\alpha][\sin\alpha - m(\lambda)\cos\alpha]$$
$$= m(\lambda) - m(\lambda').$$

On the other hand, by Lemma 2.4.1, the second term on the right in (4.3) tends to zero as $b \to \infty$. Therefore, (4.2) follows from (4.3) if we pass to the limit as $b \to \infty$ in the latter. ∎

In particular, for $\lambda' = \bar\lambda$, $\lambda = u + iv$, it follows from (4.2) that

$$\int_0^\infty |\psi(x,\lambda)|^2\,dx = -\frac{\mathrm{Im}\{m(\lambda)\}}{v}, \tag{4.4}$$

so that equality actually occurs in formula (2.7).

Lemma 2.4.3 *For fixed u_1 and u_2, as $\delta \to 0$,*

$$\int_{u_1}^{u_2} -\mathrm{Im}\{m(u+i\delta)\}\,du = O(1). \tag{4.5}$$

Proof. Let $\sin\alpha \neq 0$. By (3.7) for $x = 0$ and the Parseval equation,

$$\int_0^\infty |\psi(y,z)|^2\,dy = \int_{-\infty}^\infty \frac{d\rho(\lambda)}{(u-\lambda)^2 + v^2}, \quad z = u + iv. \tag{4.6}$$

If $\sin\alpha = 0$, it should first be proved that the Fourier transform of the function $\partial G(x,y;z)/\partial x$ is $\varphi'(x,\lambda)/(z - \lambda)$, which follows from (3.4) if the equality is differentiated throughout with respect to x, and the limit is taken as $b \to \infty$. Putting $x = 0$, the formula (4.6) is reobtained.

It follows from (4.4) and (4.6) that

$$-\mathrm{Im}\{m(u+i\delta)\} = \delta\int_{-\infty}^\infty \frac{d\rho(\lambda)}{(u-\lambda)^2 + \delta^2}.$$

Integrating the last equality with respect to u from u_1 to u_2, we have

$$\int_{u_1}^{u_2} -\mathrm{Im}\{m(u+i\delta)\}\,du = \delta\int_{u_1}^{u_2} du \int_{-\infty}^\infty \frac{d\rho(\lambda)}{(u-\lambda)^2 + \delta^2}.$$

Let (a, b) be a finite interval, with $a < u_1$ and $b > u_2$. Then

$$\delta \int_{u_1}^{u_2} du \int_{-\infty}^{a} \frac{d\rho(\lambda)}{(u - \lambda)^2 + \delta^2} = O(1), \qquad \delta \int_{u_1}^{u_2} du \int_{b}^{\infty} \frac{d\rho(\lambda)}{(u - \lambda)^2 + \delta^2} = O(1),$$

relying on (3.6).

Furthermore, we have

$$\delta \int_{u_1}^{u_2} du \int_{a}^{b} \frac{d\rho(\lambda)}{(u - \lambda)^2 + \delta^2} = \int_{a}^{b} d\rho(\lambda) \int_{(u_1 - \lambda)/\delta}^{(u_2 - \lambda)/\delta} \frac{dv}{1 + v^2} = O(1),$$

and the lemma is proved.   ∎

Theorem 2.4.1 *Let the ends of the interval $\Delta = (\lambda, \lambda + \Delta)$ be points of continuity of the function $\rho(\lambda)$. Then*

$$\rho(\lambda + \Delta) - \rho(\lambda) = \frac{1}{\pi} \lim_{\delta \to 0} \int_{\Delta} \{-\operatorname{Im}[m(u + i\delta)]\} \, du. \tag{4.7}$$

Proof. Let $f(x), g(x) \in \mathcal{L}^2(0, \infty)$ vanish outside a finite interval. Relying on (3.7'),

$$\Phi(\lambda) = \int_{0}^{\infty} R_z f \cdot g(x) \, dx = \int_{-\infty}^{\infty} \frac{F(\lambda)G(\lambda)}{z - \lambda} \, d\rho(\lambda) = \int_{-\infty}^{\infty} \frac{d\sigma(\lambda)}{z - \lambda},$$

where $\sigma(\Delta) = \int_{\Delta} F(\lambda)G(\lambda) d\rho(\lambda)$. By the Stieltjes inversion formula,

$$\sigma(\Delta) = \frac{1}{\pi} \lim_{\delta \to 0} \int_{\Delta} \{-\operatorname{Im} \Phi(u + i\delta)\} \, du. \tag{4.8}$$

Furthermore, we have

$$\operatorname{Im} \Phi(u + i\delta) = \int_{0}^{\infty} g(x) \, dx$$
$$\times \operatorname{Im} \left\{ \int_{0}^{x} [\vartheta(x, u + i\delta) + m(u + i\delta)\varphi(x, u + i\delta)]\varphi(y, u + i\delta)f(y) \, dy \right.$$
$$\left. + \int_{x}^{\infty} [\vartheta(y, u + i\delta) + m(u + i\delta)\varphi(y, u + i\delta)]\varphi(x, u + i\delta)f(y) \, dy \right\},$$

where $\varphi(x, u)$, $\vartheta(x, u)$, $f(x)$ and $g(x)$ are real-valued functions. Therefore,

$$\sigma(\Delta) = \frac{1}{\pi} \lim_{\delta \to 0} \int_{\Delta} \{-\operatorname{Im}[m(u + i\delta)]\} F(u)G(u) \, du, \tag{4.9}$$

also taking into account the relation (4.8) and Lemma 2.4.3.

Selecting $f(x)$ and $g(x)$ conveniently, we can make $F(u)$ and $G(u)$ differ as little from unity as we please in the fixed interval Δ (see the proof of Lemma 2.1.1). Therefore, (4.7) follows from (4.9) and Lemma 2.4.3.   ∎

Theorem 2.4.2 *For any nonreal z, the formula*

$$m(z) = -\cot \alpha + \int_{-\infty}^{\infty} \frac{d\rho(\lambda)}{z - \lambda} \tag{4.10}$$

holds.

Proof. It follows from (3.7) that

$$G(x, y; z) = \int_{-\infty}^{\infty} \frac{\varphi(x, \lambda)\varphi(y, \lambda)}{z - \lambda} \, d\rho(\lambda), \tag{4.11}$$

since $f(x)$ is arbitrary.

On the other hand, by definition,

$$G(x, y; z) = \begin{cases} [\vartheta(x, z) + m(z)\varphi(x, z)]\varphi(y, z), & y \leq x \\ [\vartheta(y, z) + m(z)\varphi(y, z)]\varphi(x, z), & y > x \end{cases}.$$

It then follows from the conditions (2.3) and (4.11) that

$$G(0, 0; z) = \{\cos \alpha + m(z) \sin \alpha\} \sin \alpha = \int_{-\infty}^{\infty} \frac{\sin^2 \alpha}{z - \lambda} d\rho(\lambda),$$

i.e., formula (4.10). ∎

Note that (4.7) follows from (4.10), relying on the Stieltjes inversion formula.

2.5 Proof of the Parseval equation in the case of the whole line

We again consider the equation

$$y'' + \{\lambda - q(x)\}y = 0, \tag{5.1}$$

assuming, however, that x varies in the interval $(-\infty, \infty)$ and that $q(x)$ is continuous in each finite interval.

Let $\varphi(x, \lambda)$ and $\vartheta(x, \lambda)$ be solutions of (5.1), satisfying the initial conditions

$$\varphi(0, \lambda) = 0, \qquad \varphi'_x(0, \lambda) = 1, \tag{5.2}$$
$$\vartheta(0, \lambda) = 1, \qquad \vartheta'_x(0, \lambda) = 0 \tag{5.3}$$

and let $[a, b]$ be an arbitrary finite interval. Consider the boundary-value problem determined by (5.1) and the boundary conditions

$$\begin{aligned} y(a) \sin \alpha + y'(a) \cos \alpha &= 0, \\ y(b) \sin \beta + y'(b) \cos \beta &= 0, \end{aligned} \tag{5.4}$$

where α and β are two arbitrary real numbers.

Let $\lambda_1, \lambda_2, \ldots, \lambda_n, \ldots$ be the eigenvalues, and let $y_1(x), y_2(x), \ldots, y_n(x), \ldots$ be the corresponding orthonormal eigenfunctions of the problem. Since $\varphi(x, \lambda)$ and $\vartheta(x, \lambda)$ are two linearly independent solutions of (5.1),

$$y_n(x) = \alpha_n \varphi(x, \lambda_n) + \beta_n \vartheta(x, \lambda_n).$$

Let $f(x) \in \mathcal{L}^2(a, b)$. By the Parseval equation,

$$
\begin{aligned}
\int_a^b f^2(x)\,dx &= \sum_{n=1}^{\infty} \left\{ \int_a^b f(x)[\alpha_n \varphi(x, \lambda_n) + \beta_n \vartheta(x, \lambda_n)]\,dx \right\}^2 \\
&= \sum_{n=1}^{\infty} \alpha_n^2 \left\{ \int_a^b f(x)\varphi(x, \lambda_n)\,dx \right\}^2 + \sum_{n=1}^{\infty} \beta_n^2 \left\{ \int_a^b f(x)\vartheta(x, \lambda_n)\,dx \right\}^2 \\
&\quad + 2\sum_{n=1}^{\infty} \alpha_n \beta_n \int_a^b f(x)\varphi(x, \lambda_n)\,dx \int_a^b f(x)\vartheta(x, \lambda_n)\,dx.
\end{aligned}
\tag{5.5}
$$

If we now introduce the step functions with steps at the eigenvalues of the problem (5.1), (5.4)

$$
\xi_{a,b}(\lambda) = \begin{cases} \displaystyle\sum_{0 \le \lambda_n < \lambda} \alpha_n^2 & \text{for } \lambda > 0 \\[2ex] \displaystyle -\sum_{0 > \lambda_n > \lambda} \alpha_n^2 & \text{for } \lambda \le 0 \end{cases}
$$

$$
\eta_{a,b}(\lambda) = \begin{cases} \displaystyle\sum_{0 \le \lambda_n < \lambda} \alpha_n \beta_n & \text{for } \lambda > 0 \\[2ex] \displaystyle -\sum_{0 > \lambda_n > \lambda} \alpha_n \beta_n & \text{for } \lambda \le 0 \end{cases}
$$

$$
\zeta_{a,b}(\lambda) = \begin{cases} \displaystyle\sum_{0 \le \lambda_n < \lambda} \beta_n^2 & \text{for } \lambda > 0 \\[2ex] \displaystyle -\sum_{0 > \lambda_n > \lambda} \beta_n^2 & \text{for } \lambda \le 0 \end{cases}
$$

then the equality (5.5) can be rewritten as

$$
\begin{aligned}
\int_a^b f^2(x)\,dx &= \int_{-\infty}^{\infty} \left\{ \int_a^b f(x)\varphi(x, \lambda)\,dx \right\}^2 d\xi_{a,b}(\lambda) \\
&\quad + 2\int_{-\infty}^{\infty} \left\{ \int_a^b f(x)\varphi(x, \lambda)\,dx \right\} \left\{ \int_a^b f(x)\vartheta(x, \lambda)\,dx \right\} d\eta_{a,b}(\lambda) \\
&\quad + \int_{-\infty}^{\infty} \left\{ \int_a^b f(x)\vartheta(x, \lambda)\,dx \right\}^2 d\zeta_{a,b}(\lambda).
\end{aligned}
\tag{5.5'}
$$

Lemma 2.5.1 *For any positive N, there exists a positive number $A = A(N)$ not depending on a or b, so that*

$$\bigvee_{-N}^{N} \{\rho_{ij}(\lambda)\} < A, \qquad i, j = 1, 2, \tag{5.6}$$

where

$$\rho_{11} = \xi_{a,b}, \qquad \rho_{12} = \rho_{21} = \eta_{a,b}, \qquad \rho_{22} = \zeta_{a,b}. \tag{5.7}$$

Proof. To see the validity of (5.6), it suffices to put $i = j$, because

$$\bigvee_{-N}^{N} \{\rho_{12}(\lambda)\} \leq \frac{1}{2} \{\rho_{11}(N) - \rho_{11}(-N) + \rho_{22}(N) - \rho_{22}(-N)\}. \tag{5.8}$$

Let $\varphi(x, \lambda) = \psi_1(x, \lambda)$, $\vartheta(x, \lambda) = \psi_2(x, \lambda)$. The equality (5.5′) then takes the form

$$\int_a^b f^2(x) \, dx = \int_{-\infty}^{\infty} \sum_{i,j=1}^{2} F_i(\lambda) F_j(\lambda) \, d\rho_{ij}(\lambda) \tag{5.9}$$

by (5.7), where

$$F_i(\lambda) = \int_a^b f(x)\psi_i(x, \lambda) \, dx, \quad i = 1, 2.$$

Furthermore, since the functions $\psi_i^{(k-1)}(x, \lambda)$, $(i, k = 1, 2)$ are continuous both with respect to x and λ, it follows from the initial conditions (5.2), (5.3) that $\psi_i^{(k-1)}(0, \lambda) = \delta_{ik}$ ($\delta_{ik} = 1$ for $i = k$, $\delta_{ik} = 0$ for $i \neq k$), for any $\epsilon > 0$ and given $N > 0$, there exists $h > 0$ such that

$$|\psi_i^{(k-1)}(x, \lambda) - \delta_{ik}| < \epsilon \tag{5.10}$$

where $0 \leq x \leq h$, $|\lambda| \leq N$.

Let $g_h(x)$ be a twice continuously differentiable nonnegative function on the interval $[a, b]$, vanishing along with its first derivative outside the interval $[0, h]$ and normalized, so that

$$\int_0^h g_h(x) \, dx = 1. \tag{5.11}$$

Applying the Parseval equation (5.9) to the functions $g_h^{(k-1)}(x)$, $(k = 1, 2)$, we obtain

$$\int_0^h |g_h^{(k-1)}(x)|^2 \, dx \geq \int_{-N}^{N} \sum_{i,j=1}^{2} G_{ik}(\lambda) G_{jk}(\lambda) \, d\rho_{ij}(\lambda), \tag{5.12}$$

where

$$G_{ik}(\lambda) = \int_0^h g_h^{(k-1)}(x)\psi_i(x, \lambda) \, dx = \pm \int_0^h g_h(x)\psi_i^{(k-1)}(x, \lambda) \, dx.$$

It then follows from the estimates (5.10) and from the equality (5.11) that

$$1 - \epsilon < G_{ii}(\lambda) < 1 + \epsilon, \quad i = 1, 2; \qquad |G_{12}(\lambda)|, |G_{21}(\lambda)| < \epsilon. \tag{5.13}.$$

Putting $k = 1$ in (5.12), we infer that

$$\int_0^h g_h^2(x)\, dx \ \geq \ \int_{-N}^N G_{11}^2(\lambda)\, d\rho_{11}(\lambda) + 2 \int_{-N}^N G_{11}(\lambda) G_{21}(\lambda)\, d\rho_{12}(\lambda)$$

$$+ \int_{-N}^N G_{21}^2(\lambda)\, d\rho_{22}(\lambda)$$

$$\geq \ \int_{-N}^N G_{11}^2(\lambda)\, d\rho_{11}(\lambda) - 2 \int_{-N}^N |G_{11}(\lambda)|\, |G_{21}(\lambda)|\, |d\rho_{12}(\lambda)|,$$

whence

$$\int_0^h g_h^2(x)\, dx \ \geq \ \int_{-N}^N (1 - \epsilon)^2\, d\rho_{11}(\lambda) - 2 \int_{-N}^N \epsilon(1 + \epsilon)\, |d\rho_{12}(\lambda)|$$

$$= \ (1 - \epsilon)^2 \{\rho_{11}(N) - \rho_{11}(-N)\} - 2\epsilon(1 + \epsilon) \bigvee_{-N}^N \{\rho_{12}(\lambda)\}$$

$$\geq \ (1 - \epsilon)^2 \{\rho_{11}(N) - \rho_{11}(-N)\}$$

$$- \epsilon(1 + \epsilon)\{\rho_{11}(N) - \rho_{11}(-N) + \rho_{22}(N) - \rho_{22}(-N)\}$$

$$= \ (1 - 3\epsilon)\{\rho_{11}(N) - \rho_{11}(-N)\}$$

$$- \epsilon(1 + \epsilon)\{\rho_{22}(N) - \rho_{22}(-N)\}, \tag{5.14}$$

relying on (5.13) and (5.8).

Similarly, putting $k = 2$ in (5.12) and using (5.8), (5.13), we have

$$\int_0^h |g_h'(x)|^2\, dx \geq (1 - 3\epsilon)\{\rho_{22}(N) - \rho_{22}(-N)\} - \epsilon(1 + \epsilon)\{\rho_{11}(N) - \rho_{11}(-N)\}. \tag{5.15}$$

Adding the inequalities (5.14) and (5.15) together, we obtain

$$\int_{-N}^N \{g_h^2(x) + |g_h'(x)|^2\}\, dx > (1 - 4\epsilon - \epsilon^2)\{\rho_{11}(N) - \rho_{11}(-N) + \rho_{22}(N) - \rho_{22}(-N)\},$$

which proves the estimate (5.6) for the functions $\rho_{11}(\lambda)$ and $\rho_{22}(\lambda)$, and thereby the lemma itself (if $\epsilon > 0$ is selected on the basis of the condition $1 - 4\epsilon - \epsilon^2 > 0$). ∎

We now show that, using Lemma 2.5.1 and the familiar theorems on passing to the limit under the Stieltjes integral sign, the Parseval equation can be derived from the equality (5.4) in the case of the whole line.

Theorem 2.5.1 *Let $f(x) \in \mathcal{L}^2(-\infty, \infty)$. There exist monotone functions $\xi(\lambda)$ and $\zeta(\lambda)$, bounded in each finite interval and not depending of the function $f(x)$, and* *a function $\eta(\lambda)$ with bounded variation in each finite interval, so that the Parseval equation*

$$\int_{-\infty}^{\infty} f^2(x)\,dx = \int_{-\infty}^{\infty} F^2(\lambda)d\xi(\lambda) + 2\int_{-\infty}^{\infty} F(\lambda)G(\lambda)d\eta(\lambda) + \int_{-\infty}^{\infty} G^2(\lambda)\,d\zeta(\lambda) \quad (5.16)$$

holds, where

$$F(\lambda) = \lim_{n\to\infty} \int_{-n}^{n} f(x)\varphi(x,\lambda)\,dx$$

$$G(\lambda) = \lim_{n\to\infty} \int_{-n}^{n} f(x)\vartheta(x,\lambda)\,dx.$$

The matrix $\begin{pmatrix} \xi(\lambda) & \eta(\lambda) \\ \eta(\lambda) & \xi(\lambda) \end{pmatrix}$ *is said to be spectral.*

Proof. First, let the function $f_n(x)$ vanish outside the interval $[-n, n]$, have a continuous second derivative and satisfy the boundary conditions (5.4). Applying the Parseval equation (5.5), we obtain

$$\int_{-n}^{n} f_n^2(x)\,dx = \sum_{k=1}^{\infty} \left\{ \int_{-n}^{n} f_n(x)y_k(x)\,dx \right\}^2 \quad (5.17)$$

under the assumption that $a > -n$ and $b < n$.

Since both $f_n(x)$ and $y_k(x)$ satisfy the boundary conditions (5.4) and $f(x)$ is identically zero in the neighborhood of the points a and b, integrating by parts twice, we obtain

$$\int_{-n}^{n} f_n(x)y_k(x)\,dx = -\frac{1}{\lambda_k} \int_{-n}^{n} f_n(x)\{y_k''(x) - q(x)y_k(x)\}\,dx$$

$$= -\frac{1}{\lambda_k} \int_{-n}^{n} \{f_n''(x) - q(x)f_n(x)\}y_k(x)\,dx.$$

Therefore,

$$\sum_{|\lambda_k|\geq\mu} \left\{ \int_{-n}^{n} f_n(x)y_k(x)\,dx \right\}^2 \leq \frac{1}{\mu^2} \sum_{|\lambda_k|\geq\mu} \left\{ \int_{-n}^{n} [f_n''(x) - q(x)f_n(x)]y_k(x)\,dx \right\}^2$$

$$\leq \frac{1}{\mu^2} \sum_{k=1}^{\infty} \left\{ \int_{-n}^{n} [f_n''(x) - q(x)f_n(x)]y_k(x)\,dx \right\}^2$$

$$= \frac{1}{\mu^2} \int_{-n}^{n} \{f_n''(x) - q(x)f_n(x)\}^2\,dx.$$

It follows from this estimate and from the equality (5.17) that

$$\left| \int_{n}^{n} f_n^2(x)\,dx - \sum_{-\mu\leq\lambda_k\leq\mu} \left\{ \int_{-n}^{n} f_n(x)y_k(x)\,dx \right\}^2 \right| < \frac{1}{\mu^2} \int_{-n}^{n} \{f_n''(x) - q(x)f_n(x)\}^2\,dx.$$

$$(5.18)$$

Furthermore, we have

$$\sum_{-\mu \le \lambda_k \le \mu} \left\{ \int_{-n}^{n} f_n(x) y_k(x)\, dx \right\}^2 = \sum_{-\mu \le \lambda_k \le \mu} \left\{ \int_{-n}^{n} f_n(x)[\alpha_k \varphi(x, \lambda_k) + \beta_k \vartheta(x, \lambda_k)]\, dx \right\}^2$$

$$= \int_{-\mu}^{\mu} F_n^2(\lambda)\, d\xi_{a,b}(\lambda) + 2 \int_{-\mu}^{\mu} F_n(\lambda) G_n(\lambda)\, d\eta_{a,b}(\lambda)$$

$$+ \int_{-\mu}^{\mu} G_n^2(\lambda)\, d\zeta_{a,b}(\lambda),$$

where

$$F_n(\lambda) = \int_{-n}^{n} f_n(x) \varphi(x, \lambda)\, dx, \qquad G_n(\lambda) = \int_{-n}^{n} f_n(x) \vartheta(x, \lambda)\, dx.$$

The inequality (5.18) can then be rewritten in the form

$$\left| \int_{-n}^{n} f_n^2(x)\, dx - \int_{-\mu}^{\mu} \{ F_n^2(\lambda) d\xi_{a,b}(\lambda) + 2 F_n(\lambda) G_n(\lambda)\, d\eta_{a,b}(\lambda) + G_n^2(\lambda)\, d\zeta_{a,b}(\lambda) \} \right|$$

$$< \frac{1}{\mu^2} \int_{-n}^{n} \{ f_n''(x) - q(x) f_n(x) \}^2\, dx. \quad (5.18')$$

By Lemma 2.5.1, the sets of functions $\{\xi_{a,b}(\lambda)\}$ and $\{\zeta_{a,b}(\lambda)\}$, $\mu \le \lambda \le \mu$, are bounded, while the set $\{\eta_{a,b}(\lambda)\}$, $-\mu \le \lambda \le \mu$, is bounded and has uniformly bounded variation. Therefore, we can select two sequences a_k and b_k for which the functions $\xi_{a_k,b_k}(\lambda)$, $\eta_{a_k,b_k}(\lambda)$ and $\zeta_{a_k,b_k}(\lambda)$ converge to $\xi(\lambda)$, $\eta(\lambda)$ and $\zeta(\lambda)$, respectively, as $a_k \to -\infty$ and $b_k \to \infty$. Passing to the limit in (5.18') with respect to the sequences a_k and b_k, we obtain

$$\left| \int_{-n}^{n} f_n^2(x)\, dx - \int_{-\mu}^{\mu} \{ F_n^2(\lambda) d\xi(\lambda) + 2 F_n(\lambda) G_n(\lambda)\, d\eta(\lambda) + G_n^2(\lambda)\, d\zeta(\lambda) \} \right|$$

$$\le \frac{1}{\mu^2} \int_{-n}^{n} \{ f_n''(x) - q(x) f_n(x) \}^2\, dx.$$

Finally, letting $\mu \to \infty$,

$$\int_{-n}^{n} f_n^2(x)\, dx = \int_{-\infty}^{\infty} \{ F_n^2(\lambda)\, d\xi(\lambda) + 2 F_n(\lambda) G_n(\lambda)\, d\eta(\lambda) + G_n^2(\lambda)\, d\zeta(\lambda) \},$$

i.e., the equality (5.16) for $f_n(x)$ satisfying the above conditions.

The extending the Parseval equation to arbitrary functions of the class $\mathcal{L}^2(-\infty, \infty)$ is done by the usual methods.   ∎

Let $f(x), g(x) \in \mathcal{L}^2(-\infty, \infty)$, and let $F_1(\lambda)$ and $G_1(\lambda)$ be constructed from the function $g(x)$ in the way $F(\lambda)$ and $G(\lambda)$ are from $f(x)$. It is obvious that $f(x) \pm g(x)$ have the functions $F(\lambda) \pm F_1(\lambda)$ and $G(\lambda) \pm G_1(\lambda)$ as their transforms. Therefore,

$$\int_{-\infty}^{\infty} \{ f(x) + g(x) \}^2\, dx = \int_{-\infty}^{\infty} \{ F(\lambda) + F_1(\lambda) \}^2\, d\xi(\lambda)$$

$$+ 2 \int_{-\infty}^{\infty} \{ F(\lambda) + F_1(\lambda) \} \{ G(\lambda) + G_1(\lambda) \}\, d\eta(\lambda) + \int_{-\infty}^{\infty} \{ G(\lambda) + G_1(\lambda) \}^2\, d\zeta(\lambda),$$

$$\int_{-\infty}^{\infty} \{f(x) - g(x)\}^2\, dx = \int_{-\infty}^{\infty} \{F(\lambda) - F_1(\lambda)\}^2\, d\xi(\lambda)$$

$$+ 2\int_{-\infty}^{\infty} \{F(\lambda) - F_1(\lambda)\}\{G(\lambda) - G_1(\lambda)\}d\eta(\lambda) + \int_{-\infty}^{\infty} \{G(\lambda) - G_1(\lambda)\}^2 d\zeta(\lambda).$$

Subtracting the lower equality from the upper, we have

$$\int_{-\infty}^{\infty} f(x)g(x)\, dx = \int_{-\infty}^{\infty} F(\lambda)F_1(\lambda)\, d\xi(\lambda)$$

$$+ \int_{-\infty}^{\infty} \{F(\lambda)G_1(\lambda) + F_1(\lambda)G(\lambda)\}\, d\eta(\lambda) + \int_{-\infty}^{\infty} G(\lambda)G_1(\lambda)\, d\zeta(\lambda). \tag{5.19}$$

This is called the *generalized Parseval equation.*

Theorem 2.5.2 *(Expansion Theorem.) Let $f(x) \in \mathcal{L}^2(-\infty, \infty)$ be continuous on the whole axis, and let the integrals*

$$\int_{-\infty}^{\infty} F(\lambda)\varphi(x, \lambda)\, d\xi(\lambda), \quad \int_{-\infty}^{\infty} F(\lambda)\vartheta(x, \lambda)\, d\eta(\lambda),$$

$$\int_{-\infty}^{\infty} G(\lambda)\varphi(x, \lambda)\, d\eta(\lambda), \quad \int_{-\infty}^{\infty} G(\lambda)\vartheta(x, \lambda)\, d\zeta(\lambda) \tag{5.20}$$

converge absolutely and uniformly in x over each finite interval. Then the equality

$$f(x) = \int_{-\infty}^{\infty} F(\lambda)\varphi(x, \lambda)\, d\xi(\lambda) + \int_{-\infty}^{\infty} F(\lambda)\vartheta(x, \lambda)\, d\eta(\lambda)$$

$$+ \int_{-\infty}^{\infty} G(\lambda)\varphi(x, \lambda)\, d\eta(\lambda) + \int_{-\infty}^{\infty} G(\lambda)\vartheta(x, \lambda)\, d\zeta(\lambda) \tag{5.21}$$

holds.

Proof. Let $g(x)$ be a continuous function vanishing outside a finite interval $[-n, n]$. Then the generalized Parseval equation (5.19) can be written

$$\int_{-n}^{n} f(x)g(x)\, dx = \int_{-\infty}^{\infty} F(\lambda)\left\{\int_{-n}^{n} g(x)\varphi(x, \lambda)\, dx\right\} d\xi(\lambda)$$

$$+ \int_{-\infty}^{\infty} F(\lambda)\left\{\int_{-n}^{n} g(x)\vartheta(x, \lambda)\, dx\right\} d\eta(\lambda) + \int_{-\infty}^{\infty} G(\lambda)\left\{\int_{-n}^{n} g(x)\varphi(x, \lambda)\, dx\right\} d\eta(\lambda)$$

$$+ \int_{-\infty}^{\infty} G(\lambda)\left\{\int_{-n}^{n} g(x)\vartheta(x, \lambda)\, dx\right\} d\zeta(\lambda).$$

The order of integration can be changed in the integrals on the right-hand side, since (5.20) converge absolutely. We obtain

$$\int_{-n}^{n} f(x)g(x)\, dx = \int_{-n}^{n} g(x)\left\{\int_{-\infty}^{\infty} F(\lambda)\varphi(x, \lambda)\, d\xi(\lambda) + \int_{-\infty}^{\infty} F(\lambda)\vartheta(x, \lambda)\, d\eta(\lambda)\right.$$

$$\left. + \int_{-\infty}^{\infty} G(\lambda)\varphi(x, \lambda)\, d\eta(\lambda) + \int_{-\infty}^{\infty} G(\lambda)\vartheta(x, \lambda)\, d\zeta(\lambda)\right\} dx.$$

Since $g(x)$ is arbitrary, $f(x)$ and the integrals in (5.20) are continuous (as functions of x, they are continuous thanks to the assumed uniform convergence), the theorem follows. ∎

2.6 Floquet (Bloch) solutions

Consider the equation

$$y'' + \{\lambda - q(x)\}y = 0, \tag{6.1}$$

where $q(x)$ is a real-valued and periodic function with period a, i.e., $q(x + a) = q(x)$ for any real x.

We already know (see Section 4, Chapter 1) that, because of the periodicity of $q(x)$, it is natural to consider boundary-value problems for (6.1) with the boundary conditions

$$y(0) = y(a), \qquad y'(0) = y'(a) \tag{6.2}$$

$$y(0) = -y(a), \qquad y'(0) = -y'(a). \tag{6.3}$$

The problem (6.1), (6.2) was called *periodic*, and (6.1), (6.3) was called *semi-periodic*.

Let $\varphi(x, \lambda)$ and $\vartheta(x, \lambda)$ be solutions of (6.1), satisfying the initial conditions

$$\varphi(0, \lambda) = 0, \qquad \varphi'(0, \lambda) = 1 \tag{6.4}$$

$$\vartheta(0, \lambda) = 1, \qquad \vartheta'(0, \lambda) = 0. \tag{6.5}$$

Since it follows that $W\{\vartheta, \varphi\} = 1$, $\varphi(x, \lambda)$ and $\vartheta(x, \lambda)$ are linearly independent, and, therefore, any solution of (6.1) is of the form

$$y(x, \lambda) = c_1\vartheta(x, \lambda) + c_2\varphi(x, \lambda),$$

where c_1 and c_2 are two arbitrary constants. Hence, there exists a one-to-one correspondence $y(x, \lambda) \leftrightarrow C$ between the solutions of (6.1) and the two-dimensional vectors $C = \begin{pmatrix} c_1 \\ c_2 \end{pmatrix}$.

Put $y(x + a, \lambda) = y_a(x, \lambda)$. Since $q(x)$ is a periodic function (with period a), $y_a(x, \lambda)$ is also a solution of (6.1), with

$$y_a(0, \lambda) = y(a, \lambda) = c_1\vartheta(a, \lambda) + c_2\varphi(a, \lambda),$$

$$y_a'(0, \lambda) = y'(a, \lambda) = c_1\vartheta'(a, \lambda) + c_2\varphi'(a, \lambda).$$

Therefore, $y_a(x, \lambda) = y(x + a, \lambda) \leftrightarrow T(a, \lambda)C$, with the matrix

$$T(a, \lambda) = \begin{pmatrix} \vartheta(a, \lambda) & \varphi(a, \lambda) \\ \vartheta'(a, \lambda) & \varphi'(a, \lambda) \end{pmatrix}$$

(which plays an important role in the theory of the periodic problem) called the *monodromy matrix*. It is obvious that $y(x + na, \lambda) \leftrightarrow T^nC$ for any n.

To find the eigenvalues and corresponding eigenvectors of the monodromy matrix (I being the identity 2×2 matrix), consider

$$|T(a, \lambda) - \kappa I| = \begin{vmatrix} \vartheta - \kappa & \varphi \\ \vartheta' & \varphi' - \kappa \end{vmatrix} = \kappa^2 - F(\lambda)\kappa + 1 = 0$$

$$F(\lambda) = \vartheta(a, \lambda) + \varphi'(a, \lambda).$$

We obtain the formula

$$\kappa = \frac{1}{2}\left\{F(\lambda) \pm \sqrt{F^2(\lambda) - 4}\right\} \tag{6.6}$$

for the eigenvalues of $T(a, \lambda)$.

For $\lambda \in (-\infty, \lambda_0)$, we put

$$\kappa^+(\lambda) = \frac{1}{2}\left\{F(\lambda) - \sqrt{F^2(\lambda) - 4}\right\},$$

$$\kappa^-(\lambda) = \frac{1}{2}\left\{F(\lambda) + \sqrt{F^2(\lambda) - 4}\right\}, \tag{6.6'}$$

taking the principal value of the root. For other λ, the functions $\kappa^{\pm}(\lambda)$ are determined by analytic continuation. It is obvious that, for $\lambda \in (-\infty, \lambda_0)$, $|\kappa^+(\lambda)| < 1$ and $|\kappa^-(\lambda)| > 1$. The same inequalities hold for all λ not in $[\lambda_0, \mu_0], [\lambda_1, \mu_1], [\lambda_2, \mu_2], \ldots$; this can be seen in the sequel.

Selecting the eigenvectors of $T(a, \lambda)$ in the form $e^+ = \begin{pmatrix} 1 \\ m^+ \end{pmatrix}$, $e^- = \begin{pmatrix} 1 \\ m^- \end{pmatrix}$, we obtain $(\vartheta - \kappa^+) + \varphi m^+ = 0$.

Hence,

$$m^+(\lambda) = m^+ = \frac{\kappa^+ - \vartheta}{\varphi} = \frac{\varphi' - \vartheta - \sqrt{F^2(\lambda) - 4}}{2\varphi}. \tag{6.7}$$

Similarly,

$$m^-(\lambda) = m^- = \frac{\varphi' - \vartheta + \sqrt{F^2(\lambda) - 4}}{2\varphi}, \tag{6.8}$$

$m^{\pm}(\lambda)$ coinciding with the Weyl-Titchmarsh functions (see below).

It follows from the definition of an eigenvector that $Te^{\pm} = \kappa^{\pm}e^{\pm}$; therefore,

$$T^n e^{\pm} = (\kappa^{\pm})^n e^{\pm}. \tag{6.9}$$

Consider the following solutions of the equation (4.1) :

$$\psi^+(x, \lambda) = \vartheta(x, \lambda) + m^+\varphi(x, \lambda)$$

$$\psi^-(x, \lambda) = \vartheta(x, \lambda) + m^-\varphi(x, \lambda) \tag{6.10}$$

for fixed complex λ. The functions $\psi^+(x, \lambda)$ and $\psi^-(x, \lambda)$ are called *Floquet (Bloch) solutions*.

Relying on the equation (6.9),

$$T^n \psi^\pm(x, \lambda) = \psi^\pm(x + na, \lambda) = (\kappa^\pm)^n \psi^\pm(x, \lambda).$$

Thus, for the Floquet solutions to be bounded on the whole number axis, it is necessary and sufficient that $|\kappa^\pm(\lambda)| = 1$. It can be seen from the formula (6.6) that, for real[3] λ, this holds if $|F(\lambda)| \leq 2$. However, if $|F(\lambda)| > 2$, one of $|\kappa^\pm|$ is greater, and the other one is less than 1, so the solutions are unbounded.

Put

$$\chi^\pm(x, \lambda) = (\kappa^\pm)^{-x/a} \psi^\pm(x, \lambda). \tag{6.11}$$

Then

$$\chi^\pm(x + na, \lambda) = (\kappa^\pm)^{-x/a-n} \psi^\pm(x + na, \lambda) = (\kappa^\pm)^{-x/a} \psi^\pm(x, \lambda) = \chi^\pm(x, \lambda),$$

i.e., $\chi^\pm(x, \lambda)$ are periodic functions with period a.

It follows from the representation in (6.11) that

$$\psi^\pm(x, \lambda) = (\kappa^\pm)^{x/a} \chi^\pm(x, \lambda).$$

Therefore, if λ is such that $|\kappa^+| < 1$ and, therefore, $|\kappa^-| > 1$, then $\psi^+(x, \lambda) \in \mathcal{L}^2(0, \infty)$ and $\psi^-(x, \lambda) \in \mathcal{L}^2(-\infty, 0)$, i.e., $\psi^\pm(x, \lambda)$ are Weyl solutions.

We have seen above that $|\kappa^+| < 1$ for $\lambda \in (-\infty, \lambda_0)$, and that $|\kappa^-| > 1$. Therefore, $\psi^+ \in \mathcal{L}^2(0, \infty)$ and $\psi^- \in \mathcal{L}^2(-\infty, 0)$ for these λ. For an analytic continuation of the function $\psi^\pm(x, \lambda)$, this is also valid. In particular, the above statement concerning $|\kappa^\pm|$ follows.

For real λ in the regions where $|F(\lambda)| < 2$, $\psi^+(x, \lambda) = \bar{\psi}^-(x, \lambda)$, for κ^+ and κ^- are complex conjugate. However, if $|F(\lambda)| > 2$, then $\psi^\pm(x, \lambda)$ are real-valued. Such λ cannot be associated either with periodic or semi-periodic solutions.

It now remains to discuss the case $F(\lambda) = \pm 2$. If $F(\lambda) = 2$, then $\kappa^\pm = 1$; if $F(\lambda) = -2$, $\kappa^\pm = -1$. In the former case, $\psi^+(x, \lambda) = \psi^-(x, \lambda) = \chi^\pm(x, \lambda)$; in the latter, $\psi^+(x, \lambda) = (-1)^{x/a} \chi^+(x, \lambda) = e^{i\pi x/a} \chi^+(x, \lambda) = \psi^-(x, \lambda)$, i.e., we obtain only one Floquet solution which is a periodic function.

If the corresponding value $\lambda(\mu)$ is a multiple root of the equation $F(\lambda) = 2$ (resp. $F(\lambda) = -2$), then, as has been shown above, the other solution is a periodic (resp. semi-periodic) function. However, if $\lambda(\mu)$ is a simple root, then we can show that the other solution is of the form $\psi(x, \lambda) + xp(x, \lambda)$, where $p(x, \lambda)$ is a periodic (resp. semi-periodic) function.

---

[3]For real λ, $F(\lambda)$ is also real-valued, since $q(x)$ is real-valued in the equation (6.1).

2.7 Eigenfunction expansion in the case of a periodic potential

Consider the equation

$$y'' + \{\lambda - q(x)\}y = 0 \qquad (7.1)$$

on the whole line, assuming that $q(x)$ is a real-valued periodic function with period a, i.e., $q(x + a) = q(x)$. Let $\varphi(x, \lambda)$ and $\vartheta(x, \lambda)$ be as in the preceding section. By Theorem 2.5.2, the general expansion formula

$$f(x) = \int_{-\infty}^{\infty} \{g(\lambda)\varphi(x, \lambda)\, d\xi(\lambda) + h(\lambda)\vartheta(x, \lambda)\, d\zeta(\lambda) + [g(\lambda)\vartheta(x, \lambda) + h(\lambda)\varphi(x, \lambda)]\}\, d\eta(\lambda),$$
$$(7.2)$$

where

$$g(\lambda) = \int_{-\infty}^{\infty} f(x)\varphi(x, \lambda)\, dx, \qquad h(\lambda) = \int_{-\infty}^{\infty} f(x)\vartheta(x, \lambda)\, dx,$$

is then valid.

Theorem 2.4.1 yields the formula (4.7) expressing the spectral function $\rho(\lambda)$ in terms of the Weyl-Titchmarsh function $m(\lambda)$. Similar formulas expressing the elements $\xi(\lambda)$, $\eta(\lambda)$ and $\zeta(\lambda)$ of the spectral matrix in terms of the Weyl-Titchmarsh functions $m^+(\lambda)$ and $m^-(\lambda)$ also hold (see B. M. Levitan and I. S. Sargsjan (1975, Section 8, Chapter 2)), viz.

$$\xi(\lambda) = \frac{1}{\pi} \lim_{\delta \to 0} \int_{\Delta} - \operatorname{Im}\left\{\frac{m^-(u + i\delta) \cdot m^+(u + i\delta)}{m^-(u + i\delta) - m^+(u + i\delta)}\right\} du$$

$$\eta(\lambda) = \frac{1}{2\pi} \lim_{\delta \to 0} \int_{\Delta} - \operatorname{Im}\left\{\frac{m^-(u + i\delta) + m^+(u + i\delta)}{m^-(u + i\delta) - m^+(u + i\delta)}\right\} du \qquad (7.3)$$

$$\zeta(\lambda) = \frac{1}{\pi} \lim_{\delta \to 0} \int_{\Delta} - \operatorname{Im}\left\{\frac{1}{m^-(u + i\delta) - m^+(u + i\delta)}\right\} du$$

where $\Delta = (\lambda, \lambda + \Delta)$.

We introduce the notation

$$E = (\lambda_0, \mu_0) \cup (\lambda_1, \mu_1) \cup (\lambda_2, \mu_2) \cup \dots, \qquad p(\lambda) = \sqrt{4 - F^2(\lambda)},$$

selecting the sign of the root in accordance with the analytic continuation specified in Section 6.

By the formulas (6.7) and (6.8), if follows from (7.3) that

$$\frac{d\xi(\lambda)}{d\lambda} = \begin{cases} -\vartheta'(a, \lambda)/(\pi p(\lambda)), & \lambda \in E \\ 0, & \lambda \notin E \end{cases}$$

$$\frac{d\eta(\lambda)}{d\lambda} = \begin{cases} \{\varphi'(a, \lambda) - \vartheta(a, \lambda)\}/(2\pi p(\lambda)), & \lambda \in E \\ 0, & \lambda \notin E \end{cases}$$

$$\frac{d\zeta(\lambda)}{d\lambda} = \begin{cases} \varphi(a,\lambda)/(\pi p(\lambda)), & \lambda \in E \\ 0, & \lambda \notin E \end{cases}$$

which, substituted in the formula (7.2), yields

$$f(x) = \frac{1}{\pi} \int_{\lambda_0}^{\infty *} \Big\{ h(\lambda)\vartheta(x,\lambda)\varphi'(a,\lambda) - g(\lambda)\varphi(x,\lambda)\vartheta'(a,\lambda)$$

$$+ [h(\lambda)\varphi(x,\lambda) + g(\lambda)\vartheta(x,\lambda)]\frac{\varphi'(a,\lambda) - \vartheta(a,\lambda)}{2} \Big\} \frac{d\lambda}{p(\lambda)}, \quad (7.4)$$

where the asterisk means that the integration is with respect to the spectrum, i.e., the set E, taking the positive root of $p(\lambda)$ in the interval (λ_0, μ_0) and then alternating its sign in the subsequent intervals, in accordance with the analytic continuation of the root.

We can now derive the expansion, using the basis of Floquet functions $\psi^\pm(x,\lambda)$, which often turns out to be preferable.

Without loss of generality, we assume that $\lambda_0 = 0$; otherwise, λ should be replaced by $\lambda + \lambda_0$ in (7.1), and the potential $q(x)$ should be replaced by $q(x) + \lambda_0$. Assuming that $\lambda_0 = 0$, we substitute κ^2 for λ. (7.4) then takes the form

$$f(x) = \frac{1}{\pi} \int_0^{\infty *} \Big\{ h(\kappa^2)\vartheta(x,\kappa^2)\varphi'(a,\kappa^2) - g(\lambda)\varphi(x,\kappa^2)\vartheta'(a,\kappa^2)$$

$$+ [h(\kappa^2)\varphi(x,\kappa^2) + g(\kappa^2)\vartheta(x,\kappa^2)]\frac{\varphi'(a,\kappa') - \vartheta(a,\kappa^2)}{2} \Big\} \frac{2\kappa \, d\kappa}{\sqrt{4 - F^2(\kappa^2)}}. \quad (7.5)$$

Since $\lambda = 0$ is a simple root of the equation $F(\lambda) - 2 = 0$, the function $4 - F^2(\kappa^2)$ has a root of precisely second order as $\kappa \to 0$. Therefore, the function $\sqrt{4 - F^2(\kappa^2)}$ can be naturally extended to negative κ as an odd function. Furthermore, since κ is an odd function and the other terms in the formula (7.5) are even, (7.5) can be rewritten as

$$f(x) = \frac{1}{\pi} \int_{-\infty}^{\infty *} \Big\{ h(\kappa^2)\vartheta(x,\kappa^2)\varphi'(a,\kappa^2) - g(\lambda)\varphi(x,\kappa^2)\vartheta'(a,x^2) +$$

$$[h(\kappa^2)\varphi(x,\kappa^2) + g(\kappa^2)\vartheta(x,\kappa^2)]\frac{\varphi'(a,\kappa^2) - \vartheta(a,\kappa^2)}{2} \Big\} \frac{\kappa \, d\kappa}{\sqrt{4 - F^2(\kappa^2)}}. \quad (7.6)$$

This can be written in more compact form :

$$f(x) = \frac{1}{\pi} \int_{-\infty}^{\infty *} H(x)\psi(x,\kappa)\frac{\varphi(a,\kappa^2)}{\sqrt{4 - F^2(\kappa^2)}}\kappa \, d\kappa, \quad (7.7)$$

where $H(\kappa) = \int_{-\infty}^{\infty} f(y)\bar{\psi}(y,\kappa) \, dy$, $\psi(x,\kappa) = \psi^+(x,\kappa^2)$.

In fact,

$$\frac{1}{\pi}\int_{-\infty}^{\infty} H(\kappa)\psi(x,\kappa)\frac{\varphi(a,\kappa^2)}{\sqrt{4-F^2(\kappa^2)}}\kappa\,d\kappa$$

$$=\frac{1}{\pi}\int_{-\infty}^{\infty} f(y)\left\{\int_{-\infty}^{\infty}{}^{*}\psi(x,\kappa)\bar{\psi}(y,\kappa)\frac{\varphi(a,\kappa^2)}{\sqrt{4-F^2(\kappa^2)}}\kappa\,d\kappa\right\}dy.\quad(7.8)$$

Furthermore, since $\bar{\psi}^+(x,\kappa^2)=\psi^-(x,\kappa^2)$ by the formulas (6.7), (6.8) and (6.10), using the expression (6.10), we get

$$
\begin{aligned}
\psi(x,\kappa)\bar{\psi}(y,\kappa) &= \psi^+(x,\kappa^2)\psi^-(y,\kappa^2)\\
&= [\vartheta(x,\kappa^2)+m^+(\kappa^2)\varphi(x,\kappa^2)][\vartheta(y,\kappa^2)+m^-(\kappa^2)\varphi(y,\kappa^2)]\\
&= \vartheta(x,\kappa^2)\vartheta(y,\kappa^2)+\vartheta(x,\kappa^2)\varphi(y,\kappa^2)m^-(\kappa^2)\\
&\quad +m^+(\kappa^2)\varphi(x,\kappa^2)\vartheta(x,\kappa^2)+m^+(\kappa^2)m^-(\kappa^2)\varphi(x,\kappa^2)\varphi(y,\kappa^2).
\end{aligned}
$$

Taking into account the expressions for $m^\pm(\kappa^2)$ from (6.6'), we then obtain

$$
\psi(x,\kappa)\bar{\psi}(y,\kappa)\frac{\kappa\varphi(a,\kappa^2)}{\sqrt{4-F^2(\kappa^2)}}=\vartheta(x,\kappa^2)\vartheta(y,\kappa^2)\frac{\kappa\varphi(a,\kappa^2)}{\sqrt{4-F^2(\kappa^2)}}
$$

$$
+\frac{\varphi'(a,\kappa^2)-\vartheta(a,\kappa^2)}{2}[\vartheta(x,\kappa^2)\varphi(y,\kappa^2)+\vartheta(y,\kappa^2)\varphi(x,\kappa^2)]\frac{\kappa}{\sqrt{4-F^2(\kappa^2)}}
$$

$$
-\varphi(x,\kappa^2)\varphi(y,\kappa^2)\frac{\kappa\vartheta'(a,\kappa^2)}{\sqrt{4-F^2(\kappa^2)}}
$$

$$
+\frac{\kappa i}{2}[\varphi'(a,\kappa^2)-\vartheta(a,\kappa^2)][\vartheta(x,\kappa^2)\varphi(y,\kappa^2)-\vartheta(y,\kappa^2)\varphi(x,\kappa^2)].
$$

The last addend on the right is odd; therefore, it gives no contribution to (7.8) on substituting; this shows that the formulas (7.6) and (7.7) are equivalent.

The equality (7.7) can be written in symbolic form as

$$\frac{1}{\pi}\int_{-\infty}^{\infty}{}^{*}\psi(x,\kappa)\bar{\psi}(x,\kappa)\frac{\varphi(a,\kappa^2)}{\sqrt{4-F^2(\kappa^2)}}\kappa\,d\kappa=\delta(x-y),$$

where $\delta(x)$ is the Dirac delta function.

Since φ is positive in the intervals (λ_{2n},μ_{2n}) (as a function of λ), and negative in $(\mu_{2n-1},\lambda_{2n-1})$, (7.4) can be rewritten (taking into account the domain of integration, i.e., the set E) in the form

$$
\begin{aligned}
f(x)=\frac{1}{\pi}\left(\sum_{n=0}^{\infty}\int_{\lambda_{2n}}^{\mu_{2n}}-\sum_{n=1}^{\infty}\int_{\mu_{2n-1}}^{\lambda_{2n-1}}\right)&\left\{\varphi\vartheta(x,\lambda)g(\lambda)-\vartheta'\varphi(x,\lambda)h(\lambda)\right.\\
&\left.+\frac{1}{2}(\varphi'-\vartheta)[\vartheta(x,\lambda)h(\lambda)+\varphi(x,\lambda)g(\lambda)]\right\}\frac{d\lambda}{p(\lambda)},\quad(7.9)
\end{aligned}
$$

where $\varphi(a,\lambda)$, $\varphi'(a,\lambda)$, $\vartheta(a,\lambda)$ and $\vartheta'(a,\lambda)$ are denoted by φ, φ', ϑ and ϑ', respectively.

To obtain the Parseval equation, we multiply the formula (7.9) throughout by $f(x)$, and integrate with respect to x from $-\infty$ to ∞. We obtain

$$\int_{-\infty}^{\infty} f^2(x)\,dx = \frac{1}{\pi}\left(\sum_{n=0}^{\infty}\int_{\lambda_{2n}}^{\mu_{2n}} - \sum_{n=1}^{\infty}\int_{\mu_{2n-1}}^{\lambda_{2n-1}}\right)\{\varphi g^2(\lambda) - \vartheta' h^2(\lambda) + (\varphi'-\vartheta)g(\lambda)h(\lambda)\}\frac{d\lambda}{p(\lambda)}.$$

The integrand involves the usual positive-definite quadratic form, since

$$-4\varphi\vartheta' - (\varphi' - \vartheta)^2 = 4(\vartheta\varphi' - \varphi\vartheta') - (\vartheta + \varphi')^2 = 4 - (\vartheta + \varphi')^2 \geq 0$$

in all the intervals.

Notes

Section 1

The results of this section are due to H. Weyl (1909, 1910a,b). Another proof was given by E. Titchmarsh (1962). The derivation of the Parseval equation given here is due to B. M. Levitan (1950a), and independently to N. Levinson (1951, 1954) and K. Yosida (1950).

Section 2

The circle and point result is due to H. Weyl (1910a). Theorems 2.2.2 and 2.2.3 are proved on the basis of Section 2, Chapter IX, of the monograph of E. Coddington and N. Levinson (1955). Theorem 2.2.3 is due to E. Titchmarsh (1955), while the more general result was obtained by N. Levinson (1949) and also by D. Sears and E. Titchmarsh (1950).

Section 3

The integral representation of the resolvent was first proved by H. Weyl (1910b). The proof given here is due to B. M. Levitan (1950a). Theorem 2.3.2 was obtained by B. M. Levitan (1950a). Similar theorems were proved by H. Weyl and E. Titchmarsh.

Section 4

The formula (4.7) is due to E. Titchmarsh (1962).

Section 5

The basic result of this section is due to H. Weyl (1910b). Another method for the derivation of the Parseval equation was given by E. Titchmarsh (1962). The proof obtained here is due to B. M. Levitan (1950a).

Section 6

The results of this section are classical; see, e.g., W. Magnus and W. Winkler (1966).

Section 7

See E. Titchmarsh (1962). Here, we give another proof.

Chapter 3

THE STUDY OF THE SPECTRUM

3.1 Discrete, or point, spectrum

The spectrum associated with the problem for the Sturm-Liouville operator on the half-line $[0, \infty)$ is the complement of the set of points in whose neighborhoods the spectral function $\rho(\lambda)$ is constant. In the case of the whole line $(-\infty, \infty)$, the spectrum is complementary to the set of points in whose neighborhoods all functions $\xi(\lambda)$, $\eta(\lambda)$ and $\zeta(\lambda)$ are constant. It is obvious that the spectrum is a closed set.

It is assumed that $\rho(\lambda)$, $\xi(\lambda)$, $\eta(\lambda)$ and $\zeta(\lambda)$ are unique for the corresponding problems.

A *point*, or *discrete*, spectrum is the set of all points of discontinuity of $\rho(\lambda)$ in the case of the half-line, of $(\xi(\lambda), \eta(\lambda), \zeta(\lambda))$ in the case of the whole line.

A *continuous* spectrum is the set of points of continuity of $\rho(\lambda)$ belonging to the spectrum; it is defined similarly in the case of the whole line.

The points of the discrete spectrum are also called *eigenvalues*, and the associated solutions of the problem are called *eigenfunctions*.

This chapter is devoted to the study of the spectrum.

Consider the problem

$$y'' + \{\lambda - q(x)\}y = 0 \tag{1.1}$$

$$y(0)\cos\alpha + y'(0)\sin\alpha = 0 \tag{1.2}$$

on the half-line.

We shall point out some sufficient conditions for the spectrum of the problem (1.1), (1.2) to be discrete.

Lemma 3.1.1 *For the spectrum of the problem (1.1), (1.2) to be discrete, it suffices that, for each fixed λ, any solution of the equation (1.1) should have finitely many zeros on the half-line $[0, \infty)$.*

Proof. Consider the problem posed on a finite interval $[0, b]$, i.e., subjoin the boundary condition $y(b)\cos\beta + y'(b)\sin\beta = 0$ to (1.1). It follows from the Sturm Oscillation Theorem (Theorem 1.3.3), along with the conditions expressed by the lemma, that the number of eigenvalues $N_b(\lambda)$ of the problem on $[0, b]$ remains bounded in each finite interval, uniformly with respect to b. Since the problem (1.1), (1.2), i.e., on the half-line $[0, \infty)$, is the limit of the problem on $[0, b]$, it is obvious that the spectrum of the limit problem is discrete.   ∎

Remark It follows from the Sturm Comparison Theorem (Theorem 1.3.1) that the condition of Lemma 3.1.1 necessarily holds if

$$q(x) \to +\infty \quad \text{as} \quad x \to \infty. \tag{1.3}$$

In fact, there exists, for fixed λ, $x = x_\lambda$ such that, if $x > x_\lambda$, $\lambda - q(x) < 0$. Again by the Sturm Comparison Theorem, any solution of (1.1) in the interval (x_λ, ∞) then has not more than one zero, and finitely many zeros in the interval $(0, x_\lambda)$.

It turns out that the statement of Lemma 3.1.1 holds with a considerably less restrictive condition, viz. the following.

Lemma 3.1.2 *Let $q(x)$ satisfy the conditions :*

1. *$q(x)$ is bounded from below, i.e., there exists a constant number $c > 0$ such that, for all x,*

$$q(x) > -c. \tag{1.4}$$

2. *For any $\omega > 0$,*

$$\lim_{x \to \infty} \int_x^{x+\omega} q(x) \, dx = \infty. \tag{1.5}$$

Then, for each fixed λ, any solution of the equation (1.1) has finitely many zeros on the half-line $(0, \infty)$; therefore, by Lemma 3.1.1, the spectrum of the problem (1.1), (1.2) is discrete.

Remark A. M. Molchanov has proved that, if the condition (1.4) holds, then the condition (1.5) is necessary and sufficient for the spectrum of the problem (1.1), (1.2) to be discrete. That (1.5) is sufficient follows from Lemma 3.1.2. The necessity is proved by certain methods not touched upon here (see A. M. Molchanov (1953) or I. M. Glazman (1966)).

Proof. First, notice that, without loss of generality, we can assume that

$$q(x) \geq 0, \tag{1.6}$$

and, for certain $\lambda = \lambda_0 > 0$, there is a solution of (1.1), $y = y(x)$, with infinitely many zeros $\alpha_1 < \alpha_2 < \alpha_3 < \ldots < \alpha_n < \ldots$.

Let ω be a small positive number such that $\omega < 1/(\lambda_0 + 1)$. We select N to be so large that, for $x > N$,

$$\int_x^{x+\omega} q(t) \, dt > \omega(\lambda_0 + 1), \tag{1.7}$$

which is possible because of (1.5). Furthermore, let n be so large that $\alpha_n > N$, and $m > n$ so large that $\alpha_m - \alpha_n > \omega$. If we let ω increase, then, relying on (1.6), the condition (1.7) holds for the same values of x. Therefore, without loss of generality, we can assume that $\alpha_m - \alpha_n = P\omega$, where P is a whole number.

We rewrite (1.1) in the form

$$y'' = \{q(x) - \lambda_0\}y$$

for $\lambda = \lambda_0$. Multiplying throughout by y, integrating from α_m to α_n, and integrating by parts the left-hand side, we obtain

$$-\int_{\alpha_n}^{\alpha_m} [y'(t)]^2 \, dt = \int_{\alpha_n}^{\alpha_m} q(t)y^2(t) \, dt - \lambda_0 \int_{\alpha_n}^{\alpha_m} y^2(t) \, dt. \qquad (1.8)$$

To estimate the first integral on the right-hand side, we represent it in the form

$$\int_{\alpha_n}^{\alpha_m} q(t)y^2(t) \, dt = \sum_{k=1}^{P} \int_{\alpha_n+(k-1)\omega}^{\alpha_n+k\omega} q(t)y^2(t) \, dt.$$

It follows from the mean-value theorem and from (1.7) that

$$\int_{\alpha_n+(k-1)\omega}^{\alpha_n+k\omega} q(t)y^2(t) \, dt > (\lambda_0 + 1)y^2(\xi_k)\omega, \quad \alpha_n + (k-1)\omega < \xi_k < \alpha_n + k\omega.$$

Therefore,

$$\int_{\alpha_n}^{\alpha_m} q(t)y^2(t) \, dt > (\lambda_0 + 1) \sum_{k=1}^{P} y^2(\xi_k)\omega$$

$$= (\lambda_0 + 1) \int_{\alpha_n}^{\alpha_m} y^2(t) \, dt - (\lambda_0 + 1) \sum_{k=1}^{P} \int_{\alpha_n+(k-1)\omega}^{\alpha_n+k\omega} [y^2(t) - y^2(\xi_k)] \, dt, \quad (1.9)$$

and

$$\begin{aligned}
|y^2(t) - y^2(\xi_k)| &= 2\left| \int_{\xi_k}^{t} y'(u)y(u) \, du \right| \\
&\leq \int_{\xi_k}^{t} y^2(u) \, du + \int_{\xi_k}^{t} [y'(u)]^2 \, du \\
&\leq \int_{\alpha_n+(k-1)\omega}^{\alpha_n+k\omega} y^2(u) \, du + \int_{\alpha_n+(k-1)\omega}^{\alpha_n+k\omega} [y'(u)]^2 \, du.
\end{aligned}$$

Hence, we derive from (1.9) that

$$\int_{\alpha_n}^{\alpha_m} q(t)y^2(t) \, dt > (\lambda_0 + 1) \int_{\alpha_n}^{\alpha_m} y^2(t) \, dt$$

$$-(\lambda_0 + 1) \sum_{k=1}^{P} \int_{\alpha_n+(k-1)\omega}^{\alpha_n+k\omega} \left\{ \int_{\alpha_n+(k-1)\omega}^{\alpha_n+k\omega} y^2(u) \, du \right\} dt$$

$$-(\lambda_0 + 1) \sum_{k=1}^{P} \int_{\alpha_n+(k-1)\omega}^{\alpha_n+k\omega} \left\{ \int_{\alpha_n+(k-1)\omega}^{\alpha_n+k\omega} [y'(u)]^2 \, du \right\} \, dt$$

$$= (\lambda_0 + 1) \int_{\alpha_n}^{\alpha_m} y^2(t) \, dt - (\lambda_0 + 1)\omega \int_{\alpha_n}^{\alpha_m} y^2(t) \, dt$$

$$-(\lambda_0 + 1)\omega \int_{\alpha_n}^{\alpha_m} [y'(t)]^2 \, dt, \tag{1.10}$$

and obtain from (1.8) and (1.10) that

$$-\int_{\alpha_n}^{\alpha_m} [y'(t)]^2 \, dt > \{1 - (\lambda_0 + 1)\omega\} \int_{\alpha_n}^{\alpha_m} y^2(t) \, dt - (\lambda_0 + 1)\omega \int_{\alpha_n}^{\alpha_m} [y'(t)]^2 \, dt,$$

i.e.,

$$\{1 - (\lambda_0 + 1)\omega\} \int_{\alpha_n}^{\alpha_m} \{y^2(t) + [y'(t)]^2\} \, dt < 0,$$

which is impossible, since $(\lambda_0 + 1)\omega < 1$; the proof is thus complete. ∎

Theorem 3.1.1 *If λ_0 is a point of the discrete spectrum, and $\varphi(x, \lambda_0)$ is the corresponding eigenfunction, then $\varphi(x, \lambda_0) \in \mathcal{L}^2(0, \infty)$.*

Proof. In the notation of Lemma 2.4.1,

$$\int_0^a \left\{ \int_\Delta \varphi(x, \lambda)\varphi(y, \lambda) \, d\rho_b(\lambda) \right\} dy \leq \int_\Delta \varphi^2(x, \lambda) \, d\rho_b(\lambda) \leq M,$$

where $a < b$ and M is a fixed positive number. First letting $b \to \infty$ and then $a \to \infty$, we obtain

$$\int_0^\infty \left\{ \int_\Delta \varphi(x, \lambda)\varphi(y, \lambda) \, d\rho(\lambda) \right\} dy \leq M.$$

Let an interval Δ contain only one jump of the function $\rho(\lambda)$, to be denoted by $d(\lambda_0)$. It follows from the last inequality that

$$d(\lambda_0)\varphi^2(x, \lambda_0)^2 \int_0^\infty \varphi^2(y, \lambda_0) \, dy \leq M,$$

and, since $\varphi(x, \lambda_0)$ is not identically zero, the theorem is proved. ∎

3.2 The spectrum in the case of a summable potential

3.2.1

In the sequel, we shall need the following.

Lemma 3.2.1 *Let four functions* $h_1(x)$, $h_2(x)$, $g_1(x)$, $g_2(x)$ *be nonnegative on an interval* $[0, X]$, *let* $h_1(x)$, $h_2(x)$ *be continuous, and* $g_1(x)$, $g_2(x)$ *integrable on the interval. If, for* $0 \leq x \leq X$,

$$h_1(x), h_2(x) \leq C + \int_0^x \{h_1(s)g_1(s) + h_2(s)g_2(s)\} \, ds, \qquad (2.1)$$

where C *is a constant, then*

$$h_1(x), h_2(x) \leq C \exp\left\{\int_0^x [g_1(t) + g_2(t)] \, dt\right\}, \quad 0 \leq x \leq X. \qquad (2.2)$$

Proof. Put

$$y(x) = \int_0^x \{h_1(s)g_1(s) + h_2(s)g_2(s)\} \, ds \qquad (2.3)$$

$$y'(x) = h_1(x)g_1(x) + h_2(x)g_2(x). \qquad (2.3')$$

Multiplying the first inequality in (2.1) throughout by $g_1(x)$, the second by $g_2(x)$, and adding the results, we obtain

$$y'(x) \leq C\{g_1(x) + g_2(x)\} + y(x)\{g_1(x) + g_2(x)\},$$

relying on (2.3) and (2.3′).

The last inequality can be rewritten in the form

$$\frac{d}{dx}\left\{y(x)\exp\left[-\int_0^x \{g_1(s) + g_2(s)\} \, ds\right]\right\}$$
$$\leq C[g_1(x) + g_2(x)]\exp\left\{-\int_0^x [g_1(s) + g_2(s)] \, ds\right\},$$

which, integrated from 0 to x, yields

$$y(x)\exp\left\{-\int_0^x [g_1(s) + g_2(s)] \, ds\right\} \leq C - C\exp\left\{-\int_0^x [g_1(s) + g_2(s)] \, ds\right\}.$$

Hence,

$$y(x) \leq C\exp\left\{\int_0^x [g_1(s) + g_2(s)] \, ds\right\} - C. \qquad (2.4)$$

The inequalities (2.2) directly follow from (2.1) and (2.4) because of (2.3). ∎

3.2.2

Let $\varphi(x, \lambda)$ be a solution of the equation (1.1), and let it satisfy the conditions $\varphi(0, \lambda) = \sin \alpha$, $\varphi'(0, \lambda) = -\cos \alpha$. Rewriting (1.1) as $y'' + \lambda y = q(x)y$, substituting $\lambda = s^2$ and applying the method of variation of the constants, we obtain

$$\varphi(x, \lambda) = \cos sx \cdot \sin \alpha - \frac{\sin sx}{s} \cos \alpha + \frac{1}{s}\int_0^x \sin\{s(x - t)\}q(t)\varphi(t, \lambda) \, dt. \qquad (2.5)$$

Let $s = \sigma + i\tau$, $\tau \geq 0$, and $\varphi_1(x, \lambda) = \varphi(x, \lambda)e^{-\tau x}$. It then follows from (2.5) that

$$\varphi_1(x, \lambda) = e^{-\tau x} \cos sx \cdot \sin \alpha - e^{-\tau x} \frac{\sin sx}{s} \cos \alpha$$
$$+ \frac{1}{s} \int_0^x e^{-\tau(x-t)} \sin\{s(x - t)\} q(t)\varphi_1(t, \lambda) \, dt.$$

Since $|\cos sx|, |\sin sx| \leq e^{\tau x}$, it follows from the last equality that

$$|\varphi_1(x, \lambda)| \leq 1 + \frac{1}{|s|} + \frac{1}{|s|} \int_0^x |q(t)\varphi_1(t, \lambda)| \, dt.$$

Applying Lemma 3.2.1 (putting $h_2(x) = g_2(x) \equiv 0$), we obtain

$$|\varphi_1(x, \lambda)| \leq \left(1 + \frac{1}{|s|}\right) \exp\left\{\frac{1}{|s|} \int_0^x |q(t)| \, dt\right\}. \tag{2.6}$$

Since the potential is summable, i.e., $q(x) \in \mathcal{L}(0, \infty)$, it follows that $\varphi_1(x, \lambda)$ is bounded for $0 \leq x \leq \infty$, $|s| \geq p > 0$, $\tau \geq 0$.

We first consider the real positive values of s. The function $\varphi_1(x, \lambda)$ is bounded for $s \geq p$. Therefore, it follows from the formula (2.5) that

$$\varphi(x, \lambda) = \cos sx \cdot \sin \alpha - \frac{\sin sx}{s} \cos \alpha$$
$$+ \frac{1}{s} \int_0^\infty \sin\{s(x - t)\} \cdot q(t)\varphi(t, \lambda) \, dt$$
$$- \frac{1}{s} \int_x^\infty \sin\{s(x - t)\} \cdot q(t)\varphi(t, \lambda) \, dt$$
$$= \mu(\lambda) \cos sx + \nu(\lambda) \sin sx + o(1) \tag{2.7}$$

as $x \to \infty$, where

$$\mu(\lambda) = \sin \alpha - \frac{1}{s} \int_0^\infty \sin st \cdot q(t)\varphi(t, \lambda) \, dt, \tag{2.8}$$

$$\nu(\lambda) = -\frac{\cos \alpha}{s} + \frac{1}{s} \int_0^\infty \cos st \cdot q(t)\varphi(t, \lambda) \, dt. \tag{2.9}$$

Since the integrals converge uniformly for $s \geq p > 0$, $\mu(\lambda)$ and $\nu(\lambda)$ are continuous functions of s.

Similarly, if $\vartheta(x, \lambda)$ is a solution of (1.1) and satisfies the equalities $\vartheta(0, \lambda) = \cos \alpha$, $\vartheta'(0, \lambda) = -\sin \alpha$, then

$$\vartheta(x, \lambda) = \mu_1(\lambda) \cos sx + \nu_1(\lambda) \sin sx + o(1) \tag{2.7'}$$

as $x \to \infty$, where

$$\mu_1(\lambda) = \cos \alpha - \frac{1}{s} \int_0^\infty \sin st \cdot q(t)\vartheta(t, \lambda) \, dt, \tag{2.8'}$$

$$\nu_1(\lambda) = -\frac{\sin \alpha}{s} + \frac{1}{s} \int_0^\infty \cos st \cdot q(t)\vartheta(t, \lambda)\, dt. \tag{2.9$'$}$$

Furthermore, differentiating (2.5) with respect to x, we have

$$\varphi_x'(x, \lambda) = -s \sin sx \cdot \sin \alpha - \cos sx \cdot \cos \alpha + \int_0^x \cos\{s(x - t)\} \cdot q(t)\varphi(t, \lambda)\, dt. \tag{2.10}$$

Estimating this function as above, we obtain

$$\varphi_x'(x, \lambda) = -s\mu(\lambda) \sin sx + s\nu(\lambda) \cos sx + o(1) \tag{2.10$'$}$$

as $x \to \infty$.

Similarly,

$$\vartheta_x'(x, \lambda) = -s\mu_1(\lambda) \sin sx + s\nu_1(\lambda) \cos sx + o(1).$$

Furthermore, $W_x\{\varphi, \vartheta\} = W_0\{\varphi, \vartheta\} = 1$. Therefore,

$$
\begin{aligned}
W\{\varphi, \vartheta\} &= W\{\mu(\lambda) \cos sx + \nu(\lambda) \sin sx + o(1), \mu_1(\lambda) \cos sx + \nu_1(\lambda) \sin sx + o(1)\} \\
&= s\{\mu(\lambda)\nu_1(\lambda) - \mu_1(\lambda)\nu(\lambda) + o(1)\} \\
&= 1 \tag{2.7}
\end{aligned}
$$

Since the right-hand side of the equality $s\{\mu(\lambda)\nu_1(\lambda) - \mu_1(\lambda)\nu(\lambda)\} + o(1) = 1$ does not depend on x, we obtain

$$\mu(\lambda)\nu_1(\lambda) - \mu_1(\lambda)\nu(\lambda) = \frac{1}{\sqrt{\lambda}}. \tag{2.11}$$

In particular, it follows that the functions $\mu(\lambda)$ and $\nu(\lambda)$ cannot vanish for the same values of λ.

3.2.3

We now consider the complex values of s. For fixed positive τ, it follows from (2.5) that

$$
\begin{aligned}
\varphi(x, \lambda) = \frac{1}{2}e^{-isx} \sin \alpha &+ \frac{e^{-isx}}{2is} \cos \alpha + O(e^{-\tau x}) \\
&- \frac{1}{2is} \int_0^x e^{-is(x-t)}q(t)\varphi(t, \lambda)\, dt + O\left(\int_0^x e^{-\tau(x-t)}|q(t)\varphi(t, \lambda)|\, dt\right)
\end{aligned}
$$

as $x \to \infty$.

Since $|\varphi(t, \lambda)| = O(e^{\tau t})$,

$$
\begin{aligned}
O\left(\int_0^x e^{-\tau(x-t)}|q(t)\varphi(t, \lambda)|\, dt\right) &= O\left(\int_0^x e^{\tau(2t-x)}|q(t)|\, dt\right) \\
&= O\left(e^{\tau(x-2\delta)}\int_0^{x-\delta}|q(t)|\, dt\right) + O\left(e^{\tau x}\int_{x-\delta}^x|q(t)|\, dt\right) = o(e^{\tau x})
\end{aligned}
$$

(e.g., putting $\delta = \frac{1}{2}x$).

On the other hand,

$$\int_0^\infty e^{-is(x-t)} q(t)\varphi(t,\lambda)\, dt = O\left(e^{\tau x} \int_x^\infty |q(t)|\, dt\right) = o(e^{\tau x}).$$

Therefore,

$$\varphi(x,\lambda) = e^{-isx}\{M(\lambda) + o(1)\}, \tag{2.12}$$

where

$$M(\lambda) = \frac{1}{2}\sin\alpha + \frac{1}{2is}\cos\alpha - \frac{1}{2is}\int_0^\infty e^{ist} q(t)\varphi(t,\lambda)\, dt. \tag{2.13}$$

If we use the formula (2.10), we obtain

$$\varphi_x'(x,\lambda) = -ise^{-isx}\{M(\lambda) + o(1)\}. \tag{2.14}$$

Similarly,

$$\vartheta(x,\lambda) = e^{-isx}\{M_1(\lambda) + o(1)\}, \tag{2.12'}$$

where

$$M_1(\lambda) = \frac{1}{2}\cos\alpha - \frac{1}{2is}\sin\alpha - \frac{1}{2is}\int_0^\infty e^{ist} q(t)\vartheta(t,\lambda)\, dt. \tag{2.13'}$$

3.2.4

We now have at our disposal all auxiliary means to study the Sturm-Liouville operator spectrum if $q(x) \in \mathcal{L}(0,\infty)$. Let $b > 0$. Consider the boundary-value problem

$$\begin{aligned}
y'' + \{\lambda - q(x)\}y &= 0 \\
y(0)\cos\alpha + y'(0)\sin\alpha &= 0 \\
y(b)\cos\beta + y'(b)\sin\beta &= 0
\end{aligned} \tag{2.15}$$

in the interval $[0, b]$.

We start with the negative spectrum. It follows from (2.12) and (2.13) that, for each fixed $\lambda_0 < 0$, the number of roots of the equation $\varphi(b,\lambda) = 0$ in the interval $-\infty < \lambda \leq \lambda_0$ is bounded uniformly with respect to b. Therefore, the number of points of increase of the function $\rho_b(\lambda)$ (see Section 1, Chapter 2, for the definition of $\rho_b(\lambda)$) in the interval $(-\infty, \lambda_0]$ is bounded uniformly with respect to b, and the limit function $\rho(\lambda)$ has the same property. If $q(x) \to 0$ as $x \to \infty$, this easily follows from Sturm theory (see the Remark after Lemma 3.1.1).

If there are finitely many zeros of the function $\varphi(x,0)$ on the half-line $[0,\infty)$, then there are finitely many negative eigenvalues, too. Indeed, it follows from the Sturm Oscillation Theorem that, for any $b > 0$, the problem (2.15) has a bounded number of negative eigenvalues. We have thus proved the following.

Theorem 3.2.1 *If $q(x) \in \mathcal{L}(0,\infty)$, then, for negative λ, the spectrum of the problem (1.1), (1.2) is discrete, and bounded from below.*

The following supplies a simple sufficient condition for the number of zeros of $\varphi(x, 0)$ to be finite.

Lemma 3.2.2 *If $(1+x^2)q(x) \in \mathcal{L}(0, \infty)$, then the function $\varphi(x, 0)$ has finitely many zeros on the half-line $[0, \infty)$; therefore, there are finitely many negative eigenvalues.*

Proof. In fact, it is known[1] that if $(1 + x^2)q(x) \in \mathcal{L}(0, \infty)$, then there exists a fundamental system of solutions y_1 and y_2 of the equation $y'' - q(x)y = 0$ such that $\lim_{x\to\infty} y_1(x) = 1$, $\lim_{x\to\infty}\{y_2(x) - x\} = 0$. Since $\varphi(x, 0) = C_1 y_1(x) + C_2 y_2(x)$, we have $\lim_{x\to\infty} \varphi(x, 0) = \infty$ if $C_2 \neq 0$; however, $\lim_{x\to\infty} \varphi(x, 0) = 1$ if $C_2 = 0$. In both cases, $\varphi(x, 0)$ has finitely many zeros on $[0, \infty)$, and the proof is thus complete. ∎

We now consider the positive spectrum. Since the functions $\mu(\lambda)$ and $\nu(\lambda)$ cannot vanish for the same value of λ because of (2.11), $\mu^2(\lambda) + \nu^2(\lambda) > 0$. Therefore, putting

$$\frac{\mu(\lambda)}{[\mu^2(\lambda) + \nu^2(\lambda)]^{1/2}} = \sin \delta(\lambda), \qquad \frac{\nu(\lambda)}{[\mu^2(\lambda) + \nu^2(\lambda)]^{1/2}} = \cos \delta(\lambda),$$

the formula (2.7) can be written as

$$\varphi(x, \lambda) = [\mu^2(\lambda) + \nu^2(\lambda)]^{1/2} \sin[sx + \delta(\lambda)] + o(1) \tag{2.16}$$

as $x \to \infty$.

Assuming that

$$(1 + x)q(x) \in \mathcal{L}(0, \infty) \tag{2.17}$$

(we will get rid of this condition in the sequel), differentiating (2.5) with respect to s, we obtain after simple transformations that

$$\varphi_s'(x, \lambda) = \frac{\partial}{\partial s}\{[\mu^2(\lambda) + \nu^2(\lambda)]^{1/2} \sin[sx + \delta(\lambda)]\} + o(1). \tag{2.18}$$

Let b be a large positive number, and, for simplicity, $\beta = 0$ in the boundary-value problem (2.15) (considering the case of arbitrary real β similarly). The positive eigenvalues of (2.15) are then determined by the equation

$$\sin[sb + \delta(\lambda)] = o(1). \tag{2.19}$$

Suppose that s_1 is its positive root. We notice that $\delta(\lambda) = O(1)$. Then

$$s_1 b + \delta(\lambda_1) = m\pi + o(1), \quad \lambda_1 = s_1^2. \tag{2.20}$$

If s_2 is the root of (2.19), following s_1, then either

$$s_2 b + \delta(\lambda_2) = m\pi + o(1), \quad \lambda_2 = s_2^2 \tag{2.21}$$

---

[1]See V. V. Stepanov (1959), *A course on Differential Equations*, Phismatgiz, Moscow *(Russian)*.

or

$$s_2 b + \delta(\lambda_2) = (m+1)\pi + o(1). \tag{2.22}$$

We show that (2.21) is impossible. Indeed, if (2.21) held, it would follow from Rolle's theorem that the equation $\varphi'_s(b, \lambda) = 0$ has a root s_3 between s_1 and s_2; therefore, satisfying $s_3 b + \delta(\lambda_3) = m\pi + o(1)$, which is impossible in view of the asymptotic formula (2.18).

Subtracting (2.20) from (2.22) and taking into account that the functions $\delta(\lambda)$ are continuous, we obtain the formula

$$s_2 - s_1 = \frac{\pi}{b} + o\left(\frac{1}{b}\right), \quad s_i^2 = \lambda_i, \quad i = 1, 2, \tag{2.23}$$

for two consecutive eigenvalues λ_1 and λ_2 of the problem (2.15).

Theorem 3.2.2 *If $q(x) \in \mathcal{L}(0, \infty)$, $\lambda > 0$, $\Delta > 0$, then*

$$\rho(\lambda + \Delta) - \rho(\lambda) = \frac{1}{\pi} \int_\lambda^{\lambda+\Delta} \frac{d\lambda}{\sqrt{\lambda}[\mu^2(\lambda) + \nu^2(\lambda)]}, \tag{2.24}$$

i.e., the spectrum of the problem (1.1), (1.2) is continuous in the interval $(0, \infty)$.

Proof. Let $\lambda_{1,b}, \lambda_{2,b}, \ldots, \lambda_{n,b}, \ldots$ be the eigenvalues of the problem (2.15), and let

$$\varphi(x, \lambda_{1,b}), \varphi(x, \lambda_{2,b}), \ldots, \varphi(x, \lambda_{n,b}), \ldots$$

be the corresponding eigenfunctions. We first assume that $q(x)$ satisfies the condition (2.17). By the definition of $\rho_b(\lambda)$ and by the asymptotic formulas (2.23), we have

$$
\begin{aligned}
\rho_b(\lambda + \Delta) - \rho_b(\lambda) &= \sum_{\lambda < \lambda_{n,b} \le \lambda + \Delta} \frac{1}{\int_0^b \varphi^2(x, \lambda_{n,b})\, dx} \\
&= \sum_{\lambda < \lambda_{n,b} \le \lambda + \Delta} \frac{s_{n+1,b} - s_{n,b}}{(s_{n+1,b} - s_{n,b}) \int_0^b \varphi^2(x, \lambda_{n,b})\, dx} \\
&= \frac{1}{\pi} \sum_{\lambda < \lambda_{n,b} \le \lambda + \Delta} \frac{\lambda_{n+1,b} - \lambda_{n,b}}{(s_{n+1,b} + s_{n,b}) \frac{1}{b} \int_0^b \varphi^2(x, \lambda_{n,b})\, dx + o(1)},
\end{aligned}
$$

where $s_{n,b} = \sqrt{\lambda_{n,b}}$.

It follows from the asymptotic formulas (2.7) that

$$\frac{1}{b} \int_0^b \varphi^2(x, \lambda_{n,b})\, dx = \frac{1}{2}[\mu^2(\lambda_{n,b}) + \nu^2(\lambda_{n,b})] + o(1).$$

Therefore,

$$\rho_b(\lambda + \Delta) - \rho_b(\lambda) = \frac{1}{\pi} \sum_{\lambda < \lambda_{n,b} \le \lambda + \Delta} \left\{ \frac{1}{s_{n,b}[\mu^2(\lambda_{n,b}) + \nu^2(\lambda_{n,b})]} + o(1) \right\} (\lambda_{n+1,b} - \lambda_{n,b}).$$

Passing to the limit as $b \to \infty$, we obtain the statement of the theorem in the case where $q(x)$ satisfies the condition expressed in (2.17).

Now, suppose that (2.17) does not hold. We put

$$q_n(x) = \begin{cases} q(x), & 0 \le x \le n, \\ 0, & x > n. \end{cases}$$

It is obvious that the function $q_n(x)$ satisfies (2.17). Therefore, denoting by $\rho_n(\lambda)$, $\mu_n(\lambda)$, $\nu_n(\lambda)$ the functions corresponding to $q_n(x)$, we have

$$\rho_n(\lambda + \Delta) - \rho_n(\lambda) = \frac{1}{\pi} \int_\lambda^{\lambda+\Delta} \frac{d\lambda}{\sqrt{\lambda}[\mu_n^2(\lambda) + \nu_n^2(\lambda)]}.$$

Letting $n \to \infty$, we obtain the formula (2.24) in the limit, and the proof is thus complete.   ∎

3.3   Transformation of the basic equation

In some cases, the equation (1.1) with unbounded potential $q(x)$ can be reduced to another form by a convenient substitution, admitting the derivation of asymptotic formulas.

First, we assume that λ is a real number and that the functions $q'(x)$, $q''(x)$ are continuous. We introduce the notation

$$\xi(x) = \int_0^x \{\lambda - q(t)\}^{1/2} \, dt, \qquad \eta(x) = \{\lambda - q(x)\}^{1/4} y.$$

Then

$$\frac{d\eta}{d\xi} = \frac{d\eta}{dx} \frac{dx}{d\xi} = \left[\{\lambda - q(x)\}^{1/4} \frac{dy}{dx} - \frac{1}{4} \frac{q'(x)}{(\lambda - q(x))^{3/4}} y \right] \frac{1}{(\lambda - q(x))^{1/2}}$$

$$= \{\lambda - q(x)\}^{-1/4} \frac{dy}{dx} - \frac{1}{4} \frac{q'(x)}{\{\lambda - q(x)\}^{5/4}} y$$

$$\frac{d^2\eta}{d\xi^2} = \left\{ [\lambda - q(x)]^{-1/4} \frac{d^2y}{dx^2} - \frac{1}{4} \left[\frac{q''(x)}{\{\lambda - q(x)\}^{5/4}} + \frac{5}{4} \frac{q'^2(x)}{\{\lambda - q(x)\}^{3/4}} \right] y \right\} \frac{1}{\{\lambda - q(x)\}^{1/2}}.$$

Therefore, (1.1) changes into

$$\frac{d^2\eta}{d\xi^2} + \eta + \left[\frac{1}{4} \frac{q''(x)}{\{\lambda - q(x)\}^2} + \frac{5}{16} \frac{q'^2(x)}{\{\lambda - q(x)\}^3} \right] \eta = 0, \tag{3.1}$$

which is also of the form of (1.1); however, here, the coefficient of η turns out to be small in the last term, for large λ.

It follows from (3.1) that any of its solutions $\eta = \varphi(\xi)$ satisfies the integral equation

$$\varphi(\xi) = \varphi(0)\cos\xi + \varphi'(0)\sin\xi - \int_0^\xi \sin(\xi-\tau)\left\{\frac{1}{4}\frac{q''(t)}{[\lambda-q(t)]^2} + \frac{5}{16}\frac{q'^2(t)}{[\lambda-q(t)]^3}\right\}\varphi(\tau)\,d\tau,$$

(3.2)

where $\tau = \xi(t)$; the latter can be used to obtain asymptotic formulas for $\varphi(\xi)$.

For nonreal λ or $q(x) > \lambda$, the function $\xi(x)$ assumes nonreal values; therefore, the formula (3.2) involves integrals along complex paths.

However, it is not necessary to introduce such integrals, since a similar equation with a real independent variable can be obtained, this variable turning out to be x itself. For this, we assume that $q(0) = 0$ (which can always be done by a convenient choice of the origin of the coordinates in the λ-plane). We put

$$P(x) = \{\lambda - q(x)\}^{1/4}\frac{d}{dx}\left\{[\lambda-q(x)]^{-1/2}\frac{d\eta}{dx}\right\} - \frac{d^2y}{dx^2}.$$

(3.3)

Then

$$\begin{aligned}
P(x) &= \{\lambda - q(x)\}^{1/4}\frac{d}{dx}\left\{[\lambda-q(x)]^{-1/4}\frac{dy}{dx} - \frac{1}{4}\frac{q'(x)}{[\lambda-q(x)]^{5/4}}y\right\} - \frac{d^2y}{dx^2} \\
&= -\left\{\frac{1}{4}\frac{q''(x)}{[\lambda-q(x)]} + \frac{5}{16}\frac{q'^2(x)}{[\lambda-q(x)]^2}\right\}y.
\end{aligned}$$

Relying on (1.1) and (3.3), we have

$$\begin{aligned}
I &= \int_0^x \sin\{\xi(x) - \xi(t)\}\frac{P(t)}{\{\lambda-q(t)\}^{1/4}}\,dt \\
&= \int_0^x \sin\{\xi(x) - \xi(t)\}\frac{d}{dt}\left\{[\lambda-q(t)]^{-1/2}\frac{d\eta}{dt}\right\}\,dt \\
&\quad + \int_0^x \sin\{\xi(x) - \xi(t)\}\{\lambda-q(t)\}^{3/4}y\,dt \\
&\equiv I_1 + I_2.
\end{aligned}$$

Integrating I_1 by parts twice, we obtain

$$\begin{aligned}
I_1 &= \left\{\sin[\xi(x) - \xi(t)][\lambda-q(t)]^{-1/2}\frac{d\eta}{dt}\right\}_0^x + \int_0^x \cos[\xi(x) - \xi(t)]\frac{d\eta}{dt}\,dt \\
&= -\eta'(0)\lambda^{-1/2}\sin\xi(x) + \eta(x) - \eta(0)\cos\xi(x) - I_2.
\end{aligned}$$

Therefore, $\eta(x) = \eta(0)\cos\xi(x) + \eta'(0)\lambda^{-1/2}\sin\xi(x) + I$, i.e., $\eta(x)$ satisfies the integral equation

$$\eta(x) = \eta(0)\cos\xi(x) + \frac{\eta'(0)}{\sqrt{\lambda}}\sin\xi(x) + \int_0^x \sin\{\xi(x) - \xi(t)\}R(t)\eta(t)\,dt,$$

(3.4)

where

$$R(t) = -\frac{1}{4}\frac{q''(t)}{\{\lambda - q(t)\}^{3/2}} - \frac{5}{16}\frac{q'^2(t)}{\{\lambda - q(t)\}^{5/2}}. \tag{3.5}$$

We assume in (3.4) and (3.5) that, for $\text{Im}\,\lambda > 0$, $0 < \arg\lambda < \pi$ and, for $q(t)$ varying from 0 to ∞, that $\arg\{\lambda - q(t)\}$ varies from $\arg\lambda$ to π.

Lemma 3.3.1 *Let $q(x)$ tend to infinity monotonically, and let*

$$q'(x) = O\{[q(x)]^c\}, \tag{3.6}$$

where $x \to \infty$ and c is a fixed number such that $0 < c < 3/2$, and let $q''(x)$ preserve its sign for large x. Then the integral

$$\int_0^\infty |R(x)|\,dx \tag{3.7}$$

converges uniformly with respect to λ in each domain where $\lambda - q(x) \geq \delta > 0$ for $0 \leq x < \infty$.

Proof. Indeed, relying on the estimate (3.6),

$$\int_{x_0}^x \frac{q'^2(t)}{\{q(t)\}^{5/2}}\,dt = O\left(\int_{x_0}^x \frac{q'(t)}{\{q(t)\}^{5/2-c}}\,dt\right) = O([q(t)]^{c-3/2}|_{x_0}^x) = O(1).$$

Furthermore,

$$\int_{x_0}^x \frac{q''(t)}{\{q(t)\}^{3/2}}\,dt = \frac{q'(t)}{\{q(t)\}^{3/2}}\Big|_{x_0}^x + \frac{3}{2}\int_{x_0}^x \frac{q'^2(t)}{\{q(t)\}^{5/2}}\,dt = O(1),$$

and the lemma follows from the last two estimates, because of (3.5). ∎

3.4   The study of the spectrum as $q(x) \to -\infty$

Theorem 3.4.1 *Let $q(x) \leq 0$, $q'(x) < 0$ for $x \geq 0$, $q'(x) = O(q^c(x))$ as $x \to \infty$, $q(x) \to -\infty$, and let $q''(x)$ preserve its sign for large x. If the integral*

$$\int_0^\infty |q(x)|^{-1/2}\,dx \tag{4.1}$$

diverges, then the spectrum of the problem (1.1), (1.2) is continuous and fills the whole number line $(-\infty, \infty)$.

Proof. We notice that the convergence of the integral (3.7) follows from the conditions of the theorem. Therefore, we derive from the equation (3.4) that

$$\eta(x) = \eta(0)\cos\xi(x) + \eta'(0)\lambda^{-1/2}\sin\xi(x) + \int_0^\infty \sin\{\xi(x) - \xi(t)\}R(t)\eta(t)\,dt + o(1)$$

for positive λ.

Hence, putting $\eta(x) = \{\lambda - q(x)\}^{1/4}\varphi(x, \lambda)$, where $\varphi(x, \lambda)$ is the same as in Section 2, we obtain

$$\varphi(x, \lambda) = \{\lambda - q(x)\}^{-1/4}\{\mu(\lambda)\cos\xi(x) + \nu(\lambda)\sin\xi(x) + o(1)\} \qquad (4.2)$$

for

$$\mu(\lambda) = \lambda^{1/4}\sin\alpha - \int_0^\infty \sin\xi(x)\cdot R(t)\{\lambda - q(t)\}^{1/4}\varphi(t, \lambda)\,dt \qquad (4.3)$$

$$\nu(\lambda) = -\frac{q'(0)\sin\alpha}{4\lambda^{5/4}} - \frac{\cos\alpha}{\lambda^{1/4}} + \int_0^\infty \cos\xi(t)\cdot R(t)\{\lambda - q(t)\}^{1/4}\varphi(t, \lambda)\,dt. \qquad (4.4)$$

Similarly, if $\vartheta(x, \lambda)$ is a solution of the equation (1.1), so that $\vartheta(0, \lambda) = \cos\alpha$, $\vartheta'(0, \lambda) = \sin\alpha$, then

$$\vartheta(x, \lambda) = \{\lambda - q(x)\}^{-1/4}\{\mu_1(\lambda)\cos\xi(x) + \nu_1(\lambda)\sin\xi(x) + o(1)\}, \qquad (4.5)$$

where $\mu_1(\lambda)$ and $\nu_1(\lambda)$ are obtained from $\mu(\lambda)$ and $\nu(\lambda)$ by replacing $\sin\alpha$, $-\cos\alpha$ and $\varphi(t, \lambda)$ by $\cos\alpha$, $\sin\alpha$ and $\vartheta(t, \lambda)$, respectively.

It also follows from the above that the integrals in the formulas (4.3) and (4.4) converge uniformly in λ; therefore, $\mu(\lambda)$ and $\nu(\lambda)$ are continuous functions of λ.

Furthermore, differentiating (3.4), we obtain

$$\eta'(x) = \{\lambda - q(x)\}^{1/2}$$
$$\times \left\{\lambda^{-1/2}\eta'(0)\cos\xi(x) - \eta(0)\sin\xi(x) + \int_0^x \cos\{\xi(x) - \xi(t)\}R(t)\eta(t)\,dt\right\}.$$

Proceeding as above, we obtain

$$\frac{d}{dx}\{[\lambda - q(x)]^{1/4}\varphi(x, \lambda)\} = \{\lambda - q(x)\}^{1/2}[\nu(\lambda)\cos\xi(x) - \mu(\lambda)\sin\xi(x) + o(1)];$$

hence, it follows from (4.2) and (3.6) that

$$\varphi'(x, \lambda) = \{\lambda - q(x)\}^{1/4}\{\nu(\lambda)\cos\xi(x) - \mu(\lambda)\sin\xi(x) + o(1)\}. \qquad (4.6)$$

Similarly,

$$\vartheta'(x, \lambda) = \{\lambda - q(x)\}^{1/4}\{\nu_1(\lambda)\cos\xi(x) - \mu_1(\lambda)\sin\xi(x) + o(1)\}. \qquad (4.7)$$

Thus,

$$\lim_{x\to\infty} W\{\varphi, \vartheta\} = \mu(\lambda)\nu_1(\lambda) - \mu_1(\lambda)\nu(\lambda) = W\{\varphi, \vartheta\}_{x=0} = 1.$$

Therefore, $\mu(\lambda)$ and $\nu(\lambda)$ cannot vanish for the same positive value of λ.

The above should be changed slightly if $\lambda < 0$. However, we can select x_0 so that the inequality $\lambda - q(x) > 0$ holds for $x \geq x_0$, and repeat the preceding argument now for the interval (x_0, ∞). We then arrive at the same conclusion for the behaviour of the functions $\mu(\lambda)$ and $\nu(\lambda)$ for $\lambda \leq 0$.

We first calculate the spectrum for $\lambda \geq 0$. Let $b > 0$ be arbitrary. Consider the boundary condition

$$\varphi(b, \lambda) \cos \beta + \varphi'(b, \lambda) \sin \beta = 0. \tag{4.8}$$

If $\sin \beta \neq 0$, then (4.8) can be rewritten as $b \to \infty$, using (4.2) and (4.6), as

$$\nu(\lambda) \cos \xi(b) - \mu(\lambda) \sin \xi(b) = o(1). \tag{4.9}$$

For $\sin \beta = 0$, we obtain

$$\mu(\lambda) \cos \xi(b) + \nu(\lambda) \sin \xi(b) = o(1). \tag{4.9'}$$

The analysis of the last two equations is the same. Consider (4.9′). Putting

$$\frac{\mu(\lambda)}{\{\mu^2(\lambda) + \nu^2(\lambda)\}^{1/2}} = \sin \omega(\lambda), \qquad \frac{\nu(\lambda)}{\{\mu^2(\lambda) + \nu^2(\lambda)\}^{1/2}} = \cos \omega(\lambda),$$

we rewrite the equation (4.9′) in the form

$$\sin\{\xi(b, \lambda) + \omega(\lambda)\} = o(1). \tag{4.10}$$

Let λ_1 and λ_2 be two consecutive positive roots of the equation $\varphi(b, \lambda) = 0$ (or, equivalently, of (4.10)). Using the asymptotic formula (4.2) and proceeding as in the analysis of the equation (2.19), we can show that $\xi(b, \lambda_2) - \xi(b, \lambda_1) = \pi + o(1)$. Therefore, by the definition of the function $\xi(x, \lambda)$,

$$\int_0^b \{\sqrt{\lambda_2 - q(t)} - \sqrt{\lambda_1 - q(t)}\} \, dt = \int_0^b \frac{\lambda_2 - \lambda_1}{\sqrt{\lambda_2 - q(t)} + \sqrt{\lambda_1 - q(t)}} \, dt = \pi + o(1). \tag{4.11}$$

Furthermore,

$$\int_0^b \frac{dt}{2\sqrt{\lambda_1 - q(t)}} - \int_0^b \frac{dt}{\sqrt{\lambda_2 - q(t)} + \sqrt{\lambda_1 - q(t)}}$$

$$= (\lambda_2 - \lambda_1) \int_0^b \frac{dt}{2\sqrt{\lambda_1 - q(t)}\{\sqrt{\lambda_2 - q(t)} + \sqrt{\lambda_1 - q(t)}\}}$$

which is bounded by

$$\frac{\lambda_2 - \lambda_1}{2\sqrt{\lambda_1}} \int_0^b \frac{dt}{-q(t)} = o\left(\int_0^b \frac{dt}{\sqrt{\lambda_1 - q(t)}}\right).$$

It follows, taking into account (4.11), that

$$\lambda_2 - \lambda_1 = \pi \left(\int_0^b \frac{dt}{2\sqrt{\lambda_1 - q(t)}} \right)^{-1} + o \left\{ \left(\int_0^b \frac{dt}{\sqrt{\lambda_1 - q(t)}} \right)^{-1} \right\}. \qquad (4.12)$$

We now show that

$$\int_0^b \frac{\genfrac{}{}{0pt}{}{\sin}{\cos} \{2\xi(t, \lambda)\}}{\sqrt{\lambda - q(t)}} \, dt = O(1) \qquad (4.13)$$

as $b \to \infty$.

In fact, consider the integral with the sine (the integral involving the cosine is estimated similarly). Since $\xi(t, \lambda) = \int_0^t \sqrt{\lambda - q(u)} \, du$, we have

$$\begin{aligned}
\int_0^b \frac{\sin 2\xi(t, \lambda)}{\sqrt{\lambda - q(t)}} \, dt &= \int_0^b \frac{\sin 2\xi(t, \lambda)}{\lambda - q(t)} \, d\xi(t, \lambda) \\
&= \int_0^{\xi(b, \lambda)} \frac{\sin 2\eta}{\lambda - q(t)} \, d\eta \\
&= \left\{ \int_0^{\pi/2} + \int_{\pi/2}^{\pi} + \dots \right\} \frac{\sin 2\eta}{\lambda - q(t)} \, d\eta, \quad \eta = \xi(t, \lambda). \quad (4.14)
\end{aligned}$$

As η increases, t increases, too; therefore, $\{\lambda - q(t)\}^{-1}$ decreases. Then

$$\begin{aligned}
\int_{k\pi/2}^{(k+1)\pi/2} \frac{\sin 2\eta}{\lambda - q(t)} \, dt &= \frac{1}{\lambda - q\left[(k+\vartheta)\dfrac{\pi}{2} \right]} \int_{k\pi/2}^{(k+1)\pi/2} |\sin 2\eta| \, d\eta \\
&= \frac{1}{\lambda - q\left[(k+\vartheta)\dfrac{\pi}{2} \right]}, \quad 0 < \vartheta < 1.
\end{aligned}$$

As k increases, the last expression decreases monotonically, and the estimate (4.13) follows from the expansion (4.14), an alternating sum.

It follows from the estimates (4.13), from the asymptotic formula (4.2) and from the divergence of the integral (4.1) that

$$\left(\int_0^b \frac{dt}{\sqrt{\lambda - q(t)}} \right)^{-1} \int_0^b \varphi^2(t, \lambda) \, dt = \mu^2(\lambda) + \nu^2(\lambda) + o(1).$$

We then infer, also from the formula (4.12), that, for $\lambda > 0$, $\Delta > 0$,

$$\rho_b(\lambda + \Delta) - \rho_b(\lambda) = \sum_{\lambda < \lambda_{n,b} \leq \lambda + \Delta} \left(\int_0^b \varphi^2(t, \lambda_{n,b}) \, dt \right)^{-1}$$

$$= \sum_{\lambda < \lambda_{n,b} \leq \lambda + \Delta} \frac{\lambda_{n+1,b} - \lambda_{n,b}}{(\lambda_{n+1,b} - \lambda_{n,b}) \int_0^b \varphi^2(t, \lambda_{n,b}) \, dt}$$

$$= \frac{1}{\pi} \sum_{\lambda < \lambda_{n,b} \leq \lambda + \Delta} \left\{ \frac{1}{\mu^2(\lambda_{n,b}) + \nu^2(\lambda_{n,b})} + o(1) \right\} (\lambda_{n+1,b} - \lambda_{n,b})$$

$$\rightarrow \frac{1}{\pi} \int_\lambda^{\lambda + \Delta} \frac{d\lambda}{\mu^2(\lambda) + \nu^2(\lambda)},$$

i.e., the spectrum is continuous for $\lambda \geq 0$.

If $\lambda < 0$, then, selecting x_0 so that $\lambda - q(t) > 0$ for $t > x_0$, and replacing

$$\int_0^b \{\lambda - q(t)\} \, dt \quad \text{by} \quad \int_{x_0}^b \{\lambda - q(t)\} \, dt$$

everywhere in the asymptotic formulas, we arrive at the above result. Theorem 3.4.1 is thus proved. ∎

Theorem 3.4.2 *If all conditions of Theorem 3.4.1 are fulfilled, but the integral (4.1) converges, then the spectrum of the problem (1.1), (1.2) is discrete and has a unique limit point at infinity.*

Proof. It follows from the convergence of the integral (4.1) and from the definition of the function $\xi(x, \lambda)$ that

$$\xi(x, \lambda) - \xi(x, 0) = \int_0^x \{[\lambda - q(t)]^{1/2} - [-q(t)]^{1/2}\} \, dt$$

$$= \int_0^x \frac{\lambda \, dt}{\{\lambda - q(t)\}^{1/2} + \{-q(t)\}^{1/2}} \rightarrow \int_0^\infty \frac{\lambda \, dt}{\{\lambda - q(t)\}^{1/2} + \{-q(t)\}^{1/2}} < \infty$$

as $x \rightarrow \infty$.

Therefore, the imaginary part of $\xi(x, \lambda)$ is bounded uniformly in each bounded domain of the λ-plane; in particular, in each finite interval of the real axis. Furthermore, it follows from the formula (4.2) that the integrals $\int_0^\infty \varphi^2(x, \lambda) \, dx$ are bounded uniformly in each finite interval where real λ varies.

Assume that the spectrum contains infinitely many points in a finite interval $(\lambda, \lambda + \Delta)$. It is obvious that there must be a sequence of arbitrarily large positive numbers $b_1, b_2, \ldots, b_k, \ldots$ such that the number of eigenvalues $\lambda_{n,b}$ in $(\lambda, \lambda + \Delta)$ increases indefinitely with k. Hence, relying on the boundedness of $\int_0^\infty \varphi^2(t, \lambda) \, dt$, the sum

$$\sum_{\lambda < \lambda_{n,b_k} \leq \lambda + \Delta} \left(\int_0^{b_k} \varphi^2(t, \lambda_{n,b_k}) \, dt \right)^{-1}$$

would increase indefinitely as $x \rightarrow \infty$, contrary to Lemma 2.1.1. ∎

Notes

Section 1

Lemma 3.1.1 is due to B. M. Levitan (1950a), while Lemma 3.1.2 is due to B. M. Levitan and I. S. Sargsjan (1975).

Section 2

The results here are mostly due to H. Weyl (1909); see also E. Titchmarsh (1962). Our method for calculating the function $\rho(\lambda)$ is due to B. M. Levitan (1950a).

Section 3

The results of this section are due to E. Titchmarsh (1962).

Section 4

Theorems (4.1) and (4.2) are due to E. Titchmarsh (1962). Our proof is due to B. M. Levitan (1950a).

Chapter 4

THE DISTRIBUTION OF THE EIGENVALUES

4.1 The integral equation for Green's function

4.1.1

It follows from the results of Chapter 3 that if the function $q(x)$ of the Sturm-Liouville operator

$$Ly = -y'' + q(x)y, \quad a < x < b, \tag{1.1}$$

is bounded from below, and tends to $+\infty$ as $x \to a$ or $x \to b$ (or both), then the spectrum of L is discrete (assuming that at least one of the endpoints is singular; furthermore, if at least one of them is regular, a boundary condition should be specified on it).

Here, we give another proof that the spectrum of L is discrete for the cases (1) $a = -\infty$, $b = +\infty$ and (2) $a = 0$, $b = \infty$, along with the condition that $q(x) \to +\infty$ as $x \to \pm\infty$. Additional conditions will be placed on the function $q(x)$, enabling us not only to prove that the spectrum of L is discrete, but also to obtain an asymptotic formula for the function $N(\lambda) = \sum_{\lambda_n < \lambda} 1$, or the number of eigenvalues λ_n less than λ, as $\lambda \to \infty$.

Since $q(x)$ is bounded from below by assumption, without loss of generality, we can assume that the condition $q(x) > 1$ holds for all x in the interval (a, b).

We consider the case of the whole line, i.e., where $a = -\infty$, $b = +\infty$, in detail. The changes to be introduced for the case of the half-line are not complicated; a brief account of them is given at the end of Section 6.

For fixed x and $\mu > 0$, we put

$$\kappa = \{q(x) + \mu\}^{1/2}, \qquad g(x, \eta; \mu) = \frac{1}{2\kappa} \exp(-|x - \eta|\kappa), \tag{1.2}$$

and consider the integral equation

$$G(x, \eta; \mu) = g(x, \eta; \mu) - \int_{-\infty}^{\infty} g(x, \xi; \mu)[q(\xi) - q(x)]G(\xi, \eta; \mu) \, d\xi. \tag{1.3}$$

Below, we show that (1.3) can be solved by the iteration method for sufficiently large μ, and that the solution is Green's function for the operator (1.1). That Green's function is unique follows from Theorem 2.2.3.

4.1.2

Let X be the Banach space of numerical functions $A(x, \eta)$, $(-\infty < x, \eta < +\infty)$ with the norm

$$\|A(x,\eta)\|_X^2 = \int_{-\infty}^{\infty} \left\{ \int_{-\infty}^{\infty} |A(x,\eta)|^2 \, d\eta \right\} dx,$$

and let the operator N be determined by the formula

$$NA(x,\eta) = \int_{-\infty}^{\infty} g(x,\xi;\mu)[q(\xi) - q(x)]A(\xi,\eta)\, d\xi, \tag{1.4}$$

$g(x,\xi;\mu)$ being as in (1.2).
     In the sequel, the following is important.

Lemma 4.1.1 *Let $q(x)$ satisfy the conditions*

1.

$$|q(\xi) - q(x)| \le K q^a(x)|\xi - x| \quad \text{for} \quad |x - \xi| \le 1, \tag{1.5}$$

where $K > 0$, $0 < a < 3/2$ are two constant numbers, and

2.

$$q(\xi)\exp\left(-\frac{1}{2}|x - \xi|\sqrt{q(x)}\right) \le B \quad \text{for} \quad |x - \xi| > 1, \tag{1.6}$$

where B is a constant number.

Then, for sufficiently large μ, the operator N is contracting, i.e., $\|N\| < 1$.

Proof. Put

$$
\begin{aligned}
NA(x,\eta) &\equiv M(x,\eta) \\
&= \int_{|x-\xi|\le 1} g(x,\xi;\mu)[q(\xi) - q(x)]A(\xi,\eta)\, d\xi \\
&\quad + \int_{|x-\xi|>1} g(x,\xi;\mu)[q(\xi) - q(x)]A(\xi,\eta)\, d\xi \\
&\equiv \alpha(x,\eta;\mu) + \beta(x,\eta;\mu).
\end{aligned}
\tag{1.7}
$$

     Then

$$|M(x,\eta)| \le |\alpha(x,\eta;\mu)| + |\beta(x,\eta;\mu)|$$
$$M^2(x,\eta) \le 2\{\alpha^2(x,\eta;\mu) + \beta^2(x,\eta;\mu)\}.$$

Furthermore,

$$\alpha^2(x,\eta;\mu) \le \left\{ \int_{|x-\xi|\le 1} g(x,\xi;\mu)|q(\xi) - q(x)|\,|A(\xi,\eta)|\, d\xi \right\}^2. \tag{1.8}$$

Relying on the condition (1.5), we have

$$\left(\int_{|x-\xi|\leq 1} g(x,\eta;\mu)|q(\xi)-q(x)|\,|A(\xi,\eta)|\,d\xi \right)^2$$

$$\leq K^2 \left(\int_{|x-\xi|\leq 1} q^a(x)g(x,\xi;\mu)|x-\xi|\,|A(\xi,\eta)|\,d\xi \right)^2$$

for $|x-\xi| \leq 1$. Substituting the value of $g(x,\xi;\mu)$ from (1.2), we obtain

$$\int_{|x-\xi|\leq 1} |x-\xi| q^a(x)g(x,\xi;\mu)|A(\xi,\eta)|\,d\xi$$

$$= \frac{1}{2} \int_{|x-\xi|\leq 1} |x-\xi|^{1+\epsilon}|x-\xi|^{-\epsilon} q^a(x)\kappa^{-2+\epsilon}\kappa^{1-\epsilon}\exp(-\kappa|x-\xi|)|A(\xi,\eta)|\,d\xi. \quad (1.9)$$

Since $\kappa = \{q(x)+\mu\}^{1/2}$,

$$q^a(x)\kappa^{-2+\epsilon} = O(\mu^{-\zeta}), \quad (1.10)$$

where $\mu \to +\infty$, $\zeta > 0$, for, relying on $0 < a < 3/2$, we can find such $\epsilon < 1$ that $\zeta = 1 - a + \epsilon/2$ is positive. It then follows from (1.8), (1.9) and (1.10) that

$$\alpha^2(x,\eta;\mu) \leq \frac{K^2}{\mu^{2\zeta}} \left(\int_{|x-\xi|\leq 1} |x-\xi|^{-\epsilon}|A(\xi,\eta)|\,d\xi \right)^2$$

$$= \frac{K^2}{\mu^{2\zeta}} \int_{|x-\xi|\leq 1} \frac{|A(\xi,\eta)|}{|x-\xi|^\epsilon}\,d\xi \int_{|x-\xi'|\leq 1} \frac{|A(\xi',\eta)|}{|x-\xi'|^\epsilon}\,d\xi', \quad (1.11)$$

which, integrated with respect to x from $-\infty$ to ∞, after an obvious change of variables and interchange of the integration limits (and using the Cauchy-Schwarz inequality), leads to

$$\int_{-\infty}^{\infty} \alpha^2(x,\eta;\mu)\,dx \leq \frac{K^2}{\mu^{2\zeta}} \int_{|u|\leq 1} \frac{du}{|u|^\epsilon} \int_{|u'|\leq 1} \frac{du'}{|u'|^\epsilon} \left\{ \int_{-\infty}^{\infty} |A(x+u,\eta)|\cdot|A(x+u',\eta)|\,dx \right\}$$

$$\leq \frac{K^2}{\mu^{2\zeta}} \left(\int_{|u|\leq 1} \frac{du}{|u|^\epsilon} \right)^2 \int_{-\infty}^{\infty} |A(x,\eta)|^2\,dx = \frac{K_1}{\mu^{2\zeta}} \int_{-\infty}^{\infty} |A(x,\eta)|^2\,dx.$$

Now, integrating with respect to η from $-\infty$ to ∞, we have

$$\|\alpha(x,\eta;\mu)\|_X \leq C\mu^{-2\zeta}\|A(x,\eta)\|_X, \quad (1.12)$$

where C is a constant number.

To estimate $\|\beta(x,\eta;\mu)\|_X$ as $\mu \to \infty$, we see that

$$\beta(x,\eta;\mu) = \int_{|x-\xi|>1} g(x,\xi;\mu)q(\xi)A(\xi,\eta)\,d\xi - q(x)\int_{|x-\xi|>1} g(x,\xi;\mu)A(\xi,\eta)\,d\xi \equiv \beta_1+\beta_2$$

by definition, or (1.7). We derive from the estimate (1.6) and from the formula (1.2) (proceeding as when estimating $\alpha(x, \eta; \mu)$) that

$$\beta_1^2 \leq \frac{B^2}{4\mu^2} \int_{|u|>1} e^{-(1/2)\sqrt{\mu}u} |A(x+u, \eta)| \, du \int_{|u'|\leq 1} e^{-(1/2)\sqrt{\mu}u'} |A(x+u', \eta)| \, du',$$

which, integrated with respect to x, yields

$$\int_{-\infty}^{\infty} \beta_1^2(x, \eta; \mu) \, dx \leq \frac{B^2}{4\mu^2} e^{-(1/2)\sqrt{\mu}} \int_{-\infty}^{\infty} |A(x, \eta)|^2 \, dx$$

and, by repeated integration with respect to η,

$$\|\beta_1(x, \eta; \mu)\|_X \leq C\mu^{-r} \|A(x, \eta)\|_X, \tag{1.13}$$

where $\mu \to +\infty$, C is a constant number, and r is an arbitrarily large positive number.

A similar estimate holds for $\|\beta_2(x, \eta; \mu)\|_X$. We then obtain the estimate

$$\|NA(x, \eta)\|_X \leq C\mu^{-r} \|A(x, \eta)\|_X$$

from (1.7), relying on (1.12) and (1.13) for $NA(x, \eta)$ as $\mu \to \infty$, which exactly proves the lemma. ∎

Remark Consider Banach spaces $X_2^{(p)}$, $X^{(\tau)}$, $X_1^{(\tau)}$, $p \geq 1$, τ is an arbitrary real number, whose elements are the numerical functions $A(x, \eta)$, $(-\infty < x, \eta < +\infty)$ and whose norms are

$$\|A(x, \eta)\|_{X_2^{(p)}}^p = \sup_{-\infty < x < \infty} \int_{-\infty}^{\infty} |A(x, \eta)|^p \, d\eta \tag{1.14}$$

$$\|A(x, \eta)\|_{X^{(\tau)}}^2 = \int_{-\infty}^{\infty} \left\{ \int_{-\infty}^{\infty} |A(x, \eta)|^2 q^{2\tau}(\eta) \, d\eta \right\} dx \tag{1.15}$$

$$\|A(x, \eta)\|_{X_1^{(\tau)}}^2 = \sup_{-\infty < x < \infty} \int_{-\infty}^{\infty} |A(x, \eta)|^2 q^{2\tau}(\eta) \, d\eta \tag{1.16}$$

respectively.

It is easy to see that a lemma similar to Lemma 4.1.1 also holds in all these spaces.

4.1.3

Consider the integral equation (1.3) again. It can be written, with the aid of the operator N, as

$$G(x, \eta; \mu) = g(x, \eta; \mu) - NG(\xi, \eta; \mu). \tag{1.17}$$

Since N is contracting in all the above spaces for sufficiently large μ, if the function $g(x, \eta; \mu)$ is in one of them, the equation (1.17) can be solved by the iteration method and the solution belongs to the same space.

Sufficient conditions for $g(x, \eta; \mu)$ to belong to the spaces $X_2^{(p)}$ $(p > 0)$, X, $X_1^{(1/2)}$ and $X^{(-\tau)}$ $(\tau > 0)$, respectively, are given by the following.

Lemma 4.1.2 *If $q(x) > 0$, then the functions $g(x, \eta; \mu)$ and $g'_\eta(x, \eta; \mu)$ belong to the space $X_2^{(p)}$, $(p > 0)$.*

Proof. Indeed, it follows by the definition of $g(x, \eta; \mu)$, or from the equality (1.2), that

$$\int_{-\infty}^{\infty} \{g(x, \eta; \mu)\}^p \, d\eta \leq \int_{-\infty}^{\infty} e^{-p\sqrt{\mu}|x-\eta|} \, d\eta = \int_{-\infty}^{\infty} e^{-p\sqrt{\mu}|u|} \, du = \frac{2}{p\sqrt{\mu}},$$

i.e., $g(x, \eta; \mu) \in X_2^{(p)}$. The proof for $g'_\eta(x, \eta; \mu)$ is similar. ∎

Lemma 4.1.3 *If the condition $\displaystyle\int_{-\infty}^{\infty} q^{-3/2}(x) \, dx < \infty$ is fulfilled, then $g(x, \eta; \mu)$ belongs to the space X.*

Proof. In fact, for $\mu > 0$, we have

$$\int_{-\infty}^{\infty} g^2(x, \eta; \mu) \, d\eta = \frac{1}{4\kappa^2} \int_{-\infty}^{\infty} e^{-\kappa|x-\eta|} \, d\eta = \frac{1}{2\kappa^3} \leq \frac{1}{2q^{3/2}(x)},$$

relying on (1.2); hence, $g(x, \eta; \mu) \in X$ by the lemma's conditions. ∎

Lemma 4.1.4 *If the function $q(x)$ satisfies the condition $q(\eta)q^{-1}(x) \leq C$ for $|x - \eta| \leq 1$, where C is a constant number, then $g(x, \eta; \mu) \in X_1^{(1/2)}$.*

Proof. It follows from (1.2) that

$$\int_{-\infty}^{\infty} g^2(x, \eta; \mu)q(\eta) \, d\eta = \int_{|x-\eta|\leq 1} \frac{q(\eta)}{4\kappa^2} \exp\{-2|x - \eta|\kappa\} \, d\eta$$

$$+ \int_{|x-\eta|>1} \frac{q(\eta)}{4\kappa^3} \exp\{-2|x - \eta|\kappa\} \, d\eta \equiv a(x) + b(x).$$

The conditions of the lemma imply that

$$\sup_x a(x) \leq \sup_x \int_{|x-\eta|\leq 1} e^{-2\kappa|x-\eta|} \, d\eta = \sup_x \int_{|u|\leq 1} e^{-2\kappa|u|} \, du < \infty.$$

Furthermore, we have

$$q(\eta)g^2(x, \eta; \mu) \leq C \exp\left\{-\frac{3}{2}|x - \eta|\sqrt{\mu}\right\}$$

for $|x - \eta| > 1$, relying on (1.6); therefore, as in estimating $a(x)$, we see that $\sup_x b(x) < \infty$, which completes the proof, also relying on a similar estimate for $a(x)$. ∎

Lemma 4.1.5 *Let $q(x)q^{-1}(\eta) < C$ for $|x - \eta| \leq 1$. Then $g''_{\eta\eta}(x, \eta; \mu) \in X_1^{(-1/2)}$.*

Proof. Indeed, for $\eta \neq x$,

$$\frac{\partial^2 g}{\partial \eta^2} = \kappa^2 g(x, \eta; \mu) = \frac{1}{2}\kappa \exp(-\kappa|x - \eta|).$$

The proof is therefore similar to that of Lemma 4.1.4. ∎

Lemma 4.1.6 *If $q(x)$ satisfies the conditions*

  1. $q(x)q^{-1}(\eta) < C,\ (|x - \eta| \leq 1),$
  2. $\{q(x)\}^{-3/2 - 2\tau} \in \mathcal{L}(-\infty, \infty),$

then the function $g(x, \eta; \mu)$ belongs to the space $X^{(-\tau)}$.

Proof. Put

$$\int_{-\infty}^{\infty} g^2(x, \eta; \mu)q^{-2\tau}(\eta)\, d\eta = \int_{|x-\eta|\leq 1} g^2(x, \eta; \mu)q^{-2\tau}(\eta)\, d\eta$$

$$+ \int_{|x-\eta|>1} g^2(x, \eta; \mu)q^{-2\tau}(\eta)\, d\eta \equiv a(x) + b(x).$$

It follows from Condition 1 for $\mu > 1$, by the definition of $g(x, \eta; \mu)$, that

$$g^2(x, \eta; \mu)q^{-2\tau}(\eta) \leq Cq^{-(2\tau+1)}(x)\exp\{-2|x - \eta|\sqrt{q(x)}\},$$

where $|x - \eta| \leq 1$.
  We have

$$a(x) \leq Cq^{-(2\tau+1)}(x)\int_{|x-\eta|\leq 1} \exp\{-2|x - \eta|\sqrt{q(x)}\}\, d\eta \leq C\{q(x)\}^{-(2\tau+3/2)};$$

$a(x) \in \mathcal{L}(-\infty, \infty)$, also because of Condition 2.
  Furthermore, for $|x - \eta| > 1$,

$$g^2(x, \eta; \mu)q^{-\tau}(\eta) \leq C\exp(-2|x - \eta|\kappa).$$

  Hence, $b(x) \leq C\int_1^{\infty} e^{-2u\kappa}\, du = C_1 e^{-2\kappa}$, and $b(x)$ also belongs to $\mathcal{L}(-\infty, \infty)$ with the consequence that $a(x) + b(x) \in \mathcal{L}(-\infty, \infty)$, which proves the lemma. ∎

4.2   The first derivative of the function $G(x, \eta; \mu)$

Formally differentiating the equation (1.3) with respect to η, we obtain

$$\frac{\partial G}{\partial \eta} = \frac{\partial g}{\partial \eta} - \int_{-\infty}^{\infty} g(x, \xi; \mu)[q(\xi) - q(x)]\frac{\partial}{\partial \eta}G(\xi, \eta; \mu)\, d\xi. \tag{2.1}$$

Replacing $G'_\eta(x, \eta; \mu)$ by $K(x, \eta; \mu)$, we obtain the integral equation

$$K(x, \eta; \mu) = g'_\eta(x, \eta; \mu) - \int_{-\infty}^{\infty} g(x, \xi; \mu)[q(\xi) - q(x)]K(\xi, \eta; \mu) \, d\xi. \qquad (2.2)$$

By Lemma 4.1.2, $g'_\eta(x, \eta; \mu) \in X_2^{(p)}$. Therefore, $K(x, \eta; \mu) \in X_2^{(p)}$ according to Lemma 4.1.1. It is not hard to verify that the function $K(x, \eta; \mu)$ is continuous with respect to $\eta \neq x$. Integrate (2.2) with respect to η from $-\infty$ to η, viz.

$$\int_{-\infty}^{\eta} K(x, \eta; \mu) \, d\eta = g(x, \eta; \mu) - \int_{-\infty}^{\infty} g(x, \xi; \mu)[q(\xi) - q(x)] \left\{ \int_{-\infty}^{\eta} K(\xi, \eta; \mu) \, d\eta \right\} \, d\xi. \qquad (2.3)$$

For $\eta = x$, the function $g'_\eta(x, \eta; \mu)$ and, therefore, $K(x, \lambda; \mu)$ have a discontinuity of the first kind, which is not an obstacle to apply the fundamental theorem of calculus. That the integral on the right in (2.3) exists and that the order of integration can be changed follows from $K(x, \eta; \mu) \in X_2^{(1)}$.

The equation (2.3) coincides with (1.3). Therefore, it follows from the uniqueness of the solution of (2.3) that $\int_{-\infty}^{\eta} K(x, \eta; \mu) \, d\eta = G(x, \eta; \mu)$, and we obtain by differentiating with respect to η that, for $\eta \neq x$, $K(x, \eta; \mu) = G'_\eta(x, \eta; \mu)$.

Therefore, (2.2) can be written in the form

$$G'_\eta(x, \eta; \mu) - g'_\eta(x, \eta; \mu) = - \int_{-\infty}^{\infty} g(x, \xi; \mu)[q(\xi) - q(x)]g'_\eta(\xi, \eta; \mu) \, d\xi$$
$$- \int_{-\infty}^{\infty} g(x, \xi; \mu)[q(\xi) - q(x)]|G'_\eta(\xi, \eta; \mu) - g'_\eta(\xi, \eta; \mu)| \, d\xi.$$

Applying estimates similar to those in the proof of Fundamental Lemma 4.1.1, we can show that the function

$$l(x, \eta; \mu) = - \int_{-\infty}^{\infty} g(x, \xi; \mu)[q(\xi) - q(x)]g'_\eta(\xi, \eta; \mu) \, d\xi$$

is continuous with respect to η, for which it suffices to verify that the improper integral determining $l(x, \eta; \mu)$ converges uniformly with respect to η. The latter follows from the condition (1.6).

Now, solving the equation (2.4) by iteration, it is easy to show that the function $G'_\eta(x, \eta; \mu) - g'_\eta(x, \eta; \mu)$ is continuous with respect to η. Therefore, $G'_\eta(x, \eta; \mu)$ has the same discontinuities as $g'_\eta(x, \eta; \mu)$, the latter with an unique discontinuity for $\eta = x$. Indeed,

$$g'_\eta(x, \eta; \mu) = \begin{cases} -\frac{1}{2}\exp\{-(\eta - x)\kappa\}, & \eta > x \\ \\ -\frac{1}{2}\exp\{-(x - \eta)\kappa\}, & \eta < x \end{cases} \qquad \kappa = \{q(x) + \mu\}^{1/2}.$$

It follows that

$$g'_\eta(x, \eta; \mu)|_{\eta=x+0} - g'_\eta(x, \eta; \mu)|_{\eta=x-0} = -1. \qquad (2.5)$$

Thus, $G'_\eta(x, \eta; \mu)$ is continuous with respect to η everywhere except the point $\eta = x$, where a discontinuity of the first kind occurs. Furthermore,

$$G'_\eta(x, x + 0; \mu) - G'_\eta(x, x - 0; \mu) = -1. \tag{2.6}$$

4.3 The second derivative of the function $G(x, \eta; \mu)$

It was proved in Section 2 that the function $G'_\eta(x, \eta; \mu) \in X_2^{(p)}$ is continuous with respect to η for any $p \geq 1$, $\eta \neq x$, and that it satisfies the equation (2.1) reducible to the form

$$\mathcal{L}(x, \eta; \mu) = l(x, \eta; \mu) - \int_{-\infty}^{\infty} g(x, \xi; \mu)[q(\xi) - q(x)]\mathcal{L}(\xi, \eta; \mu)\, d\xi \tag{3.1}$$

(see the equation (2.4)), where

$$\mathcal{L}(x, \eta; \mu) = G'_\eta(x, \eta; \mu) - g'_\eta(x, \eta; \mu),$$

$$l(x, \eta; \mu) = \int_{-\infty}^{\infty} g(x, \xi; \mu)[q(\xi) - q(x)]g'_\eta(\xi, \eta; \mu)\, d\xi. \tag{3.2}$$

Differentiating (3.1) formally with respect to η, we obtain

$$\mathcal{L}'_\eta(x, \eta; \mu) = l'_\eta(x, \eta; \mu) - \int_{-\infty}^{\infty} g(x, \xi; \mu)[q(\xi) - q(x)]\mathcal{L}'_\eta(\xi, \eta; \mu)\, d\xi. \tag{3.3}$$

Furthermore, differentiating (3.2) with respect to η again formally, and using (2.5), we obtain

$$
\begin{aligned}
l'_\eta(x, \eta; \mu) &= g(x, \eta; \mu)[q(\eta) - q(x)] - \int_{-\infty}^{\infty} g(x, \xi; \mu)[q(\xi) - q(x)]g''_{\eta\eta}(\xi, \eta; \mu)\, d\xi \\
&\equiv m(x, \eta; \mu),
\end{aligned}
$$

where $\eta \neq x$.

We now consider the integral equation

$$M(x, \eta; \mu) = m(x, \eta; \mu) - \int_{-\infty}^{\infty} g(x, \xi; \mu)[q(\xi) - q(x)]M(\xi, \eta; \mu)\, d\xi, \tag{3.4}$$

and show that $m(x, \eta; \mu) \in X_1^{(-1/2)}$. It will follow, also by Lemma 4.1.1, that the solution of (3.4) also belongs to $X_1^{(-1/2)}$. We have

$$
\begin{aligned}
m(x, \eta; \mu) &= g(x, \eta; \mu)[q(\eta) - q(x)] - \int_{|\xi-\eta| \leq 1} g(x, \xi; \mu)[q(\xi) - q(x)]g''_{\eta\eta}(\xi, \eta; \mu)\, d\xi \\
&\quad - \int_{|\xi-\eta| > 1} g(x, \xi; \mu)[q(\xi) - q(x)]g''_{\eta\eta}(\xi, \eta; \mu)\, d\xi \\
&\equiv m_1 + m_2 + m_3.
\end{aligned}
$$

It is not hard to derive from the conditions (1.5) and (1.6) that $m_1 \in X_1^{(1)}$; therefore, $m_1 \in X_1^{(-1/2)}$. We now consider m_2. Assuming that the condition of Lemma 4.1.5 holds, or $q(x)q^{-1}(\eta) < C$ for $|x - \eta| \leq 1$, we obtain

$$m_2 q^{-1/2}(\eta) \leq \int_{|\xi-\eta|\leq 1} g(x,\xi;\mu)|q(\xi) - q(x)|e^{-|\xi-\eta|\kappa}q^{-1/2}(\eta)\kappa \, d\xi$$

$$< C(\mu)\int_{|\xi-\eta|\leq 1} g(x,\xi;\mu)|q(\xi) - q(x)|e^{-|\xi-\eta|\kappa} \, d\xi, \quad \kappa = [q(\xi) + \mu]^{1/2},$$

where $C(\mu)$ is a constant only depending on μ.

Finally, using (1.6), we obtain the estimate

$$m_3 q^{-1/2}(\eta) \leq C(\mu)\int_{|\xi-\eta|>1} g(x,\xi;\mu)|q(\xi) - q(x)|e^{(-1/2)|\xi-\eta|\kappa} \, d\xi$$

for m_3. Therefore,

$$(m_2 + m_3)q^{-1/2}(\eta) \leq C(\mu)\int_{-\infty}^{\infty} g(x,\xi;\mu)|q(\xi) - q(\eta)|e^{-(1/2)|\xi-\eta|\kappa} \, d\xi,$$

where $\kappa = [q(\xi) + \mu]^{1/2}$.

The integral on the right can be estimated as in the proof of Lemma 4.1.1. As a result, it can easily be proved that $m(x,\eta;\mu) \in X_1^{(-1/2)}$.

We now show that $M(x,\eta;\mu)$, i.e., the solution of (3.4) (belonging to $X_1^{(-1/2)}$ as proved), is $G''_{\eta\eta}(x,\eta;\mu) - g''_{\eta\eta}(x,\eta;\mu)$ for $\eta \neq x$. Indeed, integrating (3.4) with respect to η from η_0 to η, we obtain

$$\int_{\eta_0}^{\eta} M(x,\eta;\mu) \, d\eta = \int_{\eta_0}^{\eta} m(x,\eta;\mu) \, d\eta - \int_{-\infty}^{\infty} g(x,\xi;\mu)[q(\xi) - q(x)]\left\{\int_{\eta_0}^{\eta} M(\xi,\eta;\mu) \, d\eta\right\} \, d\xi.$$
$$(3.5)$$

Furthermore, it follows from (3.1) that

$$\mathcal{L}(x,\eta;\mu) - \mathcal{L}(x,\eta_0;\mu) = l(x,\eta;\mu) - l(x,\eta_0;\mu)$$
$$- \int_{-\infty}^{\infty} g(x,\xi;\mu)[q(\xi) - q(x)]\{\mathcal{L}(\xi,\eta;\mu) - \mathcal{L}(\xi,\eta_0;\mu)\} \, d\xi. \quad (3.6)$$

If we show that

$$\int_{\eta_0}^{\eta} m(x,\eta;\mu) \, d\eta = l(x,\eta;\mu) - l(x,\eta_0;\mu), \tag{3.7}$$

then it will follow from the uniqueness of the solution of (3.5) and (3.6) that

$$\int_{\eta_0}^{\eta} M(x,\eta;\mu) \, d\eta = \mathcal{L}(x,\eta;\mu) - \mathcal{L}(x,\eta_0;\mu),$$

or

$$M(x,\eta;\mu) = \mathcal{L}'_\eta(x,\eta;\mu) = G''_{\eta\eta}(x,\eta;\mu) - g''_{\eta\eta}(x,\eta;\mu). \tag{3.8}$$

Since $g''_{\eta\eta}(x, \eta; \mu)$ exists for $\eta \neq x$, it follows from (3.8) that

$$G''_{\eta\eta}(x, \eta; \mu) \in X_1^{(-1/2)}.$$

Thus, it remains to prove (3.7).

It can easily be obtained from (1.6) that the function

$$l(x, \eta; \mu) = - \int_{-\infty}^{\infty} g(x, \xi; \mu)[q(\xi) - q(x)]g'_{\eta}(\xi, \eta; \mu) \, d\xi \tag{3.9}$$

can be differentiated under the integral sign. In fact, let n be so large that the point η is in the interval $(-n, n)$. Represent

$$l(x, \eta; \mu) = - \left\{ \int_{-\infty}^{-n} + \int_{-n}^{n} + \int_{n}^{\infty} \right\} g(x, \xi; \mu)[q(\xi) - q(x)]g'_{\eta}(\xi, \eta; \mu) \, d\xi \equiv l_1 + l_2 + l_3.$$

The integrals l_1 and l_3 can be differentiated under the integral sign, since we obtain absolutely and uniformly convergent integrals, which easily follows from (1.6). l_2 is an integral with finite limits, and its differentiation causes no difficulties. Therefore, differentiating (3.9) with respect to η and invoking (2.5), we obtain

$$\begin{aligned} l'_{\eta}(x, \eta; \mu) &= g(x, \eta; \mu)[q(\eta) - q(x)] - \int_{-\infty}^{\infty} g(x, \xi; \mu)[q(\xi) - q(x)]g''_{\eta\eta}(\xi, \eta; \mu) \, d\xi \\ &\equiv m(x, \eta; \mu) \end{aligned}$$

and

$$\int_{\eta_0}^{\eta} m(x, \eta; \mu) \, d\eta = l(x, \eta; \mu) - l(x, \eta_0; \mu),$$

which coincides with (3.7), thus proving the statement.

4.4 Further properties of the function $G(x, \eta; \mu)$

4.4.1

$G(x, \eta; \mu)$ also has some other important properties to be established here.

Lemma 4.4.1 *The function $G(x, \eta; \mu)$ satisfies the differential equation*

$$G''_{\eta\eta}(x, \eta; \mu) = \{q(x) + \mu\}G(x, \eta; \mu) \tag{4.1}$$

for $\eta \neq x$.

Proof. By the definition of the function $g(x, \eta; \mu)$, or (1.2), it can easily be seen (by straightforward calculation) that it satisfies the equation $g''_{\eta\eta} = \{q(x) + \mu\}g$. Therefore, the equation (3.4), taking into account (3.8), can be rewritten as

$$G''_{\eta\eta}(x, \eta; \mu) - [q(x) + \mu]g(x, \eta; \mu) = \{q(\eta) - q(x)\}g(x, \eta; \mu)$$
$$- \int_{-\infty}^{\infty} g(x, \xi; \mu)[q(\xi) - q(x)]G''_{\eta\eta}(\xi, \eta; \mu) \, d\xi.$$

It follows that

$$G''_{\eta\eta}\{q(\eta) + \mu\}^{-1} = g - \int_{-\infty}^{\infty} g(x, \xi; \mu)[q(\xi) - q(x)]G''_{\eta\eta}\{q(\eta) + \mu\}^{-1} d\xi.$$

Comparing with (1.3), we conclude, on the basis of the uniqueness of the solution, that $G''_{\eta\eta}\{q(\eta) + \mu\}^{-1} = G$, which coincides with (4.1). The proof is thus complete. ∎

Lemma 4.4.2 *The function* $G(x, y; \mu)$ *is symmetric with respect to the variables* x *and* y, *i.e.,*

$$G(x, y; \mu) = G(y, x; \mu). \tag{4.2}$$

Proof. Let x and y be two distinct points of the number line, with $x < y$ for definiteness. Consider the integral

$$I = \int_{-\infty}^{\infty} \{G(y, \xi; \mu)G''_{\xi\xi}(x, \xi; \mu) - G(x, \xi; \mu)G''_{\xi\xi}(y, \xi; \mu)\} d\xi$$

whose existence follows from $G \in X_1^{(1/2)}$ and $G''_{\xi\xi} \in X_1^{(-1/2)}$, noted by us in the preceding sections.

Put

$$I = \left\{ \int_{-\infty}^{x} + \int_{x}^{y} + \int_{y}^{\infty} \right\} [G(y, \xi; \mu)G''_{\xi\xi}(x, \xi; \mu) - G(x, \xi; \mu)G''_{\xi\xi}(y, \xi; \mu)] d\xi \equiv I_1 + I_2 + I_3.$$

Since

$$G(y, \xi; \mu)G''_{\xi\xi}(x, \xi; \mu) - G(x, \xi; \mu)G''_{\xi\xi}(y, \xi; \mu)$$
$$= \frac{\partial}{\partial \xi} \left[G(y, \xi; \mu)G'_{\xi}(x, \xi; \mu) - G(x, \xi; \mu)G'_{\xi}(y, \xi; \mu) \right],$$

we have

$$
\begin{aligned}
I_1 &= G(y, x; \mu)G'_{\xi}(x, \xi; \mu)|_{\xi=x-0} - G(x, x; \mu)G'_{\xi}(y, \xi; \mu)|_{\xi=x} \\
I_2 &= G(y, y; \mu)G'_{\xi}(x, \xi; \mu)|_{\xi=y} - G(x, y; \mu)G'_{\xi}(y, \xi; \mu)|_{\xi=y-0} \\
&\quad - G(y, x; \mu)G'_{\xi}(x, \xi; \mu)|_{\xi=x+0} + G(x, x, \mu)G'_{\xi}(y, \xi, \mu)|_{\xi=x} \\
I_3 &= G(x, y; \mu)G'_{\xi}(y, \xi; \mu)|_{\xi=y+0} - G(y, y; \mu)G'_{\xi}(x, \xi; \mu)|_{\xi=y}.
\end{aligned}
$$

Adding the equalities together and taking into account (2.6), we obtain

$$I = G(y, x; \mu) - G(x, y; \mu),$$

which proves (4.2), because it follows from (4.1) that $I = 0$. ∎

4.4.2

If the function $g(x)$ is bounded from below and $\lim_{|x|\to\infty} q(x) = +\infty$, then the Sturm-Liouville operator has discrete spectrum and unique Green's function for all μ outside the spectrum (see Section 1, Chapter 3).

It has been shown above that the solution of the integral equation (1.3), or the function $G(x,\eta;\mu)$, satisfies all the conditions for Green's function of the Sturm-Liouville operator. Therefore, $G(x,\eta;\mu)$ is Green's function, relying on the uniqueness. It is known from Chapter 2 that Green's function $G(x,\eta;\mu)$ can be expressed in terms of the solutions $\psi_1(x,\mu) \in \mathcal{L}^2(-\infty,0)$, $\psi_2 \in \mathcal{L}^2(0,\infty)$ determined uniquely for $q(x) \to +\infty$ as $|x| \to \infty$.

It follows from the integral representation of the resolvent (see Chapter 2, Section 3) that, for any function $f(x) \in \mathcal{L}^2(-\infty,\infty)$,

$$\int_{-\infty}^{\infty} G(x,y;\mu)f(y)\,dy = \sum_{n=1}^{\infty} \frac{a_n\varphi_n(x)}{\lambda_n + \mu},$$

$$a_n = \int_{-\infty}^{\infty} f(x)\varphi_n(x)\,dx,$$

(4.3)

where λ_n is an eigenvalue and $\varphi_n(x)$ the corresponding normalized eigenfunction of the Sturm-Liouville operator, i.e.,

$$-\varphi_n''(x) + q(x)\varphi_n(x) = \lambda_n\varphi_n(x).$$

4.5 Differentiation of Green's function with respect to its parameter

4.5.1

It follows from Lemmas 4.1.1 and 4.1.3 that if $q^{-3/2}(x) \in \mathcal{L}(-\infty,\infty)$, then the function $G(x,\eta;\mu) \in X$, i.e.,

$$\int_{-\infty}^{\infty} \left\{ \int_{-\infty}^{\infty} G^2(x,\eta;\mu)\,d\eta \right\} dx < \infty. \qquad (5.1)$$

However, if $q^{-3/2}(x) \notin \mathcal{L}(-\infty,\infty)$, then (5.1) may not hold either.

Assume that, for some natural $\tau > 0$,

$$q^{-(3/2+2\tau)}(x) \in \mathcal{L}(-\infty,\infty), \qquad (5.2)$$

in which case it is natural to consider an iteration of order τ of Green's function $G(x,\eta;\mu)$, and hope that this iteration is a Hilbert-Schmidt kernel, i.e., that (5.1) holds.

It follows from the integral representation of Green's function (see Chapter 2, Section 3) that its iterations coincide with the corresponding derivatives with respect to μ, up to the sign and a constant multiplier.

We now study $G'_\mu(x, \eta; \mu)$ in detail, according to the first iteration. The case of higher derivatives (or higher-order iterations) is studied similarly, and we confine ourselves only to stating the final result.

It follows from (1.3) that

$$G'_\mu = g'_\mu - \int_{-\infty}^{\infty} g'_\mu[q(\xi) - q(x)]G\, d\xi - \int_{-\infty}^{\infty} g[q(\xi) - q(x)]G'_\mu\, d\xi. \qquad (5.3)$$

If $q^{-7/2}(x) \in \mathcal{L}(-\infty, \infty)$, then it follows from Lemma 4.1.6 that the function $g(x, \eta; \mu)$, and therefore, by Lemma 4.1.1, $G(x, \eta; \mu)$, belong to the space $X^{(-1)}$, i.e.,

$$\int_{-\infty}^{\infty} \left\{ \int_{-\infty}^{\infty} G^2(x, \eta; \mu)q^{-2}(\eta)\, d\eta \right\} dx < \infty. \qquad (5.4)$$

Since $G(x, \eta; \mu)$ is symmetric with respect to both x and η, the condition (5.4) can be written as

$$\int_{-\infty}^{\infty} \left\{ \int_{-\infty}^{\infty} G^2(\eta, x; \mu)q^{-2}(\eta)\, d\eta \right\} dx < \infty. \qquad (5.5)$$

Moreover, it follows from Lemma 4.1.1 (and successive approximations) that

$$\left(\int_{-\infty}^{\infty} \left\{ \int_{-\infty}^{\infty} G^2(\eta, x; \mu)[q(\eta) + \mu]^{-2}\, d\eta \right\} dx \right)^{-1/2}$$
$$= O\left(\int_{-\infty}^{\infty} \left\{ \int_{-\infty}^{\infty} \frac{g^2(x, \eta; \mu)}{[q(\eta) + \mu]^2}\, d\eta \right\} dx \right) \quad (5.6)$$

as $\mu \to \infty$.

Put

$$\alpha(x, \eta; \mu) = \int_{-\infty}^{\infty} g'_\mu(x, \xi; \mu)[q(\xi) - q(x)]G(\xi, \eta; \mu)\, d\xi. \qquad (5.7)$$

Lemma 4.5.1 *Let $q(x)$ satisfy the conditions*

1. *$|q(\xi) - q(x)| \leq Aq^a(x)|\xi - x|$ for $|x - \xi| \leq 1$, where $A > 0$ and $0 < a < 3/2$ are two constant numbers,*

2. *$q(\xi)q^{-1}(x) \leq B$ for $|x - \xi| \leq 1$, where B is a constant number,*

3. *$q^2(\xi)\exp\left\{ -\frac{1}{2}|x - \xi|\sqrt{q(x)} \right\}$ for $|x - \xi| > 1$, where C is a constant number.*

4. *the function $q^{-7/2}(x) \in \mathcal{L}(-\infty, \infty)$.*

Then the function $\alpha(x, \eta; \mu)$ determined by the equality (5.7) belongs to the space X, i.e.,

$$\int_{-\infty}^{\infty} \left\{ \int_{-\infty}^{\infty} \alpha^2(x, \eta; \mu)\, d\eta \right\} dx < \infty. \qquad (5.7')$$

Proof. Indeed, if we write

$$\alpha(x,\eta;\mu) = \int_{-\infty}^{\infty} g'_{\mu}(x,\xi;\mu)[q(\xi) + \mu]\frac{q(\xi) - q(x)}{q(\xi) + \mu}G(\xi,\eta;\mu)\,d\xi,$$

and estimate the integral as in the proof of Lemma 4.1.1, we can show that (5.7′) is valid. Moreover, proceeding as in Lemma 4.1.1 and using the estimate (5.6), it is not hard to obtain the estimate

$$||\alpha(x,\eta;\mu)||_X = o(||G(x,\eta;\mu)[q(\eta) + \mu]^{-1}||_X) = o(||g(x,\eta;\mu)[q(\eta) + \mu]^{-1}||_X). \quad (5.8)$$

∎

4.5.2

It can be checked directly that if $q^{-7/2}(x) \in \mathcal{L}^2(-\infty,\infty)$, then $q'_{\mu}(x,\eta;\mu) \in X$.

Taking into account (5.7), the equation (5.3) can be rewritten as

$$G'_{\mu} = g'_{\mu} - \alpha(x,\eta;\mu) - \int_{-\infty}^{\infty} g(x,\xi;\mu)[q(\xi) - q(x)]G'_{\mu}(\xi,\eta;\mu)\,d\xi, \quad (5.3')$$

which, solved by the iteration method, yields

$$G'_{\mu}(x,\eta;\mu) = g'_{\mu}(x,\eta;\mu) - \alpha(x,\eta;\mu) + \beta(x,\eta;\mu). \quad (5.9)$$

Meanwhile, it follows from Lemma 4.1.1 that

$$||\beta(x,\eta;\mu)||_X = o(||g'_{\mu} - \alpha||_X) \quad (5.10)$$

as $\mu \to \infty$.

Since $||g'_{\mu} - \alpha||_X \leq ||g'_{\mu}||_X + ||\alpha||_X$, the estimate (5.10) can be written as

$$||\beta(x,\eta;\mu)||_X = o(||g'_{\mu}||_X + ||\alpha||_X). \quad (5.11)$$

We show that

$$||\alpha(x,\eta;\mu)||_X = o(||g'_{\mu}(x,\eta;\mu)||_X), \quad (5.12)$$

which follows from the estimate (5.8) if we prove that

$$||g(x,\eta;\mu)[q(\eta) + \mu]^{-1}||_X = o(||g'_{\mu}(x,\eta;\mu)||_X) \quad (5.13)$$

as $\mu \to \infty$.

Straightforward calculation yields

$$||g'_{\mu}(x,\eta;\mu)||_X^2 = C\int_{-\infty}^{\infty} \frac{dx}{\{q(x) + \mu\}^{7/2}}, \quad (5.14)$$

where C is a constant number.

Furthermore, using Conditions 2 and 3 of Lemma 4.5.1, we obtain

$$||g(x,\eta;\mu)[q(\eta)+\mu]^{-1}||_X = \int_{-\infty}^{\infty} \left\{ \int_{-\infty}^{\infty} g^2(x,\eta;\mu)[q(\eta)+\mu]^{-2}\,d\eta \right\}\,dx$$

$$= \int_{-\infty}^{\infty} \left\{ \int_{|x-\eta|\le 1} \frac{g^2(x,\eta;\mu)}{[q(\eta)+\mu]^2}\,d\eta + \int_{|x-\eta|>1} \frac{g^2(x,\eta;\mu)}{[q(\eta)+\mu]^2}\,d\eta \right\}\,dx$$

which is bounded by

$$C_1 \int_{-\infty}^{\infty} \left\{ \int_{-\infty}^{\infty} \frac{g^2(x,\eta;\mu)}{[q(x)+\mu]^2}\,d\eta \right\}\,dx + C_2 \int_{-\infty}^{\infty} \left\{ \int_{1}^{\infty} \exp\left[-u\sqrt{q(x)+\mu} \right]\,du \right\}\,dx$$

$$\le C_3 \int_{-\infty}^{\infty} \frac{dx}{[q(x)+\mu]^{7/2}} + C_4 \int_{-\infty}^{\infty} e^{-\sqrt{q(x)+\mu}}\,dx \le C_5 \int_{-\infty}^{\infty} \frac{dx}{[q(x)+\mu]^{7/2}}. \quad (5.15)$$

It is obvious that the estimate (5.13) follows from (5.14) and (5.15). Therefore, (5.12) is proved, and we now derive from (5.11) and (5.12) that

$$||\beta(x,\eta;\mu)||_X = o(||g'_\mu(x,\eta;\mu)||_X). \quad (5.16)$$

It follows from the equation (5.9), and because of (5.12) and (5.16), that

$$||G'_\mu(x,\eta;\mu)||_X \le ||g'_\mu(x,\eta;\mu)||_X(1+o(1)). \quad (5.17)$$

We now return to the equation (5.3′), rewriting it as

$$g'_\mu = G'_\mu + \alpha + \int_{-\infty}^{\infty} g(x,\xi;\mu)[q(\xi)-q(x)]G'_\mu(\xi,\eta;\mu)\,d\xi.$$

Taking the norm and estimating the norm of the integral according to Lemma 4.1.1, we obtain

$$||g'_\mu||_X \le ||G'_\mu||_X + o(||G'_\mu||_X) + o(||g'_\mu||_X);$$

hence,

$$||g'_\mu(x,\eta;\mu)||_X \le ||G'_\mu(x,\eta;\mu)||_X(1+o(1)). \quad (5.18)$$

We then conclude from (5.17) and (5.18) that

$$||G'_\mu('x,\eta;\mu)||_X = ||g'_\mu(x,\eta;\mu)||_X(1+o(1)) \quad (5.18')$$

as $\mu \to \infty$, i.e., $\lim_{\mu\to\infty} \dfrac{||G'_\mu(x,\eta;\mu)||_X}{||g'_\mu(x,\eta;\mu)||_X} = 1.$

4.5.3

We now state without proof the result in the general case involving higher-order iterations.

Theorem 4.5.1 *Let a function $q(x)$ fulfil Conditions 1 and 2 of Lemma 4.5.1; let also*

3. *for natural $\tau \geq 0$, $q^{-3/2-2\tau}(x) \in \mathcal{L}(-\infty, \infty)$,*

4. $q^{\tau+1}(\xi) \exp\left\{-\frac{1}{2}|x - \xi|\sqrt{q(x)}\right\} < C$ *for $|x - \xi| \geq 1$,*

where C is a constant number. Then $\dfrac{\partial^\tau}{\partial \mu^\tau} G(x, \eta; \mu) \in X$, and

$$\left\|\frac{\partial^\tau}{\partial \mu^\tau} G(x, \eta; \mu)\right\|_X = \left\|\frac{\partial^\tau}{\partial \mu^\tau} g(x, \eta; \mu)\right\|_X (1 + o(1)) \tag{5.19}$$

as $\mu \to \infty$.

Since

$$\int_{-\infty}^{\infty} \frac{\partial^\tau}{\partial \mu^\tau}\{G(x, \eta; \mu)\} \cdot \varphi_n(\eta)\, d\eta = (-1)^\tau \tau! \frac{\varphi_n(x)}{(\lambda_n + \mu)^{\tau+1}},$$

relying on the equality (4.3), it follows from the Parseval equation that

$$\int_{-\infty}^{\infty} \left\{\frac{\partial^\tau}{\partial \mu^\tau} G(x, \eta; \mu)\right\}^2 d\eta = (\tau!)^2 \sum_{n=1}^{\infty} \frac{\varphi_n^2(x)}{(\lambda_n + \mu)^{2\tau+2}}; \tag{5.20}$$

hence,

$$\int_{-\infty}^{\infty} \left\{\int_{-\infty}^{\infty} \left[\frac{\partial^\tau}{\partial \mu^\tau} G(x, \eta; \mu)\right]^2 d\eta\right\} dx = \sum_{n=1}^{\infty} \frac{(\tau!)^2}{(\lambda_n + \mu)^{2\tau+2}} \equiv (\tau!)^2 S_\tau.$$

Therefore, the asymptotic formula (5.19) can be written as

$$(\tau!)^2 S_\tau \sim \|\frac{\partial^\tau g}{\partial \mu^\tau}\|_X = \int_{-\infty}^{\infty} \left\{\int_{-\infty}^{\infty} \left[\frac{\partial^\tau}{\partial \mu^\tau} g(x, \eta; \mu)\right]^2 d\eta\right\} dx \tag{5.21}$$

when $\mu \to \infty$.

To calculate the integral on the right-hand side of (5.21), we put in the familiar equality

$$\frac{1}{\pi} \int_{-\infty}^{\infty} \frac{e^{i\alpha t}}{\alpha^2 + \beta^2}\, d\alpha = \frac{e^{-\beta t}}{\beta}, \quad t \geq 0,$$

$\beta = \kappa = \{q(x) + \mu\}^{1/2}$, $t = |x - \eta|$, and obtain

$$g(x, \eta; \mu) = \frac{1}{2\kappa} \exp(-|x - \eta|\kappa) = \frac{1}{2\pi} \int_{-\infty}^{\infty} \frac{e^{i\alpha|x-\eta|}}{\alpha^2 + q(x) + \mu} \, d\alpha$$

(see (1.2)). Now, differentiating τ times the last formula with respect to μ, we get

$$\frac{\partial^\tau}{\partial \mu^\tau} g(x, \tau; \mu) = \frac{(-1)^\tau \tau!}{2\pi} \int_{-\infty}^{\infty} \frac{e^{i\alpha|x-\eta|}}{\{\alpha^2 + q(x) + \mu\}^{\tau+1}} \, d\alpha. \tag{5.22}$$

From the Parseval equation, it follows that

$$\int_{-\infty}^{\infty} \left\{ \frac{\partial^\tau}{\partial \mu^\tau} g(x, \eta; \mu) \right\}^2 \, d\eta = \frac{(\tau!)^2}{2\pi} \int_{-\infty}^{\infty} \frac{d\alpha}{\{\alpha^2 + q(x) + \mu\}^{2\tau+2}};$$

hence,

$$\int_{-\infty}^{\infty} \left\{ \frac{\partial^\tau}{\partial \mu^\tau} g(x, \eta; \mu) \right\}^2 \, d\eta = \frac{(-1)^{2\tau+1}(\tau!)^2}{(2\tau+1)!} \frac{\partial^{2\tau+1}}{\partial \mu^{2\tau+1}} g(x, \eta; \mu)|_{\eta=x}$$

$$= \frac{(\tau!) \cdot 1 \cdot 3 \cdot 5 \cdot \ldots \cdot (4\tau + 1)}{2^{2\tau+2}(2\tau + 1)!} \frac{1}{\{q(x) + \mu\}^{2\tau+3/2}} \tag{5.23}$$

on the basis of (5.22) and (1.2).

Finally, we obtain from the asymptotic formula (5.21) and from the last equality that

$$S_\tau = \sum_{n=1}^{\infty} \frac{1}{(\lambda_n + \mu)^{2\tau+2}} \sim \frac{(4\tau + 1)!!}{2^{2\tau+2}(2\tau + 1)!} \int_{-\infty}^{\infty} \frac{dx}{\{q(x) + \mu\}^{2\tau+3/2}} \tag{5.24}$$

as $\mu \to \infty$.

4.6   Asymptotic distribution of the eigenvalues

4.6.1

Using the Tauberian theorem due to M. V. Keldysh (see B. M. Levitan and I. S. Sargs-jan (1975; Chapter 14, Theorem 4.3)), we obtain here an asymptotic formula for the distribution of the eigenvalues, deriving it from (5.24). This is contained in the following fundamental proposition of the chapter.

Theorem 4.6.1 *Let a function $q(x)$ satisfy the conditions of Theorem 4.5.1. We introduce the monotone function*

$$\sigma(\lambda) = \text{mes}\,\{q(x) < \lambda\},$$

and put

$$\psi(\lambda) = \int_0^\lambda (\lambda - \nu)^{1/2} \, d\sigma(\nu).$$

Assume that there exist two positive constants α and β such that, for sufficiently large λ, the inequalities

$$\alpha\psi(\lambda) < \lambda\psi'(\lambda) < \beta\psi(\lambda) \qquad (6.1)$$

hold. As $\lambda \to \infty$, the asymptotic formula

$$N(\lambda) = \sum_{\lambda_n < \lambda} 1 \sim \frac{1}{\pi} \int_0^\lambda (\lambda - \nu)^{1/2} \, d\sigma(\nu) = \frac{1}{\pi} \int_{q(x) < \lambda} \{\lambda - q(x)\}^{1/2} \, dx$$

then holds.

Proof. Putting $\sigma(\lambda) = \text{mes}\,\{q(x) < \lambda\}$, we have

$$\int_{-\infty}^\infty \frac{dx}{\{q(x) + \mu\}^{2\tau+3/2}} = \int_0^\infty \frac{d\sigma(\lambda)}{(\lambda + \mu)^{2\tau+3/2}}, \qquad (6.2)$$

and show that

$$\int_0^\infty \frac{d\sigma(\lambda)}{(\lambda + \mu)^{2\tau+3/2}} = \frac{\Gamma(2\tau + 2)}{\Gamma(2\tau + 3/2)\Gamma(1/2)} \int_0^\infty \frac{d\lambda}{(\lambda + \mu)^{2\tau+2}} \left\{ \int_0^\lambda \frac{d\sigma(\nu)}{(\lambda - \nu)^{1/2}} \right\}. \qquad (6.3)$$

In fact, interchanging the order of integration, we obtain

$$\int_0^\infty \frac{\lambda}{(\lambda + \mu)^{2\tau+2}} \left\{ \int_0^\lambda \frac{d\sigma(\lambda)}{(\lambda - \nu)^{1/2}} \right\} = \int_0^\infty d\sigma(\nu) \left\{ \int_\nu^\infty \frac{(\lambda - \nu)^{-1/2} \, d\lambda}{(\lambda + \mu)^{2\tau+2}} \right\}. \qquad (6.4)$$

Furthermore,

$$\int_\nu^\infty \frac{(\lambda - \nu)^{-1/2}}{(\lambda + \nu)^{2\tau+2}} \, d\lambda = \int_0^\infty \frac{t^{-1/2} \, dt}{(t + \nu + \mu)^{2\tau+2}} = \frac{1}{(\mu + \nu)^{2\tau+3/2}} \int_0^\infty \frac{z^{-1/2} \, dz}{(1 + z)^{2\tau+2}}.$$

Hence, by the familiar formulas for Eulerian integrals, we get

$$\int_\nu^\infty \frac{(\lambda - \nu)^{-1/2}}{(\lambda + \mu)^{2\tau+2}} \, d\lambda = \frac{\Gamma(2\tau + 3/2)\Gamma(1/2)}{\Gamma(2\tau + 2)(\nu + \mu)^{2\tau+3/2}},$$

which proves (6.3) along with (6.4).
    Let

$$\psi(\lambda) = \int_0^\lambda (\lambda - \nu)^{1/2} \, d\sigma(\nu).$$

Then

$$d\psi(\lambda) = \frac{1}{2} \int_0^\lambda (\lambda - \nu)^{-1/2} d\sigma(\nu) d\lambda.$$

Therefore, (6.3) can be written as

$$\int_{-\infty}^{\infty} \frac{d\sigma(\lambda)}{(\lambda + \mu)^{2\tau+3/2}} = \frac{2\Gamma(2\tau + 2)}{\Gamma(2\tau + 3/2)\Gamma(1/2)} \int_0^{\infty} \frac{d\psi(\lambda)}{(\lambda + \mu)^{2\tau+2}}. \tag{6.5}$$

Because of (6.5), it follows from the asymptotic formula (5.24) and from (6.2) that

$$S_\tau \sim \frac{(4\tau + 1)!!}{2^{2\tau+2}(2\tau + 1)!} \frac{2\Gamma(2\tau + 2)}{\Gamma(2\tau + 3/2)\Gamma(1/2)} \int_0^{\infty} \frac{d\psi(\lambda)}{(\lambda + \mu)^{2\tau+2}}$$

as $\mu \to \infty$.

Substituting the Eulerian functions $\Gamma(p)$, we obtain

$$S_\tau \sim \frac{1}{\pi} \int_0^{\infty} \frac{d\psi(\lambda)}{(\lambda + \mu)^{2\tau+2}}. \tag{6.6}$$

By definition, $S_\tau = \sum_{n=1}^{\infty}(\lambda_n + \mu)^{-2\tau-2}$. putting $N(\lambda) = \sum_{\lambda_n < \lambda} 1$, S_τ can be written as

$$S_\tau = \int_0^{\infty} \frac{dN(\lambda)}{(\lambda + \mu)^{2\tau+2}}.$$

Therefore, we finally obtain the formula

$$\int_0^{\infty} \frac{dN(\lambda)}{(\lambda + \mu)^{2\tau+2}} \sim \frac{1}{\pi} \int_0^{\infty} \frac{d\psi(\lambda)}{(\lambda + \mu)^{2\tau+2}} \tag{6.7}$$

from (6.6) as $\mu \to \infty$.

The condition (6.1) ensures that the Tauberian theorem of M. V. Keldysh can be applied to the above formula, with the consequence that, as $\lambda \to \infty$, $N(\lambda) \sim (1/\pi)\psi(\lambda)$, or, in expanded form,

$$N(\lambda) \sim \frac{1}{\pi} \int_{q(x) < \lambda} \{\lambda - q(x)\}^{1/2} dx, \tag{6.8}$$

and the proof is complete.   ∎

4.6.2

Here, we obtain a formula from which, on the one hand, (6.8) follows in a particular case and, on the other hand, we prove in the next section the theorems on the convergence and summability of expansions and differentiated eigenfunction expansions for the Sturm-Liouville operator of functions increasing polynomially at infinity.

Theorem 4.6.2 *Let τ and p be two natural numbers. Assume that the condition*

4. for $|x - \xi| > 1$,

$$q^{\tau+1}(\xi)q^p(x) \exp\left\{-\frac{1}{2}|x - \xi|\sqrt{q(x)}\right\} < B$$

holds. Let the other conditions of Theorem 4.5.1 hold in the same form. Then the asymptotic formula

$$\left\| \frac{\partial^{\tau+p}}{\partial \mu^{\tau+p}} G(x, \eta; \mu) \right\|_{X(p)} = \left\| \frac{\partial^{\tau+p}}{\partial \mu^{\tau+p}} g(x, \eta; \mu) \right\|_{X(p)} (1 + o(1)) \qquad (6.9)$$

holds together with the asymptotic formula (5.19).

Proof. We prove the formula (6.9) for $\tau = 0$, $p = 1$, the proof being quite similar in the general case, although it then relies on somewhat more tedious calculations.

We derive from the equation (5.3) that

$$\begin{aligned}
q(x) G'_\mu &= q(x) g'_\mu - \int_{-\infty}^{\infty} q(x) g'_\mu(x, \xi; \mu)[q(\xi) - q(x)] G(\xi, \eta; \mu) \, d\xi \\
&\quad - \int_{-\infty}^{\infty} q(x) g(x, \xi; \mu) \frac{q(\xi) - q(x)}{q(\xi)} q(\xi) G'_\mu(\xi, \eta; \mu) \, d\xi, \qquad (6.10)
\end{aligned}$$

multiplying throughout by $q(x)$.

Put

$$\alpha(x, \eta; \mu) = \int_{-\infty}^{\infty} q(x) g'_\mu(x, \xi; \mu)[q(x) - q(\xi)] G(\xi, \eta; \mu) \, d\xi.$$

As in the proof of Lemma 4.1.1, we can then show that

$$\|\alpha(x, \eta; \mu)\|_X = o(\|g(x, \eta; \mu)\|_X).$$

Furthermore, it follows from the same lemma and from the equation (1.3) that $\|G(x, \eta; \mu)\|_X = O(\|g(x, \eta; \mu)\|_X)$. Therefore, $\|\alpha(x, \eta; \mu)\|_X = o(\|g(x, \eta; \mu)\|_X)$. Straightforward calculation can help to establish the estimate

$$\|g(x, \eta; \mu)\|_X = O(\|q(x) g'_\mu(x, \eta; \mu)\|_X)$$

easily, which, along with the preceding estimate, leads to

$$\|\alpha(x, \eta; \mu)\|_X = o(\|q(x) g'_\mu(x, \eta; \mu)\|_X).$$

Similarly to the derivation of the equation (5.18′), it follows, also because of the equation (6.10), that

$$\|q(x) G'_\mu(x, \eta; \mu)\|_X = \|q(x) g'_\mu(x, \eta; \mu)\|_X (1 + o(1)),$$

which exactly proves the formula (6.9) for $\tau = 0$, $p = 1$, since

$$\|q^p(x) G'_\mu(x, \eta; \mu)\|_X = \|G'_\mu(x, \eta; \mu)\|_{X(p)}.$$

∎

Theorem 4.6.3 *Let $q(x)$ satisfy the conditions of Theorem 4.6.2. We introduce the monotone function $\sigma(\lambda) = \text{mes}\,\{q(x) < \lambda\}$, and put*

$$\psi_p(\lambda) = \frac{1}{\pi} \int_0^\lambda (\lambda - \nu)^{1/2} \nu^{2p}\, d\sigma(\nu). \tag{6.11}$$

Let $\psi_p(\lambda)$ satisfy the inequalities (6.1). As $\lambda \to \infty$, we then have

$$\sum_{\lambda_n < \lambda} a_n^{(p)} \sim \frac{1}{\pi} \int_{q(x) < \lambda} q^{2p}(x)\{\lambda - q(x)\}^{1/2}\, dx, \tag{6.12}$$

where

$$a_n^{(p)} = \int_{-\infty}^{\infty} q^{2p}(x)\varphi_n^2(x)\, dx \tag{6.13}$$

and $\varphi_n(x)$ is the eigenfunction corresponding to the eigenvalue λ_n.

Proof. Replacing the number τ by $\tau + p$ in the equality (5.20), we get

$$\int_{-\infty}^{\infty} \left\{ \frac{\partial^{\tau+p}}{\partial \mu^{\tau+p}} G(x, \eta; \mu) \right\}^2 d\eta = \{(\tau + p)!\}^2 \sum_{n=1}^{\infty} \frac{\varphi_n^2(x)}{(\lambda_n + \mu)^{2(\tau+p)+2}}.$$

Now, if we multiply the equality throughout by $q^{2p}(x)$, integrate it with respect to x from $-\infty$ to $+\infty$ and use the asymptotic formula (6.9) and the equality (6.13), we obtain

$$\{(\tau + p)!\}^2 \sum_{n=1}^{\infty} \frac{a_n^{(p)}}{(\lambda_n + \mu)^{2(\tau+p)+2}} \sim \int_{-\infty}^{\infty} q^{2p}(x) \left\{ \int_{-\infty}^{\infty} \left[\frac{\partial^{\tau+p}}{\partial \mu^{\tau+p}} g(x, \eta; \mu) \right]^2 d\eta \right\} dx$$

as $\mu \to \infty$.

Hence, relying on (5.23) (substituting $\tau + p$ for τ), we have

$$\sum_{n=1}^{\infty} \frac{a_n^{(p)}}{(\lambda_n + \mu)^{2(\tau+p)+2}} \sim \frac{(4\tau + 4p + 1)!!}{2^{2(\tau+p)+2}(2\tau + 2p + 1)!} \int_{-\infty}^{\infty} \frac{q^{2p}(x)}{\{q(x) + \mu\}^{2(\tau+p)+3/2}}\, dx. \tag{6.14}$$

Proceeding as in the derivation of (6.8) from (5.24), we arrive at the equality

$$\int_{-\infty}^{\infty} \frac{d\chi_p(\lambda)}{(\lambda + \mu)^{2(\tau+p)+2}} \sim \frac{1}{\pi} \int_{-\infty}^{\infty} \frac{d\psi_p(\lambda)}{(\lambda + \mu)^{2(\tau+p)+2}} \tag{6.15}$$

in this case, where $\chi_p(\lambda) = \sum_{\lambda_n < \lambda} a_n^{(p)}$ as $\mu \to \infty$ and the function $\psi_p(\lambda)$ is determined by the formula (6.11). The asymptotic formula (6.12) then follows from (6.15) by the theorem of M. V. Keldysh. The proof is thus complete. ∎

It is obvious that (6.8) is a particular case of (6.12) (associated with $p = 0$).

4.6.3   The case of the half-line

Similar results can be obtained for $[0, \infty)$. Let a function $q(x)$ be defined on $[0, \infty)$, and satisfy (1.5), (1.6) and Condition 2 of Lemma 4.5.1. Consider the operator

$$l(y) = -y'' + q(x)y, \quad x \geq 0 \tag{6.16}$$

$$y'(0) - hy(0) = 0, \tag{6.17}$$

where h is an arbitrary real number.

We first clarify what is the analog of the function $g(x, \eta; \mu)$ in this case. Let $\kappa^2 = q(x) + \mu$, $(\mu > 0)$ and

$$X(x) = \cosh \kappa x + h\kappa^{-1} \sinh \kappa x.$$

Consider the function

$$g_0(x, \eta; \mu) = \begin{cases} (\kappa + h)^{-1} X(\eta) e^{-\kappa x}, & \eta \leq x \\ \\ (\kappa + h)^{-1} X(x) e^{-\kappa \eta}, & \eta > x \end{cases} \tag{6.18}$$

which is easily verified to satisfy the equation

$$\frac{\partial^2 g_0}{\partial \eta^2} = \{q(x) + \mu\} g_0$$

and the conditions

$$\left. \left(\frac{\partial g_0}{\partial \eta} - h g_0 \right) \right|_{\eta=0} = 0, \quad \left. \frac{\partial g_0}{\partial \eta} \right|_{\eta=x+0} - \left. \frac{\partial g_0}{\partial \eta} \right|_{\eta=x-0} = -1.$$

An integral equation similar to (1.3) can be formed with the help of $g_0(x, \eta; \mu)$, and then the existence of Green's function for the operator (6.16), (6.17) can be proved.

It follows from (6.18) that

$$g_0(x, \eta; \mu) = \begin{cases} \dfrac{1}{2\kappa} \left(1 + \dfrac{h}{\kappa} \right)^{-1} e^{(\eta-x)\kappa} \left\{ 1 + O\left(\dfrac{1}{\kappa} \right) \right\}, & \eta \leq x \\ \\ \dfrac{1}{2\kappa} \left(1 + \dfrac{h}{\kappa} \right)^{-1} e^{(x-\eta)\kappa} \left\{ 1 + O\left(\dfrac{1}{\kappa} \right) \right\}, & \eta > x \end{cases} .$$

It easily follows that

$$g_0(x, \eta; \mu) = \frac{1}{2\kappa} e^{|x-\eta|\kappa} \{1 + o(1)\} = g(x, \eta; \mu)\{1 + o(1)\}$$

as $\mu \to \infty$.

This representation enables us to extend the above results to the half-line. We only note that the coefficient $1/\pi$ in the formulas (6.8) and (6.12) should then be replaced by $1/(2\pi)$.

4.7 Eigenfunction expansions with unbounded potential

Consider the problem

$$y'' + \{\lambda - q(x)\}y = 0 \tag{7.1}$$

$$y(0)\cos\alpha + y'(0)\sin\alpha = 0 \tag{7.2}$$

on the half-line $[0,\infty)$. Let $q(x)$ satisfy the conditions of Theorem 4.6.2, and the inequality

$$q(x) \geq ax^\delta \tag{7.3}$$

for large x, where a and δ are certain positive constants.

We know that if the condition (7.3) holds (more precisely, if $q(x) \to +\infty$ as $x \to +\infty$), then the eigenfunctions $\varphi_n(x)$ of the problem (7.1), (7.2) decrease rapidly; e.g., if $q(x)$ increases polynomially at infinity, then $\varphi_n(x)$ decrease more rapidly than $e^{-c|x|}$, where $c > 0$ is any constant. Therefore, there exist integrals

$$a_n^{(s)} = \int_0^\infty q^s(x)\varphi_n^2(x)\,dx < +\infty \tag{7.4}$$

for any $s > 0$, and it is not required that the function to be expanded should be square-integrable to construct the Fourier series.

Here, using the asymptotic formula (6.14), we prove certain theorems on the convergence and summation of expansions and differentiated eigenfunction expansions for the problem (7.1), (7.2), of functions increasing polynomially as $x \to \infty$.

We first prove several lemmas.

Lemma 4.7.1 *Let $q(x)$ satisfy the conditions of Theorem 4.6.2 and the condition (7.3). Then*

$$\sum_{\mu \leq \mu_n \leq \mu+1} a_n^{(s)} = O(\mu^{2s+1+2/\delta}), \quad \mu_n^2 = \lambda_n \tag{7.5}$$

as $\mu \to \infty$.

Proof. Relying on (7.3), we have

$$\sigma(\lambda) = \mathrm{mes}\,\{q(x) < \lambda\} \leq \mathrm{mes}\,\{ax^\delta < \lambda\} = (\lambda/a)^{1/\delta} = C\lambda^{1/\delta}. \tag{7.6}$$

Put

$$\sigma_s(\lambda) = \int_0^\lambda \nu^s\,d\sigma(\nu),$$

where s is a fixed number. With the notation of Section 6, the formula (6.15) can then be written as

$$\int_0^\infty \frac{d\chi_s(\lambda)}{(\lambda+\mu)^{2p}} < C \int_0^\infty \frac{d\sigma_s(\lambda)}{(\lambda+\mu)^{2p-1/2}} = C \int_0^\infty \frac{\sigma_s(\lambda)\,d\lambda}{(\lambda+\mu)^{2p+1/2}} \tag{7.7}$$

(replacing $\tau + p + 1$ by p and $2p$ by s), in the half-line case.

We derive the estimate $\sigma_s(\lambda) < C\lambda^{s+1/\delta}$ from (7.6); hence

$$\int_0^\infty \frac{d\chi_s(\lambda)}{(\lambda+\mu)^{2p}} < C \int_0^\infty \frac{\lambda^{s+1/\delta}\, d\lambda}{(\lambda+\mu)^{2p+1/2}}$$

$$= \frac{C}{\mu^{2p-s-1/\delta-1/2}} \int_0^\infty \frac{t^{s+1}\, dt}{(1+t)^{2p+1/2}}$$

$$= \frac{C}{\mu^{2p-s-1/\delta-1/2}};$$

therefore,

$$\int_{\mu^2}^{(\mu+1)^2} \frac{d\chi_s(\lambda)}{(\lambda+\mu^2)^{2p}} < C\mu^{4p-2s-2/\delta-1}. \tag{7.8}$$

On the other hand,

$$\int_{\mu^2}^{(\mu+1)^2} \frac{d\chi_s(\lambda)}{(\lambda+\mu^2)^{2p}} > \frac{C}{\mu^{4p}}\{\chi_s[(\mu+1)^2] - \chi_s(\mu^2)\}. \tag{7.9}$$

It follows from the estimates (7.8) and (7.9) that

$$\chi_s[(\mu+1)^2] - \chi_s(\mu^2) = \sum_{\mu \le \mu_n \le \mu+1} a_n^{(s)} < C\mu^{2s+2/\delta+1},$$

where $\mu_n = \sqrt{\lambda_n}$ as required.   ∎

Let a function $f(x)$ be summable in each finite interval, and let there exist a positive number s such that

$$\int_0^\infty f(x)q^{-s}(x)\, dx < \infty. \tag{7.10}$$

Put

$$S(x,\mu) = \sum_{\mu_n < \mu} c_n\varphi_n(x), \quad c_n = \int_0^\infty f(x)\varphi_n(x)\, dx.$$

Lemma 4.7.2 *If the function $f(x)$ satisfies (7.10), and if the function $q(x)$ satisfies the conditions of Lemma 4.7.1, then the estimate*

$$\bigvee_\mu^{\mu+1}\{S(x,\mu)\} = O(\mu^{s+1/2+1/\delta}) \tag{7.11}$$

is valid as $\mu \to \infty$.

Lemma 4.7.3 *Let the conditions of Lemma 4.7.2 hold, and let the function $q(x)$ have summable k^{th} derivative in each finite interval. Then the estimates*

$$\bigvee_\mu^{\mu+1}\left\{\frac{\partial^k S(x,\mu)}{\partial x^k}\right\} = O(\mu^{s+k+1/2+1/\delta}), \quad k = 1,2,\ldots \tag{7.12}$$

hold as $\mu \to \infty$.

Using (7.11) and (7.12), we can prove the following (see Chapters VII and VIII of the monograph of B. M. Levitan and I. S. Sargsjan (1975)).

Theorem 4.7.1 *Let the function $q(x)$ satisfy the conditions of Theorem 4.6.2 and the condition (7.9), and let $f(x)$ be summable in each finite interval, satisfying the condition (7.10). Then, at each point of continuity of $f(x)$, for $\tau > s + 1/2 + 1/\delta$, the equality*

$$\lim_{\mu \to \infty} \sum_{\mu_n < \mu} \left(1 - \frac{\mu_n^2}{\mu^2}\right)^\tau c_n \varphi_n(x) = f(x),$$

$$c_n = \int_0^\infty f(x) \varphi_n(x) \, dx$$

holds, i.e., the Riesz means (of order greater than $s + 1/2 + 1/\delta$) of the eigenfunction expansion of $f(x)$ for the Sturm-Liouville operator tend to $f(x)$ at each point of continuity.

Theorem 4.7.2 *Let $f(x)$ and $q(x)$ satisfy all the conditions of Theorem 4.7.1, let $q(x)$ have a summable $(k-1)^{\text{th}}$ derivative in each finite interval, and let $f(x)$ have a summable k^{th} derivative. Then, at each point of continuity of the function $f^{(k)}(x)$, for $\tau > s + k + 1/2 + 1/\delta$, the equality*

$$\lim_{\mu \to \infty} \sum_{\mu_n < \mu} \left(1 - \frac{\mu_n^2}{\mu^2}\right)^2 c_n \varphi^{(k)}(x) = f^{(k)}(x)$$

holds, i.e., the Riesz means of order greater than $s + k + 1/2 + 1/\delta$ of the kth derivative of the eigenfunction expansion of $f(x)$ for the Sturm-Liouville operator tend to $f^{(k)}(x)$ at each point of continuity of the latter.

Theorem 4.7.3 *Let the function $q(x)$ satisfy the conditions of Theorem 4.7.1, and there exist operators*

$$\mathcal{L}^0 f = f(x), \mathcal{L}f = f''(x) - q(x)f(x), \mathcal{L}^2 f, \ldots, \mathcal{L}^k f$$

in the interval $(0, \infty)$, satisfying the conditions

1. $\{\mathcal{L}^i f(x)\}_{x=0} \cos \alpha + \{\mathcal{L}^i f(x)\}'_{x=0} \sin \alpha = 0,$

2. $\mathcal{L}^i f(x) = o\left\{\exp\left[\int_0^x \sqrt{q(s)} \, ds\right]\right\}$ *as* $x \to \infty,$

3. $\{\mathcal{L}^i f(x)\}' = o\left\{\exp\left[\int_0^x \sqrt{q(s)} \, ds\right]\right\}$ *as* $x \to \infty,$

$i = 1, 2, \ldots, k - 1,$

4. there exists a positive number s such that

$$\int_0^\infty \{\mathcal{L}^k f(x)\}^2 q^{-s}(x) \, dx < \infty.$$

Then if

$$k > \left[\frac{s}{2} + \frac{1}{2\delta} + \frac{1}{4} + \frac{m}{2}\right],$$

the equality

$$f(x) = \sum_{n=1}^{\infty} c_n \varphi_n(x)$$

holds uniformly in each finite interval.

Theorem 4.7.4 *Let all the conditions of Theorem 4.7.3 hold, and let the two functions $q(x)$ and $f(x)$ have summable m^{th} derivative in each finite interval. Then if*

$$k > \left[\frac{s}{2} + \frac{1}{2\delta} + \frac{1}{4} + \frac{m}{2}\right],$$

the equality

$$f^{(m)}(x) = \sum_{n=1}^{\infty} c_n \varphi_n^{(m)}(x)$$

holds at all points of continuity of the function $f^{(m)}(x)$.

Notes

The asymptotic behaviour of $N(\lambda)$ was first obtained by J. De Wet and F. Mandl (1950), who used the Courant variational theorems. E. Titchmarsh (1962) applied the Carleman method to this problem, and obtained the asymptotic formula for $N(\lambda)$ with less stringent restrictions placed on the potential function $q(x)$ (considering the case of two variables).

The complete series for Green's function was used by B. M. Levitan (1968). Applying the estimates of the weighted trace of Green's function to the questions of an eigenfunction expansion with unbounded potential is due to B. M. Levitan (see also B. M. Levitan and I. S. Sargsjan (1975)).

Chapter 5

SHARPENING THE ASYMPTOTIC BEHAVIOUR OF THE EIGENVALUES AND THE TRACE FORMULAS

5.1 Asymptotic formulas for special solutions

Consider the equation

$$-y'' + q(x)y = \lambda y, \quad -\infty < x < \infty, \tag{1.1}$$

where $q(x)$ is an infinitely differentiable function.

We will seek its solution in the form

$$y(x, \lambda) = \exp\left\{i\mu x + \int_0^x \sigma(t, \mu) \, dt\right\}, \quad \mu = \sqrt{\lambda}, \tag{1.2}$$

which, when substituted in (1.1), yields the first-order differential equation

$$\sigma' + 2i\mu\sigma + \sigma^2 - q(x) = 0 \tag{1.3}$$

for the function $\sigma(x, \mu)$.

Let[1]

$$\sigma(x, \mu) = \sum_{k=1}^{\infty} \frac{\sigma_k(x)}{(2i\mu)^k}. \tag{1.4}$$

Substituting the expansion in (1.3) and then comparing the coefficients of the same powers of $2i\mu$, we obtain the recurrence relations

$$\sigma_1(x) = q(x), \quad \sigma_2(x) = -q'(x),$$

$$\sigma_m(x) = -\sigma'_{m-1}(x) - \sum_{j=1}^{m-2} \sigma_{m-j-1}(x)\sigma_j(x), \quad m = 3, 4, \ldots. \tag{1.5}$$

---

[1]To show that the expansion (1.4) is, in fact, asymptotic as $|\mu| \to \infty$, we should study the remainder, and refer the reader to the monograph of V. A. Marchenko (1972; Chapter I, Section 4). Note that the number of exact terms in (1.4) depends on the order of smoothness of the potential $q(x)$.

In particular, it follows that

$$\sigma_3(x) = q''(x) - q^2(x), \qquad \sigma_4(x) = -q'''(x) + 4q(x)q'(x)$$

$$\sigma_5(x) = q^{(IV)}(x) - 5q'^2(x) - 6q(x)q''(x) + 2q^3(x), \quad \text{etc.}$$

(1.6)

It can be seen from (1.5) and (1.6) that if $q(x)$ is a real-valued function, then all coefficients $\sigma_k(x)$ are also real. If $q(x)$ is periodic (resp. almost-periodic), then $\sigma_k(x)$ are also periodic (resp. almost-periodic).

For $\lambda \neq 0$, the formula (1.2) yields two linearly independent solutions

$$y_1(x, \lambda) = \exp\left\{i\mu x + \int_0^x \sigma(t, \mu)\, dt\right\}$$

(1.7)

$$y_2(x, \lambda) = \exp\left\{-i\mu x + \int_0^x \sigma(t, \mu)\, dt\right\}$$

(1.7')

where $\mu = \sqrt{\lambda}$. If $\lambda < 0$, $\sqrt{\lambda} = \mu = i\tau$, $\tau > 0$. For other λ, the root is determined by analytic continuation.

5.2 Asymptotic formulas for the eigenvalues

5.2.1

Consider the problem

$$-y'' + q(x)y = \lambda y, \qquad 0 \leq x \leq \pi,$$

(2.1)

$$y'(0) - hy(0) = 0, \quad y'(\pi) + Hy(\pi) = 0, \quad h \neq \infty, \quad H \neq \infty.$$

(2.2)

Let the eigenvalues of (2.1), (2.2) be $\lambda_0 < \lambda_1 < \lambda_2 < \ldots < \lambda_n < \ldots$. There are some familiar methods to derive the asymptotic formulas (if the potential $q(x)$ is sufficiently smooth).

Somehow or other, all these methods are based on the asymptotic formulas for the fundamental system of solutions of the equation (2.1). We also use the asymptotic formula (1.4).

To somewhat simplify the calculations, we first confine ourselves to $h = H = 0$, and then consider some other cases.

Let $y_1(x, \mu)$ and $y_2(x, \mu) = y_1(x, -\mu)$ be as in (1.7), (1.7'). Then the function

$$z(x, \mu) = y_1(x, \mu)y_2'(0, \mu) - y_2(x, \mu)y_1'(0, \mu)$$

obviously satisfies the boundary condition $z'(0, \mu) = 0$. Therefore, the eigenvalues of the problem (2.1), (2.2) (with $h = H = 0$) are the roots of

$$y_1'(\pi, \mu)y_2'(0, \mu) - y_2'(\pi, \mu)y_1'(0, \mu) = 0,$$

or, taking into account (1.7), (1.7') (after simple calculations), we have

$$\exp\left\{2i\mu\pi + \int_0^\pi [\sigma(t,\mu) - \sigma(t,-\mu)]\,dt\right\} = \tau(\mu), \tag{2.3}$$

where

$$\tau(\mu) = \frac{[-i\mu + \sigma(\pi,-\mu)][i\mu + \sigma(0,\mu)]}{[i\mu + \sigma(\pi,\mu)][-i\mu + \sigma(0,-\mu)]} \tag{2.4}$$

admits the asymptotic expansion

$$\tau(\mu) = 1 + \sum_{k=1}^\infty \frac{\tau_k}{(2i\mu)^k}, \tag{2.5}$$

where the τ_k are constant numbers.

If we take into account (1.4), we derive from (2.3), (2.5) that

$$2\pi i\mu_k - 2i\sum_{j=0}^\infty \frac{(-1)^j a_{2j+1}}{(2\mu_k)^{2j+1}} = \ln\tau(\mu_k) = 2\pi i k + \sum_{j=0}^\infty \frac{b_j}{(2i\mu_k)^j}, \tag{2.6}$$

where $a_{2j+1} = \int_0^\pi \sigma_{2j+1}(x)\,dx$, b_j are constants and[2] $\mu_k = \sqrt{\lambda_k}$.

Furthermore, since $\tau(-\mu) = 1/\tau(\mu)$, if μ_k is a root of the equation (2.3), $-\mu_k$ is also its root. Hence, μ_k is only raised to odd powers on the right-hand side of the equality (2.6). In fact, replacing μ_k by $-\mu_k$ and k by $-k$ in (2.6), we get

$$-2\pi i\mu_k + 2i\sum_{j=0}^\infty \frac{\sigma_{2j+1}}{(2\mu_k)^{2j+1}} = -2\pi i k + \sum_{j=0}^\infty \frac{(-1)^j b_j}{(2i\mu_k)^j}. \tag{2.6'}$$

Now, adding (2.6) and (2.6') together, we obtain

$$\sum_{j=0}^\infty \frac{b_{2j}}{(2i\mu_k)^{2j}} = 0;$$

hence, $b_{2j} = 0$, $j = 0, 1, 2, \dots$ as $\mu_k \to \infty$.

Therefore, (2.6) can be written in the form

$$\mu_k - \frac{1}{\pi}\sum_{j=0}^\infty \frac{a_{2j+1}}{(2\mu_k)^{2j+1}} = k - \frac{1}{2\pi}\sum_{j=0}^\infty \frac{(-1)^j b_{2j+1}}{(2\mu_k)^{2j+1}}. \tag{2.7}$$

It follows that $\mu_k = k + O(1/k)$ as $\mu_k \to \infty$, which, substituted in (2.7), yields

$$\mu_k = k + (a_1 - b_1)/k + O(1/k^2).$$

_____

[2]That the principal term in the asymptotic behaviour of μ_k is k when k is large follows, e.g., from the asymptotic formula for the solution of (2.1), satisfying the initial condition $y'(0) = 0$ (see Section 2, Chapter 1).

Substituting in (2.7) again and repeating the procedure for μ_k, we obtain the asymptotic expansion

$$\mu_k = k + \sum_{j=0}^{\infty} \frac{c_{2j+1}}{k^{2j+1}}, \tag{2.8}$$

c_{2j+1} not depending on k and eventually expressed in terms of a_{2j+1}, $\sigma_k(0)$ and $\sigma_k(\pi)$.

Remark In the case of a periodic potential $q(x)$, $\sigma(\pi, \mu) = \sigma(0, \mu)$. Therefore, it follows from (2.4) that $\tau(\mu) \equiv 1$, and that in the case of periodic $q(x)$ and $h = H = 0$, c_{2j+1} only depend on a_{2j+1} polynomially.

5.2.2

We now consider the boundary conditions $y(0) = y(\pi) = 0$. The equation for the eigenvalues ν_k then has the form

$$y_1(\pi, \nu) - y_2(\pi, \nu) = 0,$$

or, if we substitute the formulas (1.7) and (1.7'),

$$\exp\left\{2\pi i\nu_k + \int_0^{\pi} [\sigma(t, \nu_k) - \sigma(t, -\nu_k)] \, dt\right\} = 1.$$

Reasoning as above, we derive the asymptotic expansion

$$\nu_k = k + \sum_{j=0}^{\infty} \frac{c_{2j+1}}{k^{2j+1}}, \tag{2.9}$$

with the coefficients c_{2j+1} being polynomials in

$$a_{2j+1} = \int_0^{\pi} \sigma_{2j+1}(x) \, dx. \tag{2.10}$$

Now, let $q(x)$ be a smooth periodic function, and let $\nu_k(t)$ be the eigenvalues of the problem

$$-y'' + q(x + t)y = \lambda y, \quad y(0) = y(\pi) = 0,$$

$t \in \mathbf{R}^1$.

If $q(x)$ is smooth and periodic, then

$$\int_0^{\pi} \sigma_{2j+1}(x + t) \, dx = \int_0^{\pi} \sigma_{2j+1}(x) \, dx = a_{2j+1}.$$

Therefore, the right-hand side does not depend on t in the asymptotic formula (2.9), and we obtain the asymptotic expansion for the eigenvalues of the periodic and of the semi-periodic problem.

Let $\lambda_0, \lambda_2, \mu_2, \ldots$ be the eigenvalues of the periodic problem, and $\lambda_1, \mu_1, \lambda_3, \mu_3, \ldots$ those of the semi-periodic one. By Theorem 1.4.4, if t ranges over $(0, \pi)$, then $\nu_k(t)$ completely sweeps out the kth gap (λ_k, μ_k). Therefore, the asymptotic formulas

$$\lambda_{2k}, \mu_{2k} = \sum_{j=0}^{\infty} \frac{c_{2j+1}}{(2k)^{2j+1}} \tag{2.11}$$

$$\lambda_{2k-1}, \mu_{2k-1} = \sum_{j=0}^{\infty} \frac{c_{2j+1}}{(2k-1)^{2j+1}} \tag{2.12}$$

are valid, with the same coefficients c_{2j+1} as in (2.9). Hence, in the case of an infinitely differentiable periodic potential $q(x)$, the series

$$S_k(t) = \lambda_0^k + \sum_{n=1}^{\infty} \{\lambda_n^k + \mu_n^k - 2\nu_n^k(t)\} \tag{2.13}$$

converges for any natural k.

$S_k(t)$ is called the *regularized sum* of the series $\lambda_0^k + \sum_{n=1}^{\infty}(\lambda_n^k + \mu_n^k)$, or the *regularized trace* of the operator Q^k, where $Q = -d^2/dx^2 + q(x)$ (for the periodic boundary conditions).

It is obvious that $S_k(t)$ is a functional in $q(t)$ and derivatives of $q(t)$ for each fixed t. A method for calculating the functions $S_k(t)$ is given in the next section.

5.3 Calculation of the sums $S_k(t)$

There are several methods for calculating $S_k(t)$. Here, we give the formulas for $k = 1, 2, 3$ and indicate the corresponding papers. First, let the potential $q(x)$ be finite-zone. Recall that it is determined by finitely many points

$$\lambda_0 < \lambda_1 < \mu_1, \lambda_2 < \mu_2 < \ldots < \lambda_N < \mu_N,$$

the endpoints of the spectrum zones, N values $\nu_k \in [\lambda_k, \mu_k]$, $k = 1, 2, \ldots, N$ and N signs. The Weyl-Titchmarsh functions for the N-zone potential are

$$m_1(\lambda) = \frac{Q(\lambda)}{P(\lambda)} + i\frac{\sqrt{R(\lambda)}}{P(\lambda)} \tag{3.1}$$

$$m_2(\lambda) = \frac{Q(\lambda)}{P(\lambda)} - i\frac{\sqrt{R(\lambda)}}{P(\lambda)} \tag{3.1'}$$

(see B. M. Levitan (1987, Chapter 8)), where

$$R(\lambda) = (\lambda - \lambda_0) \prod_{k=1}^{N} (\lambda - \lambda_k)(\lambda - \mu_k), \qquad P(\lambda) = \prod_{k=1}^{N} (\lambda - \nu_k)$$

$$Q(\lambda) = P(\lambda) \sum_{k=1}^{N} \frac{\sigma_k \sqrt{-R(\nu_k)}}{(\lambda - \nu_k) P'(\nu_k)}.$$

The Weyl solutions are

$$\psi_1(x, \lambda) = \vartheta(x, \lambda) + m_1(\lambda)\varphi(x, \lambda) \qquad (3.2)$$

$$\psi_2(x, \lambda) = \vartheta(x, \lambda) + m_2(\lambda)\varphi(x, \lambda) \qquad (3.2')$$

where $\psi_1(x, \lambda) \in \mathcal{L}^2(0, \infty)$, $\psi_2(x, \lambda) \in \mathcal{L}^2(-\infty, 0)$ for λ in the upper half-plane, and $\vartheta(x, \lambda)$, $\varphi(x, \lambda)$ are as above solutions of (1.1), satisfying the initial conditions

$$\varphi(0, \lambda) = \vartheta'(0, \lambda) = 0, \qquad \varphi'(0, \lambda) = \vartheta(0, \lambda) = 1.$$

In the monograph of B. M. Levitan (1987; Chapter 8, Section 3), the formulas were obtained by the wave-equation method

$$\begin{aligned} \lambda_0 + \sum_{k=1}^{N} [\lambda_k + \mu_k - 2\nu_k(t)] &= q(0) \\ \lambda_0^2 + \sum_{k=1}^{N} [\lambda_k^2 + \mu_k^2 - 2\nu_k^2(t)] &= -\frac{1}{2} q''(0) + q^2(0) \end{aligned} \qquad (3.3)$$

The following formula was not written out in the monograph, but can be obtained via the same calculations :

$$\lambda_0^3 + \sum_{k=1}^{N} (\lambda_k^3 + \mu_k^3 - 2\nu_k^3) = \frac{3}{16} q^{(IV)}(0) - \frac{15}{16} q'^2(0) - \frac{3}{2} q(0)q''(0) + \frac{5}{8} q^3(0) + \frac{3}{8} q^2(0). \quad (3.3')$$

If we replace $q(x)$ by $q(x + t)$, then (λ_k, μ_k) are unaltered, ν_k should be replaced by $\nu_k(t)$ and $q(0)$ by $q(t)$. The relations (3.3), (3.3') take the form

$$\begin{aligned} \lambda_0 + \sum_{k=1}^{N} (\lambda_k + \mu_k - 2\nu_k(t)) &= q(t) \\ \lambda_0^2 + \sum_{k=1}^{N} (\lambda_k^2 + \mu_k^2 - 2\nu_k^2(t)) &= -\frac{1}{2} q''(t) + q^2(t) \end{aligned} \qquad (3.4)$$

etc.

In the case of a finite-zone and smooth potential, (3.4) remain valid for $N = \infty$, which can be shown by approximating the finite-zone potential by the finite-zone one (B. M. Levitan (1987; Chapter 9)).

Another method for deriving (3.4) can be obtained, using the results of Sections 5 and 6 of the present chapter.

5.4 Another trace regularization—auxiliary lemmas

5.4.1

Here and in the subsequent sections, another trace regularization and another method for their calculation are offered. The method is simplest to realize in the case of the periodic problem. It is based on the two subsequent lemmas.

Lemma 5.4.1 *Let λ_n, $n = 1, 2, \ldots$, have the asymptotic behaviour*

$$\sqrt{\lambda_n} = n + \frac{c_1}{n} + \frac{c_3}{n^3} + \ldots + \frac{c_{2p+1}}{n^{2p+1}} + O\left(\frac{1}{n^{2p+2}}\right), \quad p \geq 0, \tag{4.1}$$

the coefficients $c_1, c_3, \ldots$ not depending on n. Then the function

$$f(t) = \sum_{n=1}^{\infty} (\cos \sqrt{\lambda_n} t - \cos nt) \tag{4.2}$$

has continuous derivatives up to order $2p$ inclusive, on the interval $[0, \pi]$.[3]

Proof. Put

$$A_n = \frac{c_1}{n} + \frac{c_3}{n^3} + \ldots + \frac{c_{2p+1}}{n^{2p+1}}$$

$$\sqrt{\lambda_n} = n + A_n + \rho_n, \quad \rho_n = O\left(\frac{1}{n^{2p+2}}\right).$$

Then

$$\begin{aligned}
\cos \sqrt{\lambda_n} t - \cos nt &= \cos(n + A_n)t \cos \rho_n t - \sin(n + A_n)t \sin \rho_n t - \cos nt \\
&= \cos(n + A_n)t - \cos nt + r_n(t) \tag{4.3} \\
r_n(t) &= \cos(n + A_n)t[\cos \rho_n t - 1] - \sin(n + A_n)t \sin \rho_n t.
\end{aligned}$$

It is obvious that $r_n(t) = O(1/n^{2p+2})$ and the sum of the series $R(t) = \sum_{n=1}^{\infty} r_n(t)$ has continuous derivatives up to order $2p$ inclusive. It follows from (4.3) that

$$\sum_{n=1}^{\infty} (\cos \sqrt{\lambda_n} t - \cos nt) = \sum_{n=1}^{\infty} [\cos(n + A_n)t - \cos nt] + R(t) \equiv J(t) + R(t).$$

We show that the function $J(t)$ is infinitely differentiable on the interval $[0, \pi]$. We have

$$\begin{aligned}
J(t) &= \sum_{n=1}^{\infty} [\cos nt(\cos A_n t - 1) - \sin nt \sin A_n t] \\
&= \sum_{n=1}^{\infty} \cos nt \sum_{k=1}^{\infty} (-1)^k \frac{A_n^k t^{2k}}{(2k)!} - \sum_{n=1}^{\infty} \sin nt \sum_{k=0}^{\infty} (-1)^k \frac{A_n^{2k+1} t^{2k+1}}{(2k+1)!} \\
&= \sum_{k=1}^{\infty} (-1)^k \frac{t^{2k}}{(2k)!} \sum_{n=1}^{\infty} A_n^{2k} \cos nt - \sum_{k=0}^{\infty} (-1)^k \frac{t^{2k+1}}{(2k+1)!} \sum_{n=1}^{\infty} A_n^{2k+1} \sin nt \\
&= J_1(t) + J_2(t). \tag{4.4}
\end{aligned}$$

---

[3]By the derivative at the point $t = 0$ (or $t = \pi$), we mean the limit on the right (resp. on the left).

That the order of summation can be changed follows from the estimate

$$A_n = O(1/n). \tag{4.5}$$

Let N be an arbitrary natural number. We put

$$J_1(t) = \sum_{k=1}^{N}(-1)^k \frac{t^{2k}}{(2k)!} \sum_{n=1}^{\infty} A_n^{2k} \cos nt$$

$$+ \sum_{k=N+1}^{\infty} (-1)^k \frac{t^{2k}}{(2k)!} \sum_{n=1}^{\infty} A_n^{2k} \cos nt \equiv J_1^{(1)}(t) + J_1^{(2)}(t).$$

Because of (4.5), there exists a continuous derivative $d^{2N} J_1^{(2)}(t)/dt^{2N}$. The function $J_1^{(1)}(t)$ is a finite linear combination of functions

$$t^{2k} \sum_{n=1}^{\infty} \frac{\cos nt}{n^{2j}}, \quad k \geq 1, \quad j \geq 1,$$

which are infinitely differentiable on the interval $[0, \pi]$; this follows from the familiar expansion

$$\sum_{n=1}^{\infty} \frac{\sin nt}{n} = \frac{\pi - t}{2}, \quad 0 < t \leq \pi.$$

Since N has been chosen to be arbitrary, the infinite differentiability of $J_1(t)$ is thus proved. That $J_2(t)$ is infinitely differentiable can be proved similarly, in which case we must use the infinite differentiability of the functions

$$t^{2k+1} \sum_{n=1}^{\infty} \frac{\sin nt}{n^{2j+1}}.$$

∎

5.4.2

To somewhat simplify further calculations, we assume that p can be arbitrary in the asymptotic expansion (4.1), which is consistent with the infinite differentiability of the potential $q(x)$.

For arbitrary natural N, we put

$$\lambda_n^{N/2} = n^N + A_1^{(N)} n^{N-2} + \ldots + A_{N/2}^{(N)} + \alpha_n^{(N)}, \tag{4.6}$$

where N is even, and

$$\lambda_n^{N/2} = n^N + A_1^{(N)} n^{N-2} + \ldots + A_{(N+1)/2}^{(N)} n^{-1} + \alpha_n^{(N)}, \tag{4.6'}$$

where N is odd, and

$$\alpha_n^{(N)} = \begin{cases} O(n^{-2}) & \text{when } N \text{ is even} \\[2mm] O(n^{-3}) & \text{when } N \text{ is odd.} \end{cases},$$

so that the series $\lambda_0^{N/2} + \sum_{n=1}^{\infty} \alpha_n^{(N)}$ converges in any case. Its sum for even N, i.e., when $N/2$ is whole, is said to be the *regularized sum* of the series $\lambda_0^{N/2} + \sum_{n=1}^{\infty} \lambda_n^{N/2}$.

Lemma 5.4.2 *Let the asymptotic expansion (4.1) hold with arbitrary natural p, and let the function $f(t)$ be defined by the equality (4.2). Then*

1. *if N is even,*

$$f^{(N)}(+0) = (-1)^{N/2} \left\{ \sum_{n=1}^{\infty} \alpha_n^{(N)} - \frac{1}{2} A_{N/2}^{(N)} \right\}, \tag{4.7}$$

and

2. *if N is odd,*

$$f^{(N)}(+0) = (-1)^{(N-1)/2} \frac{\pi}{2} A_{(N+1)/2}^{(N)}. \tag{4.8}$$

Proof. To omit tedious calculations, we prove the lemma for $N = 4$ and $N = 3$. In the general case, the proof is similar but longer.

First, let $N = 4$. We put $\sqrt{\lambda_n} = n + \beta_n$. Then, relying on (4.6),

$$
\begin{aligned}
\lambda_n^2 \cos \sqrt{\lambda_n} t - n^4 \cos nt &= (n^4 + A_1^{(4)} n^2 + A_2^{(4)} + \alpha_n^{(4)}) \cos(n + \beta_n)t - n^4 \cos nt \\
&= (n^4 + A_1^{(4)} n^2 + A_2^{(4)} + \alpha_n^{(4)})(\cos nt \cos \beta_n t - \sin nt \sin \beta_n t) \\
&\quad -(n^4 + A_1^{(4)} n^2 + A_2^{(4)} + \alpha_n^{(4)}) \cos nt \\
&\quad +(A_1^{(4)} n^2 + A_2^{(4)} + \alpha_n^{(4)}) \cos nt \\
&= (n^4 + A_1^{(4)} n^2 + A_2^{(4)} + \alpha_n^{(4)}) \\
&\quad \times \left\{ \left[-\frac{\beta_n^2}{2!} t^2 + \frac{\beta_n^4}{4!} t^4 + O\left(\frac{t^6}{n^6}\right) \right] \cos nt - \right. \\
&\quad \left. - \left[\beta_n t - \frac{\beta_n^3}{3!} t^2 + \frac{\beta_n^5}{5!} t^5 + O\left(\frac{t^7}{n^7}\right) \right] \sin nt \right\} \\
&\quad +(A_1^{(4)} n^2 + A_2^{(4)} + \alpha_n^{(4)}) \cos nt \\
&= (A_1^{(4)} n^2 + A_2^{(4)} + \alpha_n^{(4)}) \cos nt + (B_1^{(4)} n^2 + B_2^{(4)}) t^2 \cos nt \\
&\quad + C_1^{(4)} t^4 \cos nt + (D_1^{(4)} n^3 + D_2^{(4)} n + D_3^{(4)} n^{-1}) t \sin nt \\
&\quad + (E_1^{(4)} n + E_2^{(4)} n^{-1}) t^3 \sin nt \\
&\quad + F_1^{(4)} n^{-1} t^5 \sin nt + t \varphi_n^{(4)}(t), \tag{4.9}
\end{aligned}
$$

where the coefficients A, B, C, D, E, F are polynomials in the variables $c_1, c_3, \ldots$ from the asymptotic expansion (4.1) and $\varphi_n^{(4)}(t) = O(n^{-2})$ as $n \to \infty$.

Let η and ϵ be two small positive numbers, so that $\eta < \epsilon$, $g_{\eta,\epsilon}(t)$ is an even, infinitely differentiable function vanishing both inside $(-\eta, \eta)$ and outside $(-\epsilon, \epsilon)$, arbitrary in all other respects, and j is a positive integer. Expand the function $t^j g_{\eta,\epsilon}(t)$ into Fourier cosine and sine series

$$t^j g_{\eta,\epsilon}(t) = \frac{1}{\pi} \int_\eta^\epsilon s^j g_{\eta,\epsilon}(s)\, ds + \frac{2}{\pi} \sum_{n=1}^\infty \cos nt \cdot \int_\eta^\epsilon s^j g_{\eta,\epsilon}(s) \cos ns\, ds,$$

$$t^j g_{\eta,\epsilon}(t) = \frac{2}{\pi} \sum_{n=1}^\infty \sin nt \cdot \int_\eta^\epsilon s^j g_{\eta,\epsilon}(s) \sin ns\, ds.$$

Differentiating the former $2k$ times, the latter $2k+1$ times, and putting $t = 0$, we obtain the identities

$$\sum_{n=1}^\infty (-1)^k n^{2k} \int_\eta^\epsilon s^j g_{\eta,\epsilon}(s) \cos ns\, ds = \begin{cases} 0 & \text{when } k > 0 \\ -\dfrac{1}{2} \displaystyle\int_\eta^\epsilon s^j g_{\eta,\epsilon}(s)\, ds & \text{when } k = 0 \end{cases} \tag{4.10}$$

$$\sum_{n=1}^\infty (-1)^k n^{2k+1} \int_\eta^\epsilon s^j g_{\eta,\epsilon}(s) \sin ns\, ds = 0. \tag{4.11}$$

Multiplying (4.9) throughout by $g_{\eta,\epsilon}(t)$, integrating and summing over n, we get the identity

$$\sum_{n=1}^\infty \int_\eta^\epsilon (\lambda_n^2 \cos \sqrt{\lambda_n} t - n^4 \cos nt) g_{\eta,\epsilon}(t)\, dt = \int_\eta^\epsilon f^{(IV)}(t) g_{\eta,\epsilon}(t)\, dt$$

$$= -\frac{1}{2} \int_\eta^\epsilon (A_2^{(4)} + t^2 B_2^{(4)} + t^4 C_1^{(4)}) g_{\eta,\epsilon}(t)\, dt + \int_\eta^\epsilon \left(\sum_{n=1}^\infty \alpha_n^{(4)} \cos nt \right) g_{\eta,\epsilon}(t)\, dt$$

$$+ \int_\eta^\epsilon (D_3^{(4)} t + E_2^{(4)} t^3 + F_1^{(4)} t^5) \sum_{n=1}^\infty \frac{\sin nt}{n} g_{\eta,\epsilon}(t)\, dt + \int_\eta^\epsilon t \varphi^{(4)}(t) g_{\eta,\epsilon}(t)\, dt, \tag{4.12}$$

where we have taken into account (4.10), (4.11) and (4.2), and $\varphi^{(4)}(t) = \sum_{n=1}^\infty \varphi_n^{(4)}(t)$ in the last integral on the right-hand side.

Since $g_{\eta,\epsilon}(t)$ is arbitrary in (η, ϵ), it follows from (4.12) that

$$f^{(IV)}(t) = -\frac{1}{2}(A_2^{(4)} + t^2 B_2^{(4)} + t^4 C_1^{(4)})$$

$$+ \sum_{n=1}^\infty \alpha_n^{(4)} \cos nt + (D_3^{(4)} t + E_2^{(4)} t^3 + F_1^{(4)} t^5) \frac{\pi - t}{2} + t \varphi^{(4)}(t)$$

for $t \in (\eta, \epsilon)$. Letting $t \to +0$, we get

$$f^{(IV)}(+0) = \sum_{n=1}^\infty \alpha_n^{(4)} - \frac{1}{2} A_2^{(4)},$$

i.e., the formula (4.7) for $N = 4$.

Now, let $N = 3$. Then

$$\lambda_n^{3/2} \sin \sqrt{\lambda_n} t - n^3 \sin nt = -(A_1^{(3)}n + A_2^{(3)}n^{-1}) \sin nt + (B_1^{(3)}n + B_2^{(3)}n^{-1})t^2 \sin nt$$
$$+ (C_1^{(3)}n^2 + C_2^{(3)})t \cos nt + D_1^{(3)}t^3 \cos nt + t\varphi_n^{(3)}(t), \qquad \varphi_n^{(3)}(t) = O(n^{-2}).$$

Using the identities (4.10) and (4.11), we obtain

$$f'''(+0) = -A_2^{(3)} \lim_{t \to +0} \sum_{n=1}^{\infty} \frac{\sin nt}{n} = -\frac{\pi}{2}A_2^{(3)}$$

similarly to the above, or the formula (4.8) for $N = 3$.   ∎

5.4.3

In the periodic and semi-periodic problems, the asymptotic formulas for the eigen-values are

$$\sqrt{\lambda_{2n}}, \sqrt{\mu_{2n}} = 2n + \frac{c_1}{2n} + \frac{c_3}{(2n)^3} + \cdots, \tag{4.13}$$

$$\sqrt{\lambda_{2n-1}}, \sqrt{\mu_{2n-1}} = (2n-1) + \frac{c_1}{2n-1} + \frac{c_3}{(2n-1)^3} + \cdots \tag{4.14}$$

(see the formulas (2.11), (2.12)). Using the familiar expansions

$$\sum_{n=1}^{\infty} \frac{\sin 2nt}{2n} = \frac{\pi - 2t}{4}, \qquad \sum_{n=1}^{\infty} \frac{\sin(2n-1)t}{2n-1} = \frac{\pi}{4}, \quad 0 < t \le \frac{\pi}{2}$$

and putting

$$f(t) = \sum_{n=1}^{\infty} \{\cos \sqrt{\lambda_{2n}} t + \cos \sqrt{\mu_{2n}} t - 2\cos 2nt\}$$

$$g(t) = \sum_{n=1}^{\infty} \{\cos \sqrt{\lambda_{2n-1}} t + \cos \sqrt{\mu_{2n-1}} t - 2\cos(2n-1)t\}$$

similarly to the above, we can derive the formulas :

1. For the asymptotic expansion (4.13),

$$f^{(N)}(+0) = \begin{cases} 2(-1)^{N/2} \left[\sum_{n=1}^{\infty} \alpha_{2n}^{(N)} - \frac{1}{2}A_{N/2}^{(N)} \right] & \text{when } N \text{ is even} \\ \\ (-1)^{(N+1)/2} \dfrac{\pi}{2} A_{(N+1)/2}^{(N)} & \text{when } N \text{ is odd} \end{cases} \tag{4.15}$$

and

2. for the asymptotic expansion (4.14),

$$g^{(N)}(+0) = \begin{cases} 2(-1)^{N/2} \sum_{n=1}^{\infty} \alpha_{2n-1}^{(N)} & \text{when } N \text{ is even} \\ \\ (-1)^{(N+1)/2} \dfrac{\pi}{2} A_{(N+1)/2}^{N} & \text{when } N \text{ is odd} \end{cases} \tag{4.16}$$

The numbers $A_{N/2}^{(N)}$, $A_{(N+1)/2}^{(N)}$, $\alpha_n^{(N)}$ are determined as above. The difference of the first formula in (4.15) from that in (4.16) is related to the fact that there is no absolute term in the Fourier expansion $\sum_{n=1}^{\infty} a_n \cos(2n-1)t$, in contrast to the Fourier expansion $\sum_{n=0}^{\infty} a_n \cos 2nt$.

5.5 The regularized trace formula for the periodic problem

5.5.1

Consider the equation

$$-y'' + q(x)y = \lambda y \tag{5.1}$$

in which $q(x) \in C^{\infty}(0, \pi)$, $q(x + \pi) = q(x)$ for the periodic

$$y(0) = y(\pi), \qquad y'(0) = y'(\pi) \tag{5.2}$$

and semi-periodic

$$y(0) = -y(\pi), \qquad y'(0) = -y'(\pi) \tag{5.3}$$

boundary conditions.

Let the numbers $\lambda_0, \lambda_1, \mu_1, \lambda_2, \mu_2, \ldots$, $A_k^{(N)}$, $k = 1, 2, \ldots$, $N = 1, 2, \ldots$, be the same as in the preceding section. For even N, we put

$$S_N = \lambda_0^{N/2} + \sum_{n=1}^{\infty} \{\lambda_{2n}^{N/2} + \mu_{2n}^{N/2} - 2[(2n)^N + A_1^{(N)}(2n)^{N-2} + \ldots + A_{N/2}^{(N)}]\}$$

$$R_N = \sum_{n=1}^{\infty} \{\lambda_{2n-1}^{N/2} + \mu_{2n-1}^{N/2} - 2[(2n-1)^N + A_1^{(N)}(2n-1)^{N-2} + \ldots + A_{N/2}^{(N)}]\}$$

called the *regularized traces* of the periodic (resp. semi-periodic) problems.

The problem now is to express the numbers S_N, R_N as functionals of $q(x)$ and its derivatives. We consider the semi-periodic problem in more detail. The changes to be introduced in the periodic case are obvious.

5.5.2

To use Lemma 5.4.2, we have to be able to calculate the derivative $f^{(N)}(+0)$. With this in mind, we study the auxiliary mixed problem

$$\frac{\partial^2 u}{\partial t^2} = \frac{\partial^2 u}{\partial x^2} - q(x)u \tag{5.4}$$

$$u(x,0) = f(x), \qquad \frac{\partial u}{\partial t}\Big|_{t=0} = 0 \tag{5.5}$$

$$u(0,t) = -u(\pi,t), \qquad u'_x(0,t) = -u'_x(\pi,t) \tag{5.6}$$

where $f(x)$ is a smooth semi-periodic function, i.e., $f(x+\pi) = -f(x)$.

Let $\varphi^-_{2n-1}(x)$, $\varphi^+_{2n-1}(x)$, $n = 1, 2, \ldots$ be the orthonormalized eigenfunctions of the semi-periodic problem (5.1), (5.3), corresponding to the eigenvalues λ_{2n-1}, μ_{2n-1}.

Solving the problem (5.4–6) by the Fourier method, we obtain

$$u(x,t) = \sum_{n=1}^{\infty} \left\{ \cos\sqrt{\lambda_{2n-1}}\,t \cdot \varphi^-_{2n-1}(x) \int_0^{\pi} f(s)\varphi^-_{2n-1}(s)\,ds \right.$$

$$\left. + \cos\sqrt{\mu_{2n-1}}\,t \cdot \varphi^+_{2n-1}(x) \int_0^{\pi} f(s)\varphi^+_{2n-1}(s)\,ds \right\}. \tag{5.7}$$

On the other hand, its solution can be represented in the form

$$u(x,t) = \frac{1}{2}\{f(x+t) + f(x-t)\} + \frac{1}{2}\int_{x-t}^{x+t} w(x,t,s)f(s)\,ds \tag{5.8}$$

(see B. M. Levitan and I. S. Sargsjan (1975; Chapter VI)). It is obvious that only the condition (5.6) has to be checked. Since $q(x)$ is periodic, the function $u(x+\pi,t)$, together with $u(x,t)$, is a solution of the equation (5.4), and satisfies the initial condition (5.5). Therefore,

$$u(x+\pi,0) = f(x+\pi) = -f(x),$$

and $-u(x+\pi,t)$ satisfies (5.4), (5.5). It follows from the uniqueness of a solution of the problem (5.4), (5.5) that $u(x+\pi,t) = -u(x,t)$, and the boundary condition (5.6) follows immediately.

Thus, we have obtained two expressions for the solution of the problem (5.4–6), viz. (5.7) and (5.8). Since it is unique,

$$\sum_{n=1}^{\infty} \left\{ \cos\sqrt{\lambda_{2n-1}}\,t \cdot \varphi^-_{2n-1}(x) \int_0^{\pi} f(s)\varphi^-_{2n-1}(s)\,ds \right.$$

$$\left. + \cos\sqrt{\mu_{2n-1}}\,t \cdot \varphi^+_{2n-1}(x) \int_0^{\pi} f(s)\varphi^+_{2n-1}(s)\,ds \right\}$$

$$= \frac{1}{2}\{f(x+t) - f(x-t)\} + \frac{1}{2}\int_{x-t}^{x+t} w(x,t,s)f(s)\,ds, \tag{5.9}$$

and, since $f(s)$ is arbitrary,

$$\sum_{n=1}^{\infty} \{\cos \sqrt{\lambda_{2n-1}} t \cdot \varphi_{2n-1}^-(x)\varphi_{2n-1}^-(s) + \cos \sqrt{\mu_{2n-1}} t \cdot \varphi_{2n-1}^+(x)\varphi_{2n-1}^+(s)\}$$

$$= \frac{1}{2}\{\delta(x+t-s) + \delta(x-t-s)\} + \frac{1}{2}w(x,t,s) \quad 0 \le x-t \le s \le x+t \le \pi \quad (5.10)$$

and 0 for other s in $(0, \pi)$, where δ is the Dirac delta function.

We show that the identity (5.10) remains valid in the semi-periodic case under consideration, at least for small t, irrespective of the conditions $x - t \ge 0$, $x + t \le \pi$.

The periodic (resp. semi-periodic) boundary conditions are substantially different from the others.

As an example, let $x - t < 0$, $x + t \le \pi$. We select the semi-periodic function $f(x)$ by letting it vanish on the interval $(x - t, 0)$ and not vanish (i.e., be arbitrary) on the interval $(0, \pi - (t - x))$, for which the right-hand side of (5.9) is

$$\frac{1}{2}f(x+t) + \frac{1}{2}\int_0^{x+t} w(x,t,s)f(s)\,ds.$$

Therefore, the right-hand side of (5.10) should be replaced by

$$\begin{cases} \frac{1}{2}\delta(x+t-s) + \frac{1}{2}w(x,t,s), & 0 < s < x+t < \pi \\ \\ 0 & \text{for other } s \in [0, \pi]. \end{cases} \quad (5.10')$$

If $x - t \ge 0$ and $x + t > \pi$, then the right-hand side of (5.10) should be replaced by

$$\begin{cases} \frac{1}{2}\delta(x-t-s) + \frac{1}{2}w(x,t,s), & 0 \le x-t < s < \pi \\ \\ 0 & \text{for other } s \in [0, \pi] \end{cases} \quad (5.10'')$$

If $q(x) = 0$, then $\lambda_{2n-1} = \mu_{2n-1} = 2n-1$, $\varphi_{2n-1}^-(x) = \sqrt{\frac{2}{\pi}}\sin(2n-1)x$, $\varphi_{2n-1}^+(x) = \sqrt{\frac{2}{\pi}}\cos(2n-1)x$, and the identities (5.10) assume the form

$$\frac{2}{\pi}\sum_{n=1}^{\infty}\cos(2n-1)t[\sin(2n-1)x\sin(2n-1)s + \cos(2n-1)s\cos(2n-t)x]$$

$$= \begin{cases} \frac{1}{2}[\delta(x+t-s) + \delta(x-t-s)], & 0 \le x-t < s < x+t \le \pi \\ \\ \frac{1}{2}\delta(x+t-s), & x-t < 0, 0 < s < x+t \le \pi \\ \\ \frac{1}{2}\delta(x-t-s), & x+t > \pi, 0 \le x-t < s < \pi \end{cases} \quad (5.11)$$

Subtracting (5.11) from (5.10), putting $s = x$ and integrating in $0 \le x \le \pi$ with respect to x, we obtain the identity

$$
\begin{aligned}
f(t) &= \sum_{n=1}^{\infty} \{ \cos \sqrt{\lambda_{2n-1}} t + \cos \sqrt{\mu_{2n-1}} t - 2\cos(2n-1)t \} \\
&= \frac{1}{2} \int_0^{\pi} w(x,t,s)\,dx.
\end{aligned}
\tag{5.12}
$$

It was shown in the monograph of B. M. Levitan (1987, Chapter VIII, Section 3) that the asymptotic expansion

$$
w(x,t,x) = \sum_{k=0}^{\infty} A_{2k+1}(x) t^{2k+1}
\tag{5.13}
$$

holds, where $A_{2k+1}(x)$ are polynomials in $q(x)$ and its derivatives

$$
\begin{aligned}
A_1(x) &= -\frac{1}{2}q(x), \\
A_3(x) &= -\frac{1}{48}q''(x) - \frac{1}{16}q^2(x), \\
A_5(x) &= -\frac{1}{3 \cdot 2^7}[10^{-1}q^{(IV)}(x) + q(x)q''(x) + q^3(x) - q'^2(x)],
\end{aligned}
$$

etc.

It can be seen from (5.13) that the derivatives of $w(x,t,x)$ with respect to the variable t, of even order, vanish for $t = 0$. It follows from (5.12) that

$$
f^{(N)}(+0) = \begin{cases}
0 & \text{when } N \text{ is even,} \\
\frac{1}{2}N! \int_0^{\pi} A_N(x)\,dx & \text{when } N \text{ is odd}
\end{cases}
\tag{5.14}
$$

for $f(t)$. Because of (5.14) and of the identities (4.15), (4.16), the following holds true.

Theorem 5.5.1 *Let the potential $q(x)$ be periodic (with period π), and infinitely differentiable. Then*

1. *all regularized traces R_N of the semi-periodic problem vanish, i.e.,*

$$
R_N = 0, \quad N = 1, 2, \ldots,
\tag{5.15}
$$

2. *for odd N, the formulas*

$$
A^{(N)}_{(N+1)/2} = (-1)^{(N+1)/2} \frac{N!}{\pi} \int_0^{\pi} A_N(x)\,dx
\tag{5.16}
$$

are valid, and

3. for the regularized traces S_N of the periodic problem, the identities

$$S_N = \lambda_0^{N/2} + \sum_{n=1}^{\infty}\{\lambda_{2n}^{N/2} + \mu_{2n}^{N/2} - 2[(2n)^N + A_1^{(N)}(2n)^{N-2} + \ldots + A_{N/2}^{(N)}]\}$$

$$= \lambda_0^{N/2} + \frac{1}{2}A_{N/2}^{(N)} \tag{5.17}$$

hold for even N.

Remark The coefficients $c_1, c_3, \ldots$ in the asymptotic expansion of the eigenvalues can be calculated consecutively with the help of (5.16).

5.5.3

The formulas for the regularized traces of the problem

$$-y'' + q(x)y = \lambda y, \qquad y(0) = y(\pi) = 0 \tag{5.18}$$

can be obtained by means of (3.2), (5.15) and (5.17) if $q(x)$ is a smooth periodic potential.

Let the eigenvalues of the problem (5.18) be $\nu_1, \nu_2, \nu_3, \ldots$. We already know that the coefficients in the asymptotic expansions for ν_n and λ_n, μ_n are the same. Therefore,

$$2\sum_{n=1}^{\infty}\{\nu_n^{N/2} - [n^N + A_1^{(N)}n^{N-2} + \ldots + A_{N/2}^{(N)}]\}$$

$$= 2\sum_{n=1}^{\infty}\{\nu_{2n}^{N/2} - [(2n)^N + A_1^{(N)}(2n)^{N-2} + \ldots + A_{N/2}^{(N)}]\}$$

$$+2\sum_{n=1}^{\infty}\{\nu_{2n-1}^{N/2} - [(2n-1)^N + A_1^{(N)}(2n-1)^{N-2} + \ldots + A_{N/2}^{(N)}]\}$$

$$= \sum_{n=1}^{\infty}\{\lambda_{2n}^{N/2} + \mu_{2n}^{N/2} - 2[(2n)^N + A_1^{(N)}(2n)^{N-2} + \ldots + A_{N/2}^{(N)}]\}$$

$$+ \sum_{n=1}^{\infty}\lambda_{2n-1}^{N/2} + \mu_{2n-1}^{N/2} - 2[(2n-1)^N + A_1^{(N)}(2n-1)^{N-2} + \ldots + A_{N/2}^{(N)}]\}$$

$$- \sum_{n=1}^{\infty}\{\lambda_n^{N/2} + \mu_n^{N/2} - 2\nu_n^{N/2}\} = \lambda_0^{N/2} + \frac{1}{2}A_{N/2}^{(N)} - \sum_{n=1}^{\infty}\{\lambda_n^{N/2} + \mu_n^{N/2} - 2\nu_n^{N/2}\}.$$

5.6 The regularized first trace formula in the case of separated boundary conditions

5.6.1

Consider the boundary-value problem

$$y'' + q(x)y = \lambda y, \quad 0 \le x \le \pi \tag{6.1}$$

$$y'(0) - hy(0) = 0, \qquad y'(\pi) + Hy(\pi) = 0, \tag{6.2}$$

where $q(x)$ is a smooth (non-periodic) real-valued function and h, H are two real numbers.

Let $\lambda_0 < \lambda_1 < \lambda_2 < \ldots < \lambda_n < \ldots$ be the eigenvalues of the problem (6.1), (6.2). It was shown in Section 2 that the asymptotic behaviour

$$\sqrt{\lambda_n} = n + \frac{c_1}{n} + \frac{c_3}{n^3} + \cdots$$

is valid. Therefore, we can determine the regularization of $\lambda_0^N + \sum_{n=1}^{\infty} \lambda_n^N$ similarly to the above.

Now, the problem is to express these sums as functionals of $q(x)$, its derivatives and the numbers h, H. In this case, too, we can apply the method developed in the preceding section. However, the account of (6.2) leads to considerable difficulties. There are also other methods. Discussing one of them here, we confine ourselves to calculating the first trace, i.e., the sum

$$\lambda_0 + \sum_{n=1}^{\infty} (\lambda_n - n^2 - 2c_1).$$

Let $q(x) \in C^2(0, \pi)$, and let h, H be finite. The asymptotic formula

$$\sqrt{\lambda_n} = n + \frac{c}{n} + O\left(\frac{1}{n^3}\right) \tag{6.3}$$

for the eigenvalues was obtained in Section 2 of Chapter 1 with the above assumptions, where

$$c = \frac{2}{\pi}\left(h + H + \frac{1}{2}\int_0^{\pi} q(x)\,dx\right). \tag{6.4}$$

It follows from (6.3) that

$$\lambda_n = n^2 + c + O\left(\frac{1}{n^2}\right). \tag{6.5}$$

Therefore, the series

$$\lambda_0 + \sum_{n=1}^{\infty} (\lambda_n - n^2 - c) < \infty \tag{6.6}$$

converges, and is called the *regularized first trace* for the problem (6.1), (6.2). The goal of the present section is to find its sum.

5.6.2

Let $\varphi(x, \lambda)$ be the solution of the equation (6.1), satisfying the initial conditions $\varphi(0, \lambda) = 1$, $\varphi'_x(0, \lambda) = h$. Then the eigenvalues λ_n are the roots of the entire analytic function $\varphi'_x(\pi, \lambda) + H\varphi(\pi, \lambda)$. Therefore,

$$\varphi'_x(\pi, \lambda) + H\varphi(\pi, \lambda) = A\Phi(\lambda), \tag{6.7}$$

where $\Phi(\lambda) = \prod_{n=0}^{\infty}\left(1 - \frac{\lambda}{\lambda_n}\right)$ and A is a certain constant to be determined below[4].

We now study the asymptotic behaviour of both sides of the equation (6.7) for large negative $\lambda = -\mu$. Comparing the coefficients of the same powers of μ on the left and the right, we obtain the sum of (6.6).

Put[5]

$$\Phi(-\mu) = \prod_{n=0}^{\infty}\left(1 + \frac{\mu}{\lambda_n}\right) = \left(1 + \frac{\mu}{\lambda_0}\right)\frac{\prod_{n=1}^{\infty}\left(1 + \frac{\mu}{\lambda_n}\right)}{\prod_{n=1}^{\infty}\left(1 + \frac{\mu}{n^2}\right)}\frac{\sinh\pi\sqrt{\mu}}{\pi\sqrt{\mu}}$$

$$= C_1(\lambda_0 + \mu)\psi(\mu)\frac{\sinh\pi\sqrt{\mu}}{\pi\sqrt{\mu}}, \quad (6.8)$$

where

$$C_1 = \frac{1}{\lambda_0}\prod_{n=1}^{\infty}\frac{n^2}{\lambda_n}, \qquad \psi(\mu) = \prod_{n=1}^{\infty}\left(1 - \frac{n^2 - \lambda_n}{\mu + n^2}\right). \quad (6.9)$$

To study the asymptotic behaviour of $\psi(\mu)$ for large positive μ, we consider

$$\ln\psi(\mu) = \sum_{n=1}^{\infty}\ln\left(1 - \frac{n^2 - \lambda_n}{\mu + n^2}\right) = -\sum_{n=1}^{\infty}\sum_{k=1}^{\infty}\frac{1}{k}\left(\frac{n^2 - \lambda_n}{\mu + n^2}\right)^k$$

$$= -\sum_{k=1}^{\infty}\frac{1}{k}\sum_{n=1}^{\infty}\left(\frac{n^2 - \lambda_n}{\mu + n^2}\right)^k.$$

Subsequent estimates are based on the following.

Lemma 5.6.1 *If $|n^2 - \lambda_n| \leq a$, then*

$$\sum_{n=1}^{\infty}\frac{|n^2 - \lambda_n|^k}{(\mu + n^2)^k} \leq \frac{\pi}{2}\frac{a^k}{\mu^{k-1/2}}. \quad (6.10)$$

Proof. In fact,

$$\sum_{n=1}^{\infty}\frac{|n^2 - \lambda_n|^k}{(\mu + n^2)^k} \leq a^k\sum_{n=1}^{\infty}\frac{1}{(\mu + n^2)^k} \leq a^k\int_0^{\infty}\frac{dx}{(\mu + x^2)^k}$$

$$= \frac{a^k}{\mu^{k-1/2}}\int_0^{\infty}\frac{dt}{(1 + t^2)^k} \leq \frac{\pi}{2}\frac{a^k}{\mu^{k-1/2}},$$

and the lemma is thus proved. ∎

The estimate (6.10) leads to

$$\sum_{k=2}^{\infty}\frac{1}{k}\sum_{n=1}^{\infty}\frac{|n^2 - \lambda_n|^k}{(\mu + n^2)^k} \leq \frac{\pi}{2}\sum_{k=2}^{\infty}\frac{a^k}{\mu^{k-1/2}} = \frac{\pi}{2}\frac{a^2}{\mu^{3/2}}\sum_{k=0}^{\infty}\left(\frac{a}{\mu}\right)^k = O(\mu^{-3/2}). \quad (6.11)$$

---

[4]The absence of exponential factors follows from Theorem 28, Chapter I, in the monograph of B. Ya. Levin (1980).

[5]Relying on the equality $\sin z = z\prod_{n=1}^{\infty}\left(1 - \frac{z^2}{n^2\pi^2}\right)$.

Furthermore,

$$-\sum_{n=1}^{\infty}\frac{n^2-\lambda_n}{\mu+n^2} = \sum_{n=1}^{\infty}\frac{\lambda_n-n^2-c}{\mu+n^2} + \sum_{n=1}^{\infty}\frac{c}{\mu+n^2}$$

$$= \sum_{n=1}^{\infty}\frac{c}{\mu+n^2} + \frac{1}{\mu}\sum_{n=1}^{\infty}(\lambda_n-n^2-c) - \frac{1}{\mu}\sum_{n=1}^{\infty}\frac{(\lambda_n-n^2-c)n^2}{\mu+n^2}. \quad (6.12)$$

Since $\sup_n |\lambda_n - n^2 - c|n^2 < \infty$, it follows from (6.10) that

$$\frac{1}{\mu}\sum_{n=1}^{\infty}\frac{(\lambda_n-n^2-c)n^2}{\mu+n^2} = O(\mu^{-3/2}). \quad (6.13)$$

Therefore, relying on (6.11), (6.12) and (6.13), we obtain

$$\ln\psi(\mu) = c\sum_{n=1}^{\infty}\frac{1}{\mu+n^2} + \frac{1}{\mu}\sum_{n=1}^{\infty}(\lambda-n^2-c) + O(\mu^{-3/2}). \quad (6.14)$$

As we know,

$$\sum_{n=1}^{\infty}\frac{1}{\mu+n^2} = \frac{\pi\coth\pi\sqrt{\mu}}{2\sqrt{\mu}} - \frac{1}{2\mu} = \frac{\pi}{2\sqrt{\mu}} - \frac{1}{2\mu} + O(e^{-2\pi\sqrt{\mu}}).$$

It then follows from (6.14) that

$$\ln\psi(\mu) = \frac{c\pi}{2\sqrt{\mu}} + \frac{1}{\mu}\left(s_\lambda - \lambda_0 + \frac{1}{2}c\right) + O(\mu^{-3/2}),$$

where

$$s_\lambda = \sum_{n=0}^{\infty}(\lambda_n-n^2-c). \quad (6.15)$$

Therefore,

$$\psi(\mu) = \exp\left\{\frac{c\pi}{2\sqrt{\mu}} + \frac{1}{\mu}\left(s_\lambda - \lambda_0 + \frac{c}{2}\right) + O(\mu^{-3/2})\right\}$$

$$= 1 + \frac{c\pi}{2\sqrt{\mu}} + \frac{1}{\mu}\left(s_\lambda - \lambda_0 + \frac{c}{2} + \frac{c^2\pi^2}{8}\right) + O(\mu^{-3/2}). \quad (6.16)$$

Relying on (6.16), we then derive from (6.8) that

$$\Phi(-\mu) = \frac{1}{2\pi}C_1 e^{\pi\sqrt{\mu}}\left\{\sqrt{\mu} + \frac{1}{2}c\pi + \frac{1}{\sqrt{\mu}}\left(s_\lambda + \frac{1}{2}c + \frac{c^2\pi^2}{8}\right) + O(\mu^{-1})\right\}. \quad (6.17)$$

5.6.3

We now study the asymptotic behaviour of the function $\varphi'_x(\pi, \lambda) + H\varphi(\pi, \lambda)$ for large negative $\lambda = -\mu$, using the Liouville integral equation. To simplify the calculations, we assume that $h = 0$ and that the function $q(x)$ is such that $\int_0^\pi q(x)\, dx = 0$. Then, according to the formula (2.4), Chapter 1, we have

$$\varphi(x, -\mu) = \cosh\sqrt{\mu}x + \frac{1}{\sqrt{\mu}} \int_0^x q(t) \sinh\{\sqrt{\mu}(x - t)\}\varphi(t, -\mu)\, dt,$$

and, using the estimate (2.7), Chapter 1,

$$\varphi(\pi, -\mu) = \frac{1}{2}e^{\pi\sqrt{\mu}}\{1 + O(\mu^{-1})\}. \tag{6.18}$$

Furthermore,

$$\varphi'_x(x, -\mu) = \sqrt{\mu}\sinh\sqrt{\mu}x + \int_0^x q(t) \cosh[\sqrt{\mu}(x - t)]\varphi(t, -\mu)\, dt.$$

Therefore,

$$\varphi'_x(\pi, -\mu) = \frac{1}{2}e^{\pi\sqrt{\mu}}\left\{\sqrt{\mu} + \frac{1}{4\sqrt{\mu}}[q(0) + q(\pi)] + O(\mu^{-1})\right\}. \tag{6.19}$$

Relying on (6.18) and (6.19), we have

$$\varphi'_x(\pi, -\mu) + H\varphi(\pi, -\mu) = \frac{1}{2}e^{\pi\sqrt{\mu}}\left\{\sqrt{\mu} + H + \frac{1}{4\sqrt{\mu}}[q(0) + q(\pi)] + O(\mu^{-1})\right\}. \tag{6.20}$$

It follows from the equalities (6.7), (6.17) and (6.20) that $AC_1 = \pi$, i.e., $A = \pi/C_1$. Therefore, we derive from (6.17), (6.20) that

$$H + \frac{1}{4\sqrt{\mu}}[q(0) + q(\pi)] + O(\mu^{-1}) = \frac{c\pi}{2} + \frac{1}{\sqrt{\mu}}\left(s_\lambda + \frac{c}{2} + \frac{\pi^2 c}{8}\right) + O(\mu^{-1}).$$

Comparing the coefficients of $\dfrac{1}{\sqrt{\mu}}$ and taking into account (6.4), we obtain

$$s_\lambda = \sum_{n=0}^\infty (\lambda_n - n^2 - c) = \frac{1}{4}[q(0) + q(\pi)] - \frac{H}{\pi} - \frac{H^2}{2}.$$

Notes

Section 1

The representation of the solution in the form (1.2) was used in a number of works. The study of the asymptotic expansion (1.4) was done by V. A. Marchenko (1972).

Section 2

Our method is similar to the one in the monograph of V. A. Marchenko (1972).

Section 3

The trace formulas in the case of the periodic potential, considered in this section, were first derived by H. Hochstadt (1965); see also H. McKean and P. Van Moerbeke (1975). In the nonperiodic finite-zone case, they are due to B. A. Dubrovin (1975).

Sections 4 and 5

The first trace for the periodic and semi-periodic problems was first calculated by W. Magnus and W. Winkler (1966). Our method is based on the paper of B. M. Levitan (1981).

Section 6

The regularized trace for the classical Sturm-Liouville problem was first calculated by I. M. Gelfand and B. M. Levitan (1953); see also I. M. Gelfand (1956), B. M. Levitan (1964), B. M. Levitan and I. S. Sargsjan (1975), I. A. Dikii (1955).

Chapter 6

INVERSE PROBLEMS

By inverse problems in spectral analysis, we mean those of reconstructing a linear operator from some of its spectral characteristics such as spectra (for different boundary conditions), a spectral function, scattering data, etc.

The first essential result, which provided an important impetus to the development of the theory of inverse problems, was obtained by V. A. Ambarzumian (1929).

(V. A. Ambarzumian theorem.) *Let* $\lambda_0 < \lambda_1 < \lambda_2 < \ldots$ *be the eigenvalues of the problem*

$$-y'' + q(x)y = \lambda y, \qquad y'(0) = y'(\pi) = 0, \tag{1}$$

where $q(x)$ *is a real-valued continuous function,* $0 \le x \le \pi$. *Then, if* $\lambda_n = n^2$, $n = 0, 1, 2, \ldots$, *it follows that* $q(x) \equiv 0$.

The first to pay attention to its importance was the Swedish mathematician G. Borg (1946), who systematically studied one of the most important inverse problems, viz. that for the classical Sturm-Liouville operator of the form (1) from spectra. He showed that, in the general case, the Sturm-Liouville spectrum does not determine the operator, so that Ambarzumian's result is an exception to the general rule. G. Borg also proved in that paper that two Sturm-Liouville operator spectra (for different boundary conditions) determine the operator uniquely (see the Notes at the end of the chapter concerning the development of the theory of inverse problems).

6.1 Definition and simplest properties of transformation operators

Let E be a linear topological space, let A, B be two linear (not necessarily continuous) operators from E to E, and let E_1, E_2 be two closed subspaces of E.

Definition 6.1.1 *An invertible linear operator* X *defined on the whole space* E *and acting from* E_1 *to* E_2 *is called the* transformation operator *for a pair* A *and* B *if it satisfies the conditions :*

  1. *The operator* X *and its inverse* X^{-1} *are continuous in* E.

2. The operator identity

$$AX = XB, \quad \text{or} \quad A = XBX^{-1}, \tag{1.1}$$

holds.

We now prove two simple lemmas which will be useful in the sequel.

Lemma 6.1.1 *Let $\varphi_\lambda \in E_1$ be an eigenvector of the operator B for an eigenvalue λ, i.e., $B\varphi_\lambda = \lambda\varphi_\lambda$. Then $\psi_\lambda = X\varphi_\lambda$ is an eigenvector of the operator A for the same value of λ, i.e., $A\psi_\lambda = \lambda\psi_\lambda$.*

Proof. The first equation in (1.1) implies that

$$A\psi_\lambda = AX\varphi_\lambda = XB\varphi_\lambda = X\{\lambda\varphi_\lambda\} = \lambda X\varphi_\lambda = \lambda\psi_\lambda.$$

∎

Lemma 6.1.2 *Let three linear operators A, B, C and three closed subspaces E_1, E_2, E_3 be given in the space E. Let $X_{A,B}$ be the transformation operator for the pair A and B, acting from E_2 to E_3, $X_{B,C}$ that for the pair B,C, acting from E_1 to E_2. Then the transformation operator for the pair A, C, acting from E_1 to E_3, is given by the formula*

$$X_{A,C} = X_{A,B}X_{B,C}.$$

Proof. By the definition of a transformation operator, $AX_{A,B} = X_{A,B}B$, $BX_{B,C} = X_{B,C}C$. It follows from the latter equation that $B = X_{B,C}CX_{B,C}^{-1}$. Substituting this in the former, we obtain

$$AX_{A,B} = X_{A,B}X_{B,C}CX_{B,C}^{-1}, \quad \text{or} \quad AX_{A,B}X_{B,C} = X_{A,B}X_{B,C}C.$$

The last equation means that $X_{A,B}X_{B,C}$ is the transformation operator for the pair A, C. That it acts from E_1 to E_3 is obvious.   ∎

6.2 Transformation operators with boundary condition at $x = 0$

From now onwards, E is the space of complex-valued functions $f(x)$, $0 \le x < \infty$, which are continuous and with a continuous first derivative. No restrictions are placed on the growth of the function at infinity. The topology on E is defined via the uniform convergence of the functions and their first derivatives in each finite interval.

Let $A = -d^2/dx^2 + q(x)$, $B = -d^2/dx^2 + r(x)$, where $q(x)$ and $r(x)$, $0 \le x < \infty$, are two continuous complex-valued functions.

We consider two cases.

Case 1 The space E is the same as above, E_1 is the subspace of $f(x)$ of E, satisfying the boundary condition

$$f'(0) = h_1 f(0), \tag{2.1}$$

where h_1 is an arbitrary finite complex number, E_2 is the subspace of E of functions $f(x)$, so that the boundary condition

$$f'(0) = h_2 f(0) \tag{2.1'}$$

is fulfilled, h_2 being an arbitrary finite complex number, too.

Theorem 6.2.1 *The transformation operator $X = X_{A,B}$ mapping E_1 to E_2 can be realized in the form*

$$Xf(x) = f(x) + \int_0^x K(x,t)f(t)\,dt. \tag{2.2}$$

Its kernel $K(x,t)$ is the solution of the problem

$$K''_{xx}(x,t) - q(x)K(x,t) = K''_{tt}(x,t) - r(t)K(x,t), \tag{2.3}$$

$$K(x,x) = h_2 - h_1 + \frac{1}{2}\int_0^x [q(s) - r(s)]\,ds, \tag{2.4}$$

$$[K'_t(x,t) - h_1 K(x,t)]_{t=0} = 0. \tag{2.5}$$

Conversely, if the function $K(x,t)$ is the solution of (2.3), (2.4), (2.5), then X determined by formula (2.2) is the transformation operator for the pair of operators A, B, and acts from E_1 to E_2.

Proof. Differentiating (2.2), we obtain

$$(Xf)' = f'(x) + K(x,x)f(x) + \int_0^x K'_x(x,t)f(t)\,dt. \tag{2.6}$$

Since $Xf(x) \in E_2$ and $f(x) \in E_1$ (given), putting $x = 0$ in (2.6), we have that

$$(Xf)'_{x=0} = h_2(Xf)_{x=0} = h_2 f(0) = f'(0) + K(0,0)f(0) = \{h_1 + K(0,0)\}f(0).$$

Hence,

$$K(0,0) = h_2 - h_1. \tag{2.7}$$

Now, differentiating (2.6) once again, we get

$$(Xf)'' = f''(x) + \frac{dK(x,x)}{dx}f(x) + K(x,x)f'(x) + K'_x(x,t)|_{t=x}f(x) + \int_0^x K''_{xx}(x,t)f(t)\,dt.$$

Therefore,

$$\begin{aligned} A(Xf) &= -(Xf)'' + q(x)(Xf) \\ &= -f''(x) - f(x)\frac{d}{dx}K(x,x) - K(x,x)f'(x) - K'_x(x,t)|_{t=x}f(x) \\ &\quad + q(x)f(x) - \int_0^x [K''_{xx}(x,t) - q(x)K(x,t)]f(t)\,dt. \end{aligned} \tag{2.8}$$

We now transform $X(Bf)$ by partial integration. We have

$$
\begin{aligned}
X(Bf) &= -f''(x) + r(x)f(x) + \int_0^x K(x,t)[-f''(t) + r(t)f(t)]\,dt \\
&= -f''(x) + r(x)f(x) - K(x,x)f'(x) + K(x,0)f'(0) \\
&\quad + K_t'(x,t)|_{t=x}f(x) - K_t'(x,t)|_{t=0}f(0) \\
&\quad - \int_0^x [K_{tt}''(x,t) - r(t)K(x,t)]f(t)\,dt. \qquad (2.9)
\end{aligned}
$$

Equalizing (2.8), (2.9) and since $f(x)$ is arbitrary, we obtain

$$
K_{xx}''(x,t) - q(x)K(x,t) = K_{tt}''(x,t) - r(t)K(x,t)
$$

$$
[K_t'(x,t) - h_1 K(x,t)]_{t=0} = 0
$$

$$
2\frac{d}{dx}K(x,x) = q(x) - r(x). \qquad (2.10)
$$

(2.4) then follows, also from (2.7).

We have thus shown that if a transformation operator X can be realized in the form of (2.2), then its kernel $K(x,t)$ is the solution of the problem (2.3), (2.4), (2.5). Conversely, if $K(x,t)$ is the solution, then the operator X constructed from it by the formula (2.2) is the transformation operator for the two operators A, B and maps E_1 onto E_2; to see this, it suffices to repeat the above calculations in inverse order.

Thus, to complete the proof, we should show that (2.3), (2.4), (2.5) is solvable. First of all, we can confine ourselves to considering the cases $r(x) \equiv 0$, $h_1 = 0$ or $h_2 = 0$. We invoke Lemma 6.1.2. In fact, let $A = -d^2/dx^2 + q(x)$, $B = -d^2/dx^2$, $C = -d^2/dx^2 + r(x)$, and let E_1 be the subspace of E, containing $f(x)$ which satisfy the boundary condition $f'(0) = h_1 f(0)$, E_2 the subspace of functions satisfying the equality $f'(0) = 0$, and, finally, E_3 the subspace of functions such that $f'(0) = h_2 f(0)$. Suppose that the transformation operator $X_{A,B}$ mapping E_2 onto E_3, and the transformation operator $X_{B,C}$ mapping E_1 onto E_2, are of the form

$$
X_{A,B}f(x) = f(x) + \int_0^x K_1(x,t)f(t)\,dt \qquad (2.11)
$$

$$
X_{B,C}f(x) = f(x) + \int_0^x K_2(x,t)f(t)\,dt. \qquad (2.12)
$$

By Lemma 6.1.2, $X_{A,C} = X_{A,B}X_{B,C}$ is then the transformation operator for the pair of operators A, C, and maps E_1 onto E_2. We show that $X_{A,C}$ can also be represented in the form of (2.2). In fact, it follows from (2.11) and (2.12) that

$$
\begin{aligned}
X_{A,C}f(x) &= X_{A,B}X_{B,C}f(x) \\
&= X_{A,B}\left[f(x) + \int_0^x K_2(x,s)f(s)\,ds\right] \\
&= f(x) + \int_0^x K_2(x,s)f(s)\,ds
\end{aligned}
$$

$$+ \int_0^x K_1(x,t) \left[f(t) + \int_0^t K_2(t,s) f(s) \, ds \right] dt$$

$$= f(x) + \int_0^x \left[K_2(x,t) + K_1(x,t) + \int_t^x K_1(x,s) K_2(s,t) \, ds \right] f(t) \, dt.$$

Thus, let $A = -d^2/dx^2 + q(x)$, $B = -d^2/dx^2$, h_1 or h_2 is zero. We show that $X_{A,B}$ can be represented as in (2.2). Consider the case $h_2 = 0$, that of $h_1 = 0$ being similar and even simpler.

Due to the above, we now have to show the solvability of the problem

$$K''_{xx}(x,t) - q(x) K(x,t) = K''_{tt}(x,t) \tag{2.13}$$

$$K(x,x) = -h_1 + \frac{1}{2} \int_0^x q(s) \, ds \tag{2.14}$$

$$[K'_t(x,t) - h_1 K(x,t)]_{t=0} = 0. \tag{2.15}$$

However, we first assume that $q(x)$ is differentiable. Changing the variables $\xi = x + t$, $\eta = x - t$, we obtain the problem

$$N''_{\xi\eta}(\xi,\eta) = -\frac{1}{4} q \left(\frac{\xi + \eta}{2} \right) N(\xi,\eta) \tag{2.16}$$

$$N(\xi,0) = -h_1 + \frac{1}{2} \int_0^{\xi/2} q(s) \, ds \tag{2.17}$$

$$[N'_\xi(\xi,\eta) - N'_\eta(\xi,\eta) - h_1 N(\xi,\eta)]_{\xi=\eta} = 0 \tag{2.18}$$

for the function $N(\xi,\eta) \equiv K \left(\frac{\xi+\eta}{2}, \frac{\xi-\eta}{2} \right)$.

Now, integrating with respect to η from 0 to η, we obtain

$$N'_\xi(\xi,\eta) - N'_\xi(\xi,\eta)|_{\eta=0} = -\frac{1}{4} \int_0^\eta q \left(\frac{\xi + \alpha}{2} \right) N(\xi,\alpha) \, d\alpha.$$

It follows from the condition (2.17) that $N'_\xi(\xi,\eta)|_{\eta=0} = \frac{1}{4} q(\xi/2)$. Therefore,

$$N'_\xi(\xi,\eta) = -\frac{1}{4} \int_0^\eta q \left(\frac{\xi + \alpha}{2} \right) N(\xi,\alpha) \, d\alpha + \frac{1}{4} q \left(\frac{\xi}{2} \right), \tag{2.19}$$

which, integrated with respect to ξ from η to ξ, yields

$$N(\xi,\eta) = -\frac{1}{4} \int_\eta^\xi d\beta \int_0^\eta q \left(\frac{\beta + \alpha}{2} \right) N(\beta,\alpha) \, d\alpha + \frac{1}{4} \int_\eta^\xi q \left(\frac{\alpha}{2} \right) d\alpha + N(\eta,\eta).$$

To calculate $N(\eta,\eta)$, we see that

$$2N'_\xi(\xi,\eta)|_{\xi=\eta} = [h_1 N(\xi,\eta) + N'_\xi(\xi,\eta) + N'_\eta(\xi,\eta)]_{\xi=\eta}$$

$$= -\frac{1}{2} \int_0^\eta q \left(\frac{\eta + \alpha}{2} \right) N(\eta,\alpha) \, d\alpha + \frac{1}{2} q \left(\frac{\eta}{2} \right),$$

relying on (2.18) and (2.19); hence,

$$[e^{h_1\eta}N(\eta,\eta)]' = \frac{1}{2}e^{h_1\eta}\left[q\left(\frac{\eta}{2}\right) - \int_0^\eta q\left(\frac{\eta+\alpha}{2}\right)N(\eta,\alpha)\,d\alpha\right],$$

which, integrated from 0 to η, taking into account (2.17), yields

$$N(\eta,\eta) = -h_1 e^{-h_1\eta} + \frac{1}{2}e^{-h_1\eta}\int_0^\eta e^{h_1\beta}\left[q\left(\frac{\beta}{2}\right) - \int_0^\beta q\left(\frac{\alpha+\beta}{2}\right)N(\beta,\alpha)\,d\alpha\right]d\beta.$$

The function $N(\xi,\eta)$ thus should satisfy the integral equation

$$N(\xi,\eta) = -\frac{1}{4}\int_\eta^\xi d\beta \int_0^\eta q\left(\frac{\beta+\alpha}{2}\right)N(\beta,\alpha)\,d\alpha + \frac{1}{4}\int_0^\xi q\left(\frac{\alpha}{2}\right)d\alpha - h_1 e^{-h_1\eta}$$

$$+\frac{1}{2}e^{-h_1\eta}\int_0^\eta e^{h_1\beta}\left[q\left(\frac{\beta}{2}\right) - \int_0^\beta q\left(\frac{\alpha+\beta}{2}\right)N(\beta,\alpha)\,d\alpha\right]d\beta. \quad (2.20)$$

Conversely, if $N(\xi,\eta)$ satisfies (2.20) and $q(x)$ is once differentiable, then we can directly verify that $N(\xi,\eta)$ is the solution of the problem (2.16), (2.17), (2.18). ∎

(2.20) is a Volterra integral equation. Therefore, its solution is unique, with the existence possibly proved by the method of successive approximations. The differentiability assumption is inessential, since a continuous function can be approximated by smooth ones, and then the limit can be taken. The limit function $K(x,t)$ so formed is the transformation operator kernel.

Case 2 The space E is the same as in Case 1, and $E_1 = E_2$ the subspace of E, containing the functions $f(x)$ which satisfy the boundary condition $f(0) = 0$. Formally, the latter condition is associated with (2.1) for $h_1 = \infty$. It is obvious that the case $h_1 = \infty$ is not embraced by the above.

Let A and B be two operators of the above form. Similarly to Case 1, without loss of generality, we can assume that $B = -d^2/dx^2$. We shall look for the transformation operator $X = X_{A,B}$ in the form

$$Xf(x) = f(x) + \int_0^x N(x,t)f(t)\,dt$$

again.

Reasoning similarly to the above, we can easily obtain the problem

$$N''_{xx}(x,t) - q(x)N(x,t) = N''_{tt}(x,t)$$

$$N(x,0) = 0$$

$$N(x,x) = C + \frac{1}{2}\int_0^x q(s)\,ds$$

for the kernel $N(x,t)$, where $C = N(0,0)$ is arbitrary.

The solvability and the uniqueness of the solution for this problem can be proved as in Case 1, reducing it to an integral equation.

6.3   Derivation of the basic integral equation

6.3.1

Consider the problem

$$-y'' + q(x)y = \lambda y \tag{3.1}$$

$$y(0) = 1, \qquad y'(0) = h \tag{3.2}$$

on the half-line $0 \le x < \infty$, where $q(x)$ and h are real, $h \ne \infty$, and also the unperturbed problem

$$-y'' = \lambda y, \qquad y(0) = 1, \qquad y'(0) = 0. \tag{3.3}$$

Let $\rho(\lambda)$ be the spectral function of (3.1), (3.2), $\rho_0(\lambda)$ that of (3.3), which is explicitly computable, viz. $\rho_0(\lambda) = \frac{2}{\pi}\sqrt{\lambda}$, $\lambda \ge 0$.

Denote the solution of (3.1), (3.2) by $\varphi(x, \lambda)$. The solution of (3.3) is $\cos\sqrt{\lambda}x$. Therefore, we obtain the formulas

$$\varphi(x, \lambda) = \cos\sqrt{\lambda}x + \int_0^x K(x, t)\cos\sqrt{\lambda}t\, dt \tag{3.4}$$

$$\cos\sqrt{\lambda}x = \varphi(x, \lambda) + \int_0^x H(x, t)\varphi(t, \lambda)\, dt \tag{3.5}$$

via the transformation operator for the pair of operators $-d^2/dx^2 + q(x)$ and $-d^2/dx^2$, where $K(x, t)$ and $H(x, t)$ are two continuously differentiable functions.

6.3.2

We first give a non-rigorous discussion of the fundamental integral equation of the inverse problem, and justify the derivation in the sequel. The procedure is based on the symbolic equalities

$$\int_{-\infty}^{\infty} \varphi(x, \lambda)\varphi(s, \lambda)\, d\rho(\lambda) = \delta(x - s) \tag{3.6}$$

$$\int_0^{\infty} \cos\sqrt{\lambda}x \cdot \cos\sqrt{\lambda}s\, d\rho_0(\lambda) = \delta(x - s), \tag{3.7}$$

where $\delta(x)$ is the Dirac delta function.

Note that (3.6), (3.7) are the symbolic notation of the generalized Parseval equations

$$\int_0^{\infty} f(x)g(x)\, dx = \int_{-\infty}^{\infty} F(\lambda)G(\lambda)\, d\rho(\lambda)$$

for the problems (3.1), (3.2) and (3.3), respectively (see the formula (1.11), Chapter 2).

It follows from the formulas (3.6) and (3.5) that, for $0 < s < x$,

$$\int_{-\infty}^{\infty} \varphi(x, \lambda) \cos \sqrt{\lambda} s \, d\rho(\lambda) = 0.$$

Substituting the right-hand side of the equality (3.4) for $\varphi(x, \lambda)$, we obtain for $0 < s < x$

$$\begin{aligned}
0 &= \int_{-\infty}^{\infty} \left[\cos \sqrt{\lambda} x \cos \sqrt{\lambda} s + \int_0^x K(x, t) \cos \sqrt{\lambda} t \cos \sqrt{\lambda} s \, dt \right] d\rho(\lambda) \\
&= \int_{-\infty}^{\infty} \cos \sqrt{\lambda} x \cos \sqrt{\lambda} s \, d\sigma(\lambda) + \int_0^{\infty} \cos \sqrt{\lambda} x \cos \sqrt{\lambda} s \, d\rho_0(\lambda) \\
&\quad + \int_0^x K(x, t) \left[\int_{-\infty}^{\infty} \cos \sqrt{\lambda} t \cos \sqrt{\lambda} s \, d\sigma(\lambda) \right] dt \\
&\quad + \int_0^x K(x, t) \left[\int_0^{\infty} \cos \sqrt{\lambda} t \cos \sqrt{\lambda} s \, d\rho_0(\lambda) \right] dt \\
&= F(x, s) + \delta(x - s) + \int_0^x K(x, t) F(t, s) \, dt + K(x, s) \\
&= F(x, s) + K(x, s) + \int_0^x K(x, t) F(t, s) \, dt,
\end{aligned}$$

where

$$\sigma(\lambda) = \begin{cases} \rho(\lambda) - \rho_0(\lambda), & \lambda \geq 0 \\ \rho(\lambda), & \lambda < 0 \end{cases}$$

$$F(x, s) = \int_{-\infty}^{\infty} \cos \sqrt{\lambda} x \cdot \cos \sqrt{\lambda} s \, d\sigma(\lambda). \tag{3.8}$$

We have thus shown (not rigorously for the present) that the kernel $K(x, t)$ of the transformation operator satisfies the integral equation

$$F(x, s) + K(x, s) + \int_0^x K(x, t) F(t, s) \, dt = 0, \quad 0 \leq s \leq x, \tag{3.9}$$

where $F(x, s)$ is determined by the formula (3.8). (3.9) is a Fredholm integral equation for each fixed x, in which $K(x, s)$ is an unknown function, and x a parameter.

The above derivation of (3.9) is, of course, non-rigorous, at least because it is not clear for the time being whether the integral (3.8) exists in any sense at all.

Lemma 6.3.1 *Let*

$$F_N(x, y) = \int_{-\infty}^{N} \cos \sqrt{\lambda} x \cos \sqrt{\lambda} y \, d\sigma(\lambda).$$

Fix x, and denote by $g(y)$ a smooth function with bounded support contained in the interval $(0, x)$. Then

$$\lim_{N \to \infty} \int_0^x F_N(x, y) g(y) \, dy + \int_0^x K(x, y) g(y) \, dy$$

$$+ \lim_{N \to \infty} \int_0^{\infty} g(y) \left\{ \int_0^x K(x, t) F_N(t, y) \, dt \right\} dy = 0.$$

Proof. For each smooth function with bounded support, by Theorem 2.1.2,

$$\lim_{N \to \infty} \int_{-\infty}^{N} \varphi(x, \lambda) \left\{ \int_{0}^{\infty} g(y) \varphi(y, \lambda) \, dy \right\} d\rho(\lambda)$$

$$= \lim_{N \to \infty} \int_{0}^{\infty} g(y) \left\{ \int_{-\infty}^{N} \varphi(x, \lambda) \varphi(y, \lambda) \, d\rho(\lambda) \right\} dy = g(x).$$

Since the support of $g(y)$ is contained in $(0, x)$ for fixed x, and $g(y)$ is continuous, $g(x) = 0$. Therefore,

$$\lim_{N \to \infty} \int_{0}^{\infty} g(y) \left\{ \int_{-\infty}^{N} \varphi(x, \lambda) \varphi(y, \lambda) \, d\rho(\lambda) \right\} dy = 0,$$

and, relying on the formula (3.5),

$$\lim_{N \to \infty} \int_{0}^{\infty} g(y) \left\{ \int_{-\infty}^{N} \varphi(x, \lambda) \cos \sqrt{\lambda} y \, d\rho(\lambda) \right\} dy$$

$$= \lim_{N \to \infty} \int_{0}^{\infty} g(y) \left\{ \int_{-\infty}^{N} \varphi(x, \lambda) \varphi(y, \lambda) \, d\rho(\lambda) \right\} dy$$

$$+ \lim_{N \to \infty} \int_{0}^{\infty} g(y) \left\{ \int_{0}^{y} H(y, t) \left[\int_{-\infty}^{N} \varphi(x, \lambda) \varphi(t, \lambda) \, d\rho(\lambda) \right] dt \right\} dy$$

$$+ \lim_{N \to \infty} \int_{0}^{\infty} \left\{ \int_{t}^{\infty} H(y, t) g(y) \, dy \int_{-\infty}^{N} \varphi(x, \lambda) \varphi(t, \lambda) \, d\rho(\lambda) \right\} dt$$

$$= \int_{x}^{\infty} H(y, x) g(x) \, dy = 0,$$

since $H(x, y) = 0$ for $y > x$.

Furthermore, using (3.4), we have that

$$\int_{-\infty}^{N} \varphi(x, \lambda) \cos \sqrt{\lambda} y \, d\rho(\lambda) = \int_{-\infty}^{N} \cos \sqrt{\lambda} x \cos \sqrt{\lambda} y \, d\rho(\lambda)$$

$$+ \int_{0}^{x} K(x, t) \left\{ \int_{-\infty}^{N} \cos \sqrt{\lambda} t \cos \sqrt{\lambda} y \, d\rho(\lambda) \right\} dt$$

$$= \int_{-\infty}^{N} \cos \sqrt{\lambda} x \cos \sqrt{\lambda} y \, d\rho_0(\lambda)$$

$$+ \int_{-\infty}^{N} \cos \sqrt{\lambda} x \cos \sqrt{\lambda} y \, d\sigma(\lambda)$$

$$+ \int_{0}^{x} K(x, t) \left\{ \int_{-\infty}^{N} \cos \sqrt{\lambda} x \cos \sqrt{\lambda} y \, d\rho_0(\lambda) \right\} dt$$

$$+ \int_{0}^{x} K(x, t) \left\{ \int_{-\infty}^{N} \cos \sqrt{\lambda} x \cos \sqrt{\lambda} y \, d\sigma(\lambda) \right\} dt.$$

Multiplying throughout by $g(y)$, integrating by y and letting $N \to \infty$, we have

$$\lim_{N \to \infty} \int_0^\infty g(y) F_N(x,y)\, dy + \int_0^x K(x,y) g(y)\, dy$$

$$+ \lim_{N \to \infty} \int_0^\infty g(y) \left\{ \int_0^x K(x,t) F_N(t,y)\, dt \right\} dy = 0$$

as required.   ∎

It follows from the lemma that (3.9) necessarily holds if the sequence of functions $F_N(x,y)$ tends to the limit $F(x,y)$ as $N \to \infty$ at each finite point (x,y), remaining bounded in each bounded domain of the (x,y)-plane (i.e., converges boundedly). As we shall see below, this is always so. Note for the present that examples of the functions $\rho(\lambda)$ can be easily found, with uniform convergence (and, therefore, boundedly convergent) in each bounded domain of the (x,y)-plane.

Lemma 6.3.2 *The sequence of $F_N(x,y)$ converges boundedly.*

Proof. It is known (see B. M. Levitan and I. S. Sargsjan (1975; Chapter 7, Section 4)) that the sequence of functions

$$\Phi_N(x,y) = \int_{-\infty}^N \varphi(x,\lambda)\varphi(y,\lambda)\, d\rho(\lambda) - \int_0^N \cos\sqrt{\lambda}x \cos\sqrt{\lambda}y\, d\rho_0(\lambda) \qquad (3.10)$$

converges boundedly. Writing the formula (3.5) as $\cos\sqrt{\lambda}x = (I_x + H_x)\varphi$, applying the operator $(I_x + H_x)(I_y + H_y)$, we obtain

$$\int_{-\infty}^N \cos\sqrt{\lambda}x \cos\sqrt{\lambda}y\, d\rho(\lambda) - \int_0^N (I_x + H_x)(I_y + H_y)\cos\sqrt{\lambda}x \cos\sqrt{\lambda}y\, d\rho_0(\lambda)$$

$$= F_N(x,y) - \int_0^N \cos\sqrt{\lambda}x \left\{ \int_0^y H(y,s)\cos\sqrt{\lambda}s\, ds \right\} d\rho_0(\lambda)$$

$$- \int_0^N \cos\sqrt{\lambda}y \left\{ \int_0^x H(x,s)\cos\sqrt{\lambda}s\, ds \right\} d\rho_0(\lambda)$$

$$- \int_0^N \left\{ \int_0^x H(x,s)\cos\sqrt{\lambda}s\, ds \right\} \left\{ \int_0^y H(y,s)\cos\sqrt{\lambda}s\, ds \right\} d\rho_0(\lambda). \quad (3.11)$$

We have already noted that the left-hand side of the above identity converges boundedly. Therefore, the same also holds for the right-hand side, whose second, third and fourth addends, however, converge boundedly to

$$H(y,x), \qquad H(x,y), \qquad \int_0^{\min(x,y)} H(x,s)H(y,s)\, ds,$$

respectively, as the usual Fourier integrals of differentiable functions. Consequently, the first addend on the right in (3.11), i.e., the sequence of $F_N(x,y)$, converges boundedly. The proof is now complete.   ∎

Theorem 6.3.1 *The kernel of the transformation operator $K(x,y)$ satisfies the integral equation*

$$K(x,y) + F(x,y) + \int_0^x K(x,t)F(t,y)\,dt = 0, \quad 0 \le y \le x, \qquad (3.12)$$

where $F(x,y)$ is the limit of

$$F_N(x,y) = \int_{-\infty}^N \cos\sqrt{\lambda}x \cos\sqrt{\lambda}y\, d\sigma(\lambda)$$

as $N \to \infty$, and

$$\sigma(\lambda) = \begin{cases} \rho(\lambda) - (2/\pi)\sqrt{\lambda}, & \lambda \ge 0 \\ \rho(\lambda), & \lambda < 0. \end{cases}$$

If $q(x)$ has continuous derivatives up to order n on the interval $[0,\infty)$, then the function $F(x,y)$ has continuous partial derivatives up to order $n+1$ in the domain $x \ge y \ge 0$.

Proof. The first part has already been proved. It now remains to prove the differentiability proposition.

It is not hard to derive from the integral equation (2.20) that if $q(x)$ has derivatives up to order n, then the function $K(x,y)$ has partial derivatives up to order $n+1$ for $x \ge y$. We now assume that $K(x,y)$ is known in (3.12), a Volterra integral equation with respect to $F(x,y)$, which can be solved by the method of successive approximations. The series for $F(x,y)$, so formed, can be differentiated $n+1$ times. ∎

Remark Put

$$\Phi_N(x) = \int_{-\infty}^N \cos\sqrt{\lambda}x\, d\sigma(x), \qquad \Phi(x) = \lim_{N\to\infty} \Phi_N(x).$$

Then, for $x \ge y \ge 0$,

$$F(x,y) = \frac{1}{2}\{\Phi(x+y) - \Phi(x-y)\}, \qquad (3.13)$$

$$F(x,x) = \frac{1}{2}\{\Phi(2x) + \Phi(0)\}. \qquad (3.14)$$

It follows from the latter that if $F(x,y)$ has continuous partial derivatives up to order $n+1$ for $x \ge y$, then $\Phi(x)$ has continuous derivatives up to order $n+1$ for $x \ge 0$. The converse follows from (3.13).

6.3.3

The following simple property of the spectral function $\rho(\lambda)$ will be useful in the sequel.

Theorem 6.3.2 *Let $f(x)$ be an arbitrary function with bounded support, and of class $\mathcal{L}^2(0, \infty)$. Put*

$$E(\lambda) = \int_0^\infty f(x) \cos \sqrt{\lambda} x \, dx.$$

If

$$\int_{-\infty}^\infty E^2(\lambda) \, d\rho(\lambda) = 0, \tag{3.15}$$

then $f(x) \equiv 0$.

Proof. It follows from the formula (3.5) that

$$E(\lambda) = \int_0^\infty f(x) \left\{ \varphi(x, \lambda) + \int_0^x H(x, t)\varphi(t, \lambda) \, dt \right\} dx = \int_0^\infty g(x)\varphi(x, \lambda) \, dx,$$

$$g(x) = f(x) + \int_x^\infty H(t, x)f(t) \, dt.$$

Therefore, $E(\lambda)$ is the Fourier φ-transform of the function $g(x)$. It follows from the Parseval equation

$$\int_0^\infty g^2(x) \, dx = \int_{-\infty}^\infty E^2(\lambda) \, d\rho(\lambda)$$

and (3.15) that $\int_0^\infty g^2(x) \, dx = 0$; hence, $g(x) \equiv 0$, or

$$f(x) + \int_x^\infty H(t, x)f(t) \, dt = 0. \tag{3.16}$$

Because $f(x)$ is a function with bounded support, the integral is, in fact, with finite limits; therefore, (3.16) is a Volterra integral equation, and $f(x) \equiv 0$.   ∎

6.4   Solvability of the basic integral equation

6.4.1

It has been shown in the preceding section that if we want a monotonically increasing function $\rho(\lambda)$ to be the spectral function of the problem (3.1), (3.2) with real-valued and n times differentiable $q(x)$, it is necessary that the following conditions should hold :

1. Put

$$\sigma(\lambda) = \begin{cases} \rho(\lambda) - (2/\pi)\sqrt{\lambda}, & \lambda \geq 0 \\ \rho(\lambda), & \lambda < 0 \end{cases}$$

$$\Phi_N(x) = \int_{-\infty}^{N} \cos \sqrt{\lambda} x \, d\sigma(\lambda).$$

The sequence of the functions $\Phi_N(x)$ as $N \to \infty$ converges boundedly to a function $\Phi(x)$ with continuous derivatives up to order $n + 1$ on the interval $[0, \infty]$.

2. Let $f(x)$ be an arbitrary function with bounded support, of class $\mathcal{L}^2(0, \infty)$, and

$$E(\lambda) = \int_0^{\infty} f(x) \cos \sqrt{\lambda} x \, dx.$$

If $\int_{-\infty}^{\infty} E^2(\lambda) \, d\rho(\lambda) = 0$, then $f(x) \equiv 0$.

We show here and in the subsequent section that Conditions 1 and 2 are not only necessary, but also sufficient for a boundary-value problem of the form (3.1), (3.2) to exist with an n times continuously differentiable function $q(x)$.

Thus, let $\rho(\lambda)$ be a monotonically increasing function fulfilling Conditions 1 and 2. Put

$$\Phi(x) = \lim_{N \to \infty} \Phi_N(x), \qquad F(x, y) = \frac{1}{2}\{\Phi(x + y) + \Phi(x - y)\},$$

and consider the family of integral equations

$$K(x, y) + F(x, y) + \int_0^x K(x, t)F(t, y) \, dt = 0, \quad 0 \leq y \leq x < \infty. \tag{4.1}$$

Theorem 6.4.1 *For any fixed $x > 0$, the integral equation (4.1) has a unique solution $K(x, y) \equiv K_x(y)$.*

Proof. It should be shown that, for each fixed $x > 0$, the homogeneous equation

$$g(y) + \int_0^x F(t, y)g(t) \, dt = 0 \tag{4.2}$$

only has the trivial solution. Assume the contrary, i.e., that there exists a function $g(y) \in \mathcal{L}^2(0, \infty)$, $g(y) \not\equiv 0$, satisfying (4.2). Multiplying throughout by $g(y)$ and integrating with respect to y from 0 to x, we obtain

$$\int_0^x g^2(y) \, dy + \int_0^x \int_0^x F(t, y)g(t)g(y) \, dt \, dy = 0.$$

Substituting the corresponding limit for $F(t, y)$ and interchanging the integrals and the limit (which can easily be justified), we get

$$\int_0^x g^2(y)\, dy + \int_{-\infty}^{\infty} \left\{ \int_0^x \int_0^x g(t)g(y) \cos \sqrt{\lambda}t \cos \sqrt{\lambda}y \, dt\, dy \right\} d\rho(\lambda)$$
$$- \int_0^{\infty} \left\{ \int_0^x \int_0^x g(t)g(y) \cos \sqrt{\lambda}t \cos \sqrt{\lambda}y \, dt\, dy \right\} d\left(\frac{2}{\pi}\sqrt{\lambda} \right) = 0. \quad (4.3)$$

By the Parseval equation for the usual Fourier integrals, the last integral is $\int_0^x g^2(y)\, dy$. Therefore, (4.3) can be written as

$$\int_{-\infty}^{\infty} \left\{ \int_0^x g(y) \cos \sqrt{\lambda}y \, dy \right\}^2 d\rho(\lambda) = 0,$$

and, by Condition 2, $g(y) \equiv 0$ as required. ∎

6.4.2

It follows immediately from the integral equation (4.1) that the function $K(x, y)$ has as many partial derivatives with respect to y as $F(x, y)$. To examine the differentiability of $K(x, y)$ with respect to x is more complicated, but can be done with the help of the following.

Lemma 6.4.1 *Given the integral equation*

$$h(t, a) + \int_0^1 H(t, s; a)h(s, a)\, ds = g(t; a), \quad (4.4)$$

where the kernel $H(t, s; a)$ and the absolute term $g(t; a)$ are continuous functions of the parameter a and of the independent variable t.

Then, if, for $a = a_0$, the homogeneous equation only has the trivial solution, the solution $h(t, a)$ of (4.4) is a continuous function of t and a in a certain neighborhood of a_0. If $H(t, s; a)$ and $g(t; a)$ have m continuous derivatives with respect to a, then $h(t, a)$ also has as many continuous derivatives with respect to a.

Proof. Put

$$H(t, s; a) = H(t, s; a_0) + H_1(t, s; a) \equiv H_0 + H_1,$$

with $|H_1(t, s; a)| < \epsilon$ if a is in a small neighborhood of the point a_0. (4.4) can be written symbolically as

$$g = h + H_0 h + H_1 h.$$

Applying the operator $(I + H_0)^{-1}$ to both sides, we obtain

$$(I + H_0)^{-1} g = h + (I + H_0)^{-1} H_1 h. \quad (4.5)$$

Since the norm of $(I + H_0)^{-1} H_1$ can be made as small as we please, the equation (4.5) can be solved by the method of successive approximations; hence, the continuity of the function $h(t, a)$ follows. To prove the differentiability, a Taylor series segment should be distinguished from the function $H(t, s; a)$, or $H = H_0 + H_1$ put, where

$$H_0 = H(t, s; a_0) + \left.\frac{\partial H}{\partial a}\right|_{a=a_0} (a - a_0) + \ldots + \frac{1}{m!} \left.\frac{\partial^m H}{\partial a^m}\right|_{a=a_0} (a - a_0)^m.$$

We now apply the proved lemma to (4.1), replace y with yx, t with xt, and obtain

$$K(x, xy) + F(x, xy) + x \int_0^1 K(x, xt) F(xt, xy)\, dt = 0,$$

i.e., the integral equation with kernel $xF(xt, xy)$ and absolute term $F(x, xy)$. Since $F(x, y)$ is a continuous function, $K(x, y)$ is continuous as a function of two variables. It also follows that $K(x, y)$ has as many partial derivatives as $F(x, y)$. ∎

6.5   Derivation of the differential equation

The function $K(x, y)$ was determined in the previous section from a given monotone function $\rho(\lambda)$ satisfying Conditions 1 and 2. Similarly to the formula (3.4), we put

$$\varphi(x, \lambda) = \cos\sqrt{\lambda}x + \int_0^x K(x, t) \cos\sqrt{\lambda}t\, dt, \tag{5.1}$$

which satisfies the initial conditions

$$\varphi(0, \lambda) = 1, \qquad \varphi'(0, \lambda) = K(0, 0) = -F(0, 0) = h. \tag{5.2}$$

The following contains an essential part of the solution of the inverse problem from the spectral function.

Theorem 6.5.1 *Let $K(x, y)$ satisfy (4.1). Then the function $\varphi(x, \lambda)$, determined by the formula (5.1), satisfies the differential equation*

$$-y'' + q(x)y = \lambda y, \quad 0 \le x < \infty, \tag{5.3}$$

where

$$q(x) = 2\frac{d}{dx} K(x, x). \tag{5.4}$$

Proof. We first assume that the function $F(x, y)$ has continuous second partial derivatives (which is associated with the existence of the first derivative of the function $q(x)$ in the direct problem). Then the function $K(x, y)$ also has continuous second derivatives. Differentiating the identity

$$J(x, y) = F(x, y) + K(x, y) + \int_0^x K(x, t) F(t, y)\, dt \tag{5.5}$$

twice with respect to x, $0 \le y \le x < \infty$, we obtain

$$J''_{xx} = F''_{xx} + K''_{xx} + \frac{dK(x,x)}{dx}F + K(x,x)F'_x$$

$$+ \frac{\partial K(x,t)}{\partial x}\Big|_{t=x}F + \int_0^x K''_{xx}(x,t)F(t,y)\,dt = 0. \quad (5.6)$$

Differentiating (5.5) with respect to y, we get

$$J'_y = F'_y + K'_y + \int_0^x K(x,t)F'_y(t,y)\,dt = 0, \quad (5.7)$$

$$J''_{yy} = F''_{yy} + K''_{yy} + \int_0^x K(x,t)F''_{yy}(t,y)\,dt = 0. \quad (5.8)$$

Since $F(x,y) = \frac{1}{2}\{\Phi(x+y) + \Phi(x-y)\}$, we have $F'_y(x,0) = 0$, $F''_{yy} = F''_{xx}$. It then follows from (5.7) for $y = 0$ that

$$K'_y(x,0) = 0. \quad (5.9)$$

Replacing F''_{yy} with F''_{tt} in the identity (5.8) and then integrating by parts twice, we get

$$J''_{yy} = F''_{yy} + K''_{yy} + K(x,x)F'_x - \frac{\partial K(x,t)}{\partial t}\Big|_{t=x} F + \int_0^x K''_{tt}(x,t)F(t,y)\,dt. \quad (5.10)$$

After obvious eliminations, it follows from (5.5), (5.6), (5.10) and (5.4) that

$$J''_{xx} - J''_{yy} - q(x)J = K''_{xx} - K''_{yy} - q(x)K$$

$$+ \int_0^x \{K''_{xx}(x,t) - K''_{tt}(x,t) - q(x)K(x,t)\}F(t,y)\,dt = 0 \quad (5.11)$$

for $0 \le y \le x < \infty$, i.e., the function $K''_{xx}(x,y) - K''_{yy}(x,y) - q(x)K(x,y)$ satisfies the homogeneous integral equation. By Theorem 6.4.1,

$$K''_{xx} - K''_{yy} - qK = 0, \quad 0 \le y \le x < \infty. \quad (5.12)$$

Since

$$\frac{d}{dx}K(x,x) = \frac{1}{2}q(x), \qquad K(0,0) = h, \qquad \frac{\partial K}{\partial y}\Big|_{y=0} = 0, \quad (5.13)$$

$Xf(x) = f(x) + \int_0^x K(x,t)f(t)\,dt$ is the transformation operator for the pair $-d^2/dx^2 + q(x)$ and $-d^2/dx^2$ (see Section 2). Therefore, the function $\varphi(x,\lambda)$ determined by (5.1) satisfies (5.3) and the initial conditions $y(0) = 1$, $y'(0) = h = -F(0,0)$. Thus, if $F(x,y)$ has continuous second derivatives, then Theorem 6.5.1 is proved.

We now assume that only the first derivatives of $F(x,y)$ are continuous. Then only the first derivatives of $K(x,y)$ are continuous; therefore, the identity (5.5) can be differentiated only once.

To avoid this difficulty, we approximate the function $\Phi(x)$ uniformly in each finite interval by twice continuously differentiable functions $\Phi_n(x)$, so that, for any $N < \infty$,

$$\lim_{n \to \infty} \int_0^N |\Phi'_n(x) - \Phi'(x)|\, dx = 0.$$

We put

$$F_n(x, y) = \frac{1}{2}\{\Phi_n(x + y) + \Phi_n(x - y)\}.$$

Then, for each fixed x and sufficiently large n, the integral equation

$$F_n(x, y) + K_n(x, y) + \int_0^x K_n(x, t) F_n(t, y)\, dt = 0, \quad 0 \leq y \leq x < \infty,$$

is solvable, while

$$K_n(x, y) \to K(x, y), \quad \frac{\partial}{\partial x} K_n(x, y) \to \frac{\partial}{\partial x} K(x, y), \quad \frac{\partial}{\partial y} K_n(x, y) \to \frac{\partial}{\partial y} K(x, y)$$

(see the proof of Lemma 6.4.1).

On the other hand, it has already been shown that the function

$$\varphi_n(x, \lambda) = \cos \sqrt{\lambda} x + \int_0^x K_n(x, t) \cos \sqrt{\lambda} t\, dt \tag{5.14}$$

satisfies the equation

$$y'' + \left\{\lambda - 2\frac{d}{dx} K_n(x, x)\right\} y = 0 \tag{5.15}$$

and the initial conditions

$$\varphi_n(0, \lambda) = 1, \qquad \varphi'_n(0, \lambda) = K_n(0, 0). \tag{5.16}$$

Passing to the limit in (5.14), (5.15) and (5.16), we obtain that the function (5.1) satisfies the equation

$$y'' + \left\{\lambda - 2\frac{d}{dx} K(x, x)\right\} y = 0$$

and the initial conditions $y(0) = 1$, $y'(0) = K(0, 0)$. ∎

6.6   Derivation of the Parseval equation

6.6.1

In Sections 4 and 5, we determined the equation of the form (5.3) and the number h from the boundary condition and given monotone function $\rho(\lambda)$ satisfying Conditions 1 and 2 of Section 4.

It now remains to see that the original function $\rho(\lambda)$ is, in fact, spectral for the boundary-value problem constructed; it suffices to show that the Parseval equation

$$\int_0^\infty f^2(x)\,dx = \int_{-\infty}^\infty F^2(\lambda)\,d\rho(\lambda) \tag{6.1}$$

is valid for any smooth function $f(x)$ with bounded support, where

$$F(\lambda) = \int_0^\infty f(x)\varphi(x,\lambda)\,dx. \tag{6.2}$$

6.6.2

Let $F(x,y)$ be determined by the formula (3.8), let $K(x,y)$ be the solution of the equation (4.1), and

$$\varphi(x,\lambda) = \cos\sqrt{\lambda}x + \int_0^x K(x,t)\cos\sqrt{\lambda}t\,dt,$$

$$\cos\sqrt{\lambda}x = \varphi(x,\lambda) + \int_0^x H(x,t)\varphi(x,t)\,dt.$$

Lemma 6.6.1 *The identity*

$$H(x,t) = F(x,t) + \int_0^t K(t,u)F(x,u)\,du, \quad 0 \le t \le x, \tag{6.3}$$

holds.

Proof. We first assume that $F(x,t)$ has continuous second derivatives. Then $K(x,t)$ also has continuous second derivatives, and satisfies the equation (5.12)

$$K''_{xx} - q(x)K = K''_{tt} \tag{6.4}$$

and the conditions (5.13)

$$\left.\frac{\partial K}{\partial t}\right|_{t=0} = 0, \qquad K(x,x) = h + \frac{1}{2}\int_0^x q(s)ds. \tag{6.5}$$

The function $H(x,t)$ is the kernel of the transformation operator $X_{B,A}$, where

$$Bf = -\frac{d^2 f}{dx^2}, \qquad f'(0) = 0,$$

$$Af = -\frac{d^2 f}{dx^2} + q(x)f, \qquad f'(0) = hf(0).$$

Therefore, $H(x,t)$ satisfies the equation

$$H''_{xx} = H''_{tt} - q(t)H \tag{6.6}$$

(see Section 2) and the conditions

$$(H'_t - hH)_{t=0} = 0, \qquad H(x,x) = -h - \frac{1}{2}\int_0^x q(s)\,ds = -K(x,x). \qquad (6.7)$$

Thus, to prove the lemma, it should be shown that the right-hand side of (6.3) is a solution of the problem (6.6), (6.7). Put

$$N(x,t) = F(x,t) + \int_0^1 K(t,u)F(x,u)\,du.$$

Then $N(x,0) = F(x,0)$; since $F'_t(x,0) = 0$,

$$N'_t(x,0) = F'_t(x,0) + K(0,0)F(x,0) = hN(x,0).$$

Therefore, $N(x,t)$ satisfies the first condition in (6.7). We show that $N(x,t)$ also satisfies the second. In fact, by (4.1),

$$N(x,x) = F(x,x) + \int_0^x K(x,u)F(x,u)\,du = -K(x,x).$$

It remains to show that $N(x,t)$ satisfies the equation (6.6). We have

$$N''_{tt} = F''_{tt} + \frac{dK(t,t)}{dt}F + K(t,t)F'_t + K'_t|_{u=t}F + \int_0^t K''_{tt}(t,u)F(x,u)\,du,$$

$$N''_{xx} = F''_{xx} + K(t,t)F'_u|_{u=t} - K'_u|_{u=t}F + \int_0^t K''_{uu}(t,u)F(x,u)\,du.$$

Subtracting the upper equation from the lower and taking (5.4) into account, we obtain

$$
\begin{aligned}
N''_{xx} - N''_{tt} &= -q(t)F(x,t) + \int_0^t (K''_{uu} - K''_{tt})F(x,u)\,du \\
&= -q(t)N(x,t) + q(t)\int_0^t K(t,u)F(x,u)\,du + \int_0^t (K''_{uu} - K''_{tt})F(x,u)\,du,
\end{aligned}
$$

or, by (6.4),

$$N''_{xx} - N''_{yy} + q(t)N = \int_0^t [K''_{uu} - K''_{tt} + q(t)K]\,F(x,u)\,du = 0.$$

Thus, in the case where (x,t) has second derivatives, the lemma is proved. The case of only first derivatives can be examined by a limiting procedure (as in Section 5). ∎

6.6.3

We now prove the Parseval equation. Let $f(x)$ be an arbitrary function with bounded support, and

$$F(\lambda) = \int_0^\infty f(x)\varphi(x,\lambda)\,dx = \int_0^\infty f(t)\left\{\cos\sqrt{\lambda}t + \int_0^t K(t,s)\cos\sqrt{\lambda}s\,ds\right\}dt$$

$$= \int_0^\infty g(t)\cos\sqrt{\lambda}t\,dt, \quad (6.8)$$

where

$$g(t) = f(t) + \int_t^\infty K(s,t)f(s)\,ds. \quad (6.9)$$

The function $g(t)$ also has a bounded support, and is smooth. The equality (6.8) means that $F(\lambda)$ is the Fourier cosine transform of $g(t)$. Therefore, by (3.8),

$$\int_{-\infty}^\infty F^2(\lambda)\,d\rho(\lambda) = \int_0^\infty F^2(\lambda)d\left(\frac{2}{\pi}\sqrt{\lambda}\right) + \int_{-\infty}^\infty F^2(\lambda)\,d\sigma(\lambda)$$

$$= \int_0^\infty g^2(t)\,dt$$

$$+ \int_{-\infty}^\infty d\sigma(\lambda)\left\{\int_0^\infty g(t)\cos\sqrt{\lambda}t\,dt\right\}\left\{\int_0^\infty g(x)\cos\sqrt{\lambda}x\,dx\right\}$$

$$= \int_0^\infty g^2(t)\,dt + \int_0^\infty\int_0^\infty F(x,t)g(x)g(t)\,dx\,dt.$$

Furthermore, it follows from the equations (6.3) and (4.1) that

$$\int_0^\infty F(x,t)g(t)\,dt = \int_0^\infty F(x,t)\left\{f(t) + \int_t^\infty K(u,t)f(u)\,du\right\}dt$$

$$= \int_0^\infty f(t)\left\{F(x,t) + \int_0^t K(t,u)F(x,u)\,du\right\}dt$$

$$= \int_0^x f(t)\left\{F(x,t) + \int_0^t K(t,u)F(x,u)\,du\right\}dt$$

$$+ \int_x^\infty f(t)\left\{F(x,t) + \int_0^t K(t,u)F(x,u)\,du\right\}dt$$

$$= \int_0^x f(t)H(x,t)\,dt - \int_x^\infty f(t)K(t,x)\,dt.$$

Therefore,

$$\int_{-\infty}^\infty F^2(\lambda)\,d\rho(\lambda) = \int_0^\infty g^2(t)\,dt + \int_0^\infty g(x)\left\{\int_0^x H(x,t)f(t)\,dt\right\}dx$$

$$- \int_0^\infty g(x)\left\{\int_x^\infty K(t,x)f(t)\,dt\right\}dx$$

$$= \int_0^\infty g^2(t)\,dt + \int_0^\infty f(t)\left\{\int_t^\infty H(x,t)g(x)\,dx\right\}dt$$

$$- \int_0^\infty g(x)\left\{\int_x^\infty K(t,x)f(t)\,dt\right\}dx. \quad (6.10)$$

By (6.9),

$$\int_x^\infty K(x,t)f(t)\,dt = g(x) - f(x). \tag{6.11}$$

To show that

$$\int_0^\infty H(x,t)g(x)\,dx = f(t) - g(t), \tag{6.12}$$

we put

$$f(t) + \int_0^t K(t,s)f(s)\,ds = (I+K)f.$$

By definition, $(I+K)^{-1} = I+H$; therefore, $(I+K^*)^{-1} = I+H^*$, where K^*, H^* are two adjoint operators. The equality (6.9) can be written in the form $g = f + K^*f$; therefore, $f = (I+K^*)^{-1}g = g + H^*g$, which coincides with (6.12). It follows from (6.10), (6.11) and (6.12) that

$$\int_{-\infty}^\infty F^2(\lambda)\,d\rho(\lambda) = \int_0^\infty g^2(t)\,dt + \int_0^\infty f(t)\{f(t) - g(t)\}\,dt$$

$$- \int_0^\infty g(x)\{g(x) - f(x)\}\,dx = \int_0^\infty f^2(t)\,dt,$$

i.e., the Parseval equation as required.

The main results of Sections 4 to 6, i.e., the solution of the inverse problem for the Sturm-Liouville operator from the spectral function $\rho(\lambda)$, are formulated as the following.

Theorem 6.6.1 *For the monotonically increasing function $\rho(\lambda)$ to be the spectral function of a problem of the form (3.1), (3.2) with an n times continuously differentiable function $q(x)$, it is necessary and sufficient that :*

1. *The sequence of*

$$\Phi_N(x) = \int_{-\infty}^N \cos\sqrt{\lambda}x\,d\sigma(\lambda),$$

   where

$$\sigma(\lambda) = \begin{cases} \rho(\lambda) - 2\sqrt{\lambda}/\pi, & \lambda \geq 0 \\ \rho(\lambda), & \lambda < 0 \end{cases}$$

   should converge boundedly (in each finite interval) to the limit function $\Phi(x)$, with continuous derivatives up to order $n+1$ on the interval $[0,\infty)$, and $\Phi(+0) = -h$.

2. *It follows from the equality*

$$\int_{-\infty}^\infty E^2(\lambda)d\rho(\lambda) = 0$$

   that $f(x) \equiv 0$, where $E(\lambda)$ is the Fourier cosine transform of an arbitrary smooth function $f(x)$ with bounded support.

Remark It is obvious that Condition 2 holds if the points of increase of the function $\rho(\lambda)$ have at least one finite limit point.

For the discrete spectrum, Condition 2 can be replaced by an efficient sufficient condition. Let $\lambda_1 < \lambda_2 < \ldots < \lambda_n < \ldots \to \infty$ be the points of increase of $\rho(\lambda)$, and let $N(\lambda)$ the number of those λ_n which are less than λ. It can be shown that Condition 2 holds if

$$\lim_{k \to \infty} \sup \left(N(\lambda)/\lambda \right) = 0.$$

6.7 Generalization of the basic integral equation

6.7.1

Until now, while studying the inverse problem, we assumed that the simplest Sturm-Liouville operator

$$ly = -y'', \qquad y'(0) = 0, \quad 0 \le x < \infty,$$

is unperturbed. However, the results obtained in the preceding sections can be generalized easily by taking an arbitrary Sturm-Liouville operator A_0,

$$A_0 y = -y'' + q_0(x)y, \qquad y'(0) = h_0 y(0), \quad 0 \le x < \infty, \tag{7.1}$$

as unperturbed, where $q_0(x)$ is a fixed real-valued continuous function (or having continuous derivatives), and h_0 is a real number.

Let A be another real-valued Sturm-Liouville operator

$$Ay = -y'' + q(x)y, \qquad y'(0) = hy(0), \quad 0 \le x < \infty. \tag{7.2}$$

Let the solution of the problem

$$-y'' + q_0(x)y = \lambda y, \qquad \varphi_0(0, \lambda) = 1, \qquad \varphi_0'(0, \lambda) = h_0,$$

be $\varphi_0(x, \lambda)$, and that of the problem

$$-y'' + q(x)y = \lambda y, \qquad \varphi(0, \lambda) = 1, \qquad \varphi'(0, \lambda) = h,$$

$\varphi(x, \lambda)$.

It was shown in Section 2 that the formulas

$$\varphi(x, \lambda) = \varphi_0(x, \lambda) + \int_0^x A(x, t)\varphi_0(t, \lambda) \, dt \tag{7.3}$$

$$\varphi_0(x, \lambda) = \varphi(x, \lambda) + \int_0^x A_0(x, t)\varphi(t, \lambda) \, dt \tag{7.4}$$

are valid. Suppose that $\rho_0(\lambda)$ and $\rho(\lambda)$ are the spectral functions of the operators A_0 and A. The Parseval equations for A_0 and A are equivalent to the symbolic formulas

$$\int_{-\infty}^{\infty} \varphi_0(x, \lambda)\varphi_0(y, \lambda) \, d\rho_0(\lambda) = \delta(x - y) \tag{7.5}$$

$$\int_{-\infty}^{\infty} \varphi(x, \lambda)\varphi(y, \lambda) \, d\rho(\lambda) = \delta(x - y), \tag{7.6}$$

where $\delta(x)$ is the Dirac delta function.

Using (7.3), (7.4), (7.5), (7.6) and reasoning as in Section 3 (first formally), it is not hard to derive the integral equation

$$A(x, y) + F(x, y) + \int_0^x A(x, t)F(t, y) \, dt = 0 \tag{7.7}$$

for the kernel $A(x, y)$, $0 \le y \le x < \infty$, where

$$F(x, y) = \int_{-\infty}^{\infty} \varphi_0(x, \lambda)\varphi_0(y, \lambda) d\{\rho(\lambda) - \rho_0(\lambda)\}. \tag{7.8}$$

The rigorous proof of (7.7) follows from the bounded convergence of the sequence of functions

$$F_N(x) = \int_{-\infty}^{\infty} \varphi_0(x, \lambda)\varphi_0(y, \lambda) \, d\{\rho(\lambda) - \rho_0(\lambda)\}, \tag{7.9}$$

which, in turn, follows from the identity

$$F_N(x, y) = \int_{-\infty}^{N} \varphi_n(x, \lambda)\varphi_0(y, \lambda) d\left\{\rho(\lambda) - \frac{2}{\pi}\sqrt{\lambda}\right\}$$
$$- \int_{-\infty}^{N} \varphi_0(x, \lambda)\varphi_0(y, \lambda) \, d\left\{\rho_0(\lambda) - \frac{2}{\pi}\sqrt{\lambda}\right\}$$

and from the formula

$$\varphi_0(x, \lambda) = \cos\sqrt{\lambda}x + \int_0^x K_0(x, t) \cos\sqrt{\lambda}t \, dt.$$

6.7.2

It is now clear how to solve the inverse problem in case (7.1) is taken as unperturbed, i.e., an arbitrary Sturm-Liouville operator, on the half-line. We formulate the corresponding result.

Let a monotonically increasing function $\rho(\lambda)$ be given, satisfying the conditions :

1. *The sequence (7.9) converges boundedly, and the limit function $F(x, y)$ has $n + 1$ continuous partial derivatives for $x \ge y \ge 0$.*

2. *It follows from the equality*

$$\int_{-\infty}^{\infty} E^2(\lambda) \, d\rho(\lambda) = 0,$$

where

$$E(\lambda) = \int_0^{\infty} f(x)\varphi_0(x, \lambda) \, dx$$

and $f(x)$ is an arbitrary smooth function with bounded support, that $f(x) \equiv 0$.

Consider (7.7), where $F(x,y)$ was determined in Condition 1. It follows from Condition 2 that (7.7) is solvable for each $x < \infty$.

Furthermore, it can be shown similarly to Section 5 that the solution of (7.7), the function $A(x,y)$, satisfies the differential equation

$$A''_{xx} - q(x)A = A''_{yy} - q_0(y)A, \tag{7.10}$$

where

$$q(x) = q_0(x) + 2dA(x,x)/dx. \tag{7.11}$$

For $x = y = 0$, we derive from (7.7) that $A(0,0) = -F(0,0)$; therefore,

$$A(x,x) = -F(0,0) + \frac{1}{2}\int_0^x \{q(s) - q_0(s)\}\, ds.$$

Another consequence of (7.7) is the condition

$$(A'_t - h_0 A)_{t=0} = 0. \tag{7.12}$$

Put

$$\varphi(x,\lambda) = \varphi_0(x,\lambda) + \int_0^x A(x,t)\varphi_0(x,t)\, dt.$$

It follows from (7.10) and (7.12) that $\varphi(x,\lambda)$ satisfies the equation $-y'' + q(x)y = \lambda y$ and the boundary condition $\varphi'(0,\lambda) = h\varphi(0,\lambda)$ where $h = h_0 - F(0,0)$.

That $\rho(\lambda)$ is, in fact, the spectral function of the problem constructed follows from the Parseval equation which can be derived on the basis of

$$B(x,t) = F(x,t) + \int_0^t A(t,u)F(x,u)du$$

by an argument similar to Section 6.

6.8   The case of the zero boundary condition

The zero boundary condition is not covered by the preceding results. Consider the problem

$$-y'' + q(x)y = \lambda y \tag{8.1}$$
$$y(0) = 0 \tag{8.2}$$

on the half-line $[0, \infty)$, where $q(x)$ is a real-valued continuous function.

Let $s(x,\lambda)$ be the solution of the equation (8.1), satisfying the initial conditions $s(0,\lambda) = 0$, $s'(0,\lambda) = 1$. Denote the spectral function associated with the problem (8.1), (8.2) by $\rho(\lambda)$, $-\infty < \lambda < \infty$, i.e., a monotonically increasing function such that, for any smooth function $f(x)$ with bounded support, the Parseval equation

$$\int_0^\infty f^2(x)\, dx = \int_{-\infty}^\infty F^2(\lambda)\, d\rho(\lambda), \qquad F(\lambda) = \int_0^\infty f(x)s(x,\lambda)\, dx$$

holds. In this case, the unperturbed problem is of the form

$$-y'' = \lambda y, \qquad y(0) = 0,$$

for which

$$s_0(x, \lambda) = \frac{\sin \sqrt{\lambda} x}{\sqrt{\lambda}}, \qquad \rho_0(\lambda) = 2\lambda^{3/2}/(3\pi), \quad \lambda \geq 0, \qquad \rho_0(\lambda) = 0, \quad \lambda < 0.$$

The transformation operator for the zero boundary condition was considered in Section 2 (Case 2). It follows from its properties that

$$s(x, \lambda) = \frac{\sin \sqrt{\lambda} x}{\sqrt{\lambda}} + \int_0^x N(x, t) \frac{\sin \sqrt{\lambda} t}{\sqrt{\lambda}} \, dt.$$

Repeating the formal derivation of Section 3, it is not difficult to obtain the integral equation

$$F(x, y) + N(x, y) + \int_0^\infty N(x, t) F(t, y) \, dt = 0, \quad 0 \leq y \leq x < \infty \qquad (8.3)$$

for the kernel $N(x, t)$, where

$$F(x, y) = \int_{-\infty}^\infty \frac{\sin \sqrt{\lambda} x}{\sqrt{\lambda}} \frac{\sin \sqrt{\lambda} y}{\sqrt{\lambda}} \, d\sigma(\lambda)$$

$$\sigma(\lambda) = \begin{cases} \rho(\lambda) - 2\lambda^{3/2}/(3\pi), & \lambda \geq 0 \\ \rho(\lambda), & \lambda < 0. \end{cases} \qquad (8.4)$$

If the sequence of functions

$$\Phi_N(x) = \int_{-\infty}^N \frac{\cos \sqrt{\lambda} x - 1}{\lambda} \, d\sigma(\lambda) \qquad (8.5)$$

converges boundedly, then (8.3) is, in fact, valid.

Conversely, if a monotonically increasing function $\rho(\lambda)$ is given with the conditions :

1. *The sequence in (8.5) converges boundedly*

2. *it follows from the equality* $\int_{-\infty}^\infty E^2(\lambda) d\rho(\lambda) = 0$, *where*

$$E(\lambda) = \int_0^\infty f(x) \frac{\sin \sqrt{\lambda} x}{\sqrt{\lambda}} \, dx$$

and f is a smooth function with compact support, that $f(x) \equiv 0$,

then the results of the preceding section on the solvability of the inverse problem can be extended obviously to the case of the zero boundary condition.

6.9   Reconstructing the classical problem

Let $q(x)$ be a real-valued function, continuous on the interval $[0, \pi]$, and let h, H be two real numbers.

The *classical Sturm-Liouville problem* (with separated boundary conditions) is

$$
\begin{aligned}
-y'' + q(x)y &= \lambda y \quad (0 \leq x \leq \pi) & (9.1) \\
y'(0) - hy(0) &= 0 & (9.2) \\
y'(\pi) + Hy(\pi) &= 0. & (9.3)
\end{aligned}
$$

Let $\varphi(x, \lambda)$ be the solution of the equation (9.1), satisfying the initial conditions $\varphi(0, \lambda) = 1$, $\varphi'(0, \lambda) = h$. It is obvious that $\varphi(x, \lambda)$ satisfies the boundary condition (9.2) for any λ. Therefore, the eigenvalues of our problem are the roots λ_n of the equation

$$
\varphi'(\pi, \lambda) + H\varphi(\pi, \lambda) = 0,
$$

while the corresponding eigenfunction $\psi_n(x)$ is $\varphi(x, \lambda_n)$, $(n = 0, 1, 2, \ldots)$.

Put

$$
\alpha_n = \int_0^\pi \varphi^2(x, \lambda_n) \, dx.
$$

The values α_n are called *norming constants*.

It was proved in Chapter 1 that the eigenfunctions corresponding to different eigenvalues are orthogonal and that the Parseval equation

$$
\int_0^\pi f(x)g(x) \, dx = \sum_{n=0}^\infty \frac{1}{\alpha_n} \int_0^\pi f(x)\varphi(x, \lambda_n) \, dx \int_0^\pi g(x)\varphi(x, \lambda_n) \, dx \qquad (9.4)
$$

holds for arbitrary functions $f(x) \in \mathcal{L}^2(0, \pi)$ and $g(x) \in \mathcal{L}^2(0, \pi)$.

It can be seen that $\rho(\lambda)$ is a step function for the classical Sturm-Liouville problem, determined by the equality

$$
\rho(\lambda) = \sum_{\lambda_n < \lambda} \frac{1}{\alpha_n}. \qquad (9.5)
$$

Thus, the spectral function of the classical Sturm-Liouville problem is determined by two sequences of numbers $\{\lambda_n\}$ and $\{\alpha_n\}$. The latter consists of positive numbers, while only finitely many terms of the former can be negative. The set of quantities $\{\lambda_n, \alpha_n\}_{n=0}^\infty$ determining the spectral function $\rho(\lambda)$ in the formula (9.5) are called the *spectral characteristics* of the problem (9.1), (9.2), (9.3).

Repeating the argument of Sections 3 to 6, it is easy to obtain necessary and sufficient conditions for the classical inverse problem to be solvable. It is obvious that the spectral function of the simplest classical Sturm-Liouville problem

$$
-y'' = \lambda y, \quad y'(0) = y'(\pi) = 0, \quad 0 \leq x \leq \pi,
$$

then should be taken as the unperturbed spectral function $\rho_0(\lambda)$, and not $2\sqrt{\lambda}/\pi$.

For the above problem, $\rho_0(\lambda)$ is determined by

$$\rho_0(\lambda) = \begin{cases} \dfrac{1}{\pi} + \displaystyle\sum_{\lambda_n < \lambda} \dfrac{2}{\pi}, & \lambda \geq 0 \\[2mm] 0, & \lambda < 0 \end{cases}.$$

We now state, without proof, three theorems which can be proved in a similar way as those in Sections 3 to 6.

Theorem 6.9.1 *Let $\{\lambda_n, \alpha_n\}_{n=0}^{\infty}$ be the spectral characteristics of the problem (9.1), (9.2), (9.3), and let the function $q(x)$ possess m continuous derivatives on the interval $[0, \pi]$. We put*

$$F(x, y) = \frac{1}{\alpha_0} \cos \sqrt{\lambda_0} x \cos \sqrt{\lambda_0} y$$

$$+ \sum_{n=1}^{\infty} \left\{ \frac{1}{\alpha_n} \cos \sqrt{\lambda_n} x \cos \sqrt{\lambda_n} y - \frac{2}{\pi} \cos nx \cos ny \right\} - \frac{1}{\pi}, \quad (9.6)$$

which has continuos partial derivatives for $0 \leq y \leq x \leq \pi$ up to order $m+1$ inclusive.

Theorem 6.9.2 *Let the sequence $\{\lambda_n, \alpha_n\}_{n=0}^{\infty}$, $\lambda_n \to \infty$, $\alpha_n > 0$, possess the properties :*

1. *$F(x, y)$ has continuous partial derivatives up to order $m + 1$ inclusive*

2. *it follows from the equalities*

$$\int_0^{\pi} f(x) \cos \sqrt{\lambda_n} x \, dx = 0, \quad n = 0, 1, 2, \ldots,$$

   where $f(x)$ is an arbitrary smooth function, that $f(x) \equiv 0$.

If Conditions 1 and 2 are fulfilled, then the integral equation

$$F(x, y) + K(x, y) + \int_0^x K(x, t) F(t, y) \, dt = 0, \quad 0 \leq y \leq x \leq \pi, \quad (9.7)$$

has a unique solution, and the function $K(x, y)$ has continuous partial derivatives up to order $m + 1$ inclusive for $0 \leq y \leq x \leq \pi$. Furthermore, the function

$$\varphi(x, \lambda) = \cos \sqrt{\lambda} x + \int_0^x K(x, t) \cos \sqrt{\lambda} t \, dt \quad (9.8)$$

satisfies the equation

$$-y'' + 2\frac{d}{dx} K(x, x) y = \lambda y, \quad 0 \leq x \leq \pi,$$

and the initial conditions

$$y(0) = 1, \quad y'(0) = h = K(0, 0) = -F(0, 0).$$

Remark We know (see Chapter 1, the formula (2.12)) that if $q(x) \in \mathcal{L}(0, \pi)$, then $\sqrt{\lambda_n} = n + o(1)$. Therefore, the set of functions $\{\cos \sqrt{\lambda_n} x\}_{n=0}^{\infty}$ is known[1] to be complete in the space $C(0, \pi)$ and, as a consequence, Condition 2 is necessarily fulfilled.

Theorem 6.9.3 *With the conditions of Theorem 6.9.2 for arbitrary $f(x) \in \mathcal{L}^2(0, \pi)$, $g(x) \in \mathcal{L}^2(0, \pi)$, the Parseval equation (9.4) holds. The function $\varphi(x, \lambda)$ is determined by the equation (9.8).*

Theorem 6.9.4 *Let $\sqrt{\lambda_n} = n + \delta_n$, $\sum_{n=0}^{\infty} \delta_n^2 < \infty$, $\alpha_n = (2/\pi) + o(1)$, and let Conditions 1 and 2 of Theorem 6.9.2 be fulfilled. Then the functions $\varphi(x, \lambda_n)$ are pairwise orthogonal and*

$$\int_0^{\pi} \varphi^2(x, \lambda_n)\, dx = \alpha_n^2.$$

Proof. Let the series

$$\sum_{n=0}^{\infty} c_n \varphi(x, \lambda_n),$$

where

$$c_n = \frac{1}{\alpha_n} \int_0^{\pi} f(x) \varphi(x, \lambda_n)\, dx,$$

converge uniformly. We then show that its sum is $f(x)$.

Let it be $s(x)$. It follows from the Parseval equation (9.4) and the assumed uniform convergence that

$$\int_0^{\pi} f(x) g(x)\, dx = \int_0^{\infty} s(x) g(x)\, dx,$$

and, since $g(x)$ is an arbitrary function, $s(x) = f(x)$.

In particular, let $f(x) = \varphi(x, \lambda_k)$ for fixed k. Since $\varphi(x, \lambda)$ satisfies (9.1) and the boundary condition (9.2), it follows from Green's formula that

$$
\begin{aligned}
c_n(k) &= \frac{1}{\alpha_n} \int_0^{\pi} \varphi(x, \lambda_k) \varphi(x, \lambda_n)\, dx \\
&= \frac{\varphi'(\pi, \lambda_n) \varphi(\pi, \lambda_k) - \varphi'(\pi, \lambda_k) \varphi(\pi, \lambda_n)}{\alpha_n (\lambda_k - \lambda_n)}.
\end{aligned}
$$

It follows from (9.8) that

$$\varphi(\pi, \lambda_n) = O(1), \qquad \varphi'(\pi, \lambda_n) = O(\sqrt{\lambda_n} \delta_n).$$

Therefore,

$$c_n(k) = O\left(\frac{\delta_n}{\sqrt{\lambda_n}}\right) + O(\lambda_n^{-1});$$

---

[1] B. Ya. Levin (1980; p. 547).

hence, the series $\sum_{n=0}^{\infty} c_n(k)\varphi(x,\lambda_n)$ converges absolutely and uniformly. We have proved that

$$\varphi(x,\lambda_k) = \sum_{n=0}^{\infty} c_n(k)\varphi(x,\lambda_n). \qquad (9.9)$$

The set of functions $\cos\sqrt{\lambda_n}x$ is known to be minimal (strongly linearly independent)[2]. By the formula (9.8), the set of $\varphi(x,\lambda_n)$ is also minimal. It therefore follows from (9.9) that

$$\int_0^\pi \varphi(x,\lambda_k)\varphi(x,\lambda_n)\,dx = 0, \quad n \neq k$$
$$\int_0^\pi \varphi^2(x,\lambda_k)\,dx = \alpha_k$$

as required.   ∎

Theorem 6.9.5 *The ratio $\varphi'(\pi,\lambda_n)/\varphi(\pi,\lambda_n)$, $n = 0,1,2,\ldots$, does not depend on n with the conditions of Theorem 6.9.4, i.e., the functions $\varphi(x,\lambda_n)$ satisfy a classical boundary condition of the form (9.3).*

Proof. It follows from the established orthogonality of the functions $\varphi(x,\lambda_n)$ and Green's identity that

$$\varphi'(\pi,\lambda_n)\varphi(\pi,\lambda_m) - \varphi'(\pi,\lambda_m)\varphi(\pi,\lambda_n) = 0 \qquad (9.10)$$

for $n \neq m$. If, for certain n, $\varphi(\pi,\lambda_n) = 0$, it follows from the above equality that $\varphi(\pi,\lambda_m) = 0$ for all m. Let $\varphi(\pi,\lambda_n) \neq 0$. It then follows from (9.10) that

$$\frac{\varphi'(\pi,\lambda_n)}{\varphi(\pi,\lambda_n)} = \frac{\varphi'(\pi,\lambda_m)}{\varphi(\pi,\lambda_m)} = C,$$

which proves the lemma.   ∎

Theorem 6.9.6 *For the sequences $\{\lambda_n\}_{n=0}^{\infty}$ and $\{\alpha_n\}_{n=0}^{\infty}$ to be the spectral characteristics of a boundary-value problem of the form (9.1), (9.2), (9.3) with infinitely differentiable function $q(x)$, it is necessary and sufficient that the infinite classical asymptotic expansions*

$$\sqrt{\lambda_n} = n + \frac{a_0}{n} + \frac{a_1}{n^3} + \ldots, \quad \alpha_n = \frac{\pi}{2} + \frac{b_0}{n^2} + \frac{b_1}{n^4} + \ldots \qquad (9.11)$$

hold.

---

[2] B. Ya. Levin (1980; p. 547).

Proof. The necessity easily follows from the results of Section 2, Chapter 1.

By Theorem 6.9.2, to prove the sufficiency, we must show that the function

$$s(x) = \sum_{n=1}^{\infty} \left\{ \frac{1}{\alpha_n} \cos \sqrt{\lambda_n} x - \frac{2}{\pi} \cos nx \right\}$$

is infinitely differentiable on the interval $0 \le x \le 2\pi$.

Let N be an arbitrary natural number. If we substitute the asymptotic expansions (9.11) in the expression for $s(x)$, series of functions are distinguished, which can be N times differentiated termwise, and, on the other hand, products of polynomials in x, i.e., series of the form

$$\sum_{n=1}^{\infty} \frac{\sin nx}{n^{2k+1}} \qquad (2k + 1 \le N + 2)$$

$$\sum_{n=1}^{\infty} \frac{\cos nx}{n^{2k}} \qquad (2k < N + 2).$$

Thus, it remains to show that the sums of these series are infinitely differentiable on $0 \le x \le 2\pi$.

Differentiating the first series $2k$ times, and the second $2k - 1$ times, we obtain $\sum_{n=1}^{\infty} \frac{\sin nx}{n}$, up to the sign, whose sum is well known to be $(\pi - x)/2$ on the interval $0 < x < 2\pi$. However, this function is obviously infinitely differentiable on $0 \le x \le 2\pi$. ∎

6.10   Inverse periodic problem

6.10.1

Here, we assume that the function $q(x)$ in the equation

$$-y'' + q(x)y = \lambda y, \quad 0 \le x \le \pi, \tag{10.1}$$

is real-valued, smooth and periodic (with period π). Consider the three boundary-value problems

A. $y(0) = y(\pi)$, $y'(0) = y'(\pi)$,

B. $y(0) = -y(\pi)$, $y'(0) = -y'(\pi)$,

C. $y(0) = y(\pi) = 0$,

for (10.1).

The first, $\{(10.1) + A\}$, is said to be *periodic*; the second, $\{(10.1) + B\}$, *semi-periodic*; and, finally, the third, $\{(10.1) + C\}$, a *fixed-endpoint* problem.

Let $\varphi(x, \lambda)$ and $\vartheta(x, \lambda)$ be the solutions of (10.1), satisfying the initial conditions

$$\varphi(0, \lambda) = \vartheta'(0, \lambda) = 0, \qquad \varphi'(0, \lambda) = \vartheta(0, \lambda) = 1.$$

Furthermore, put

$$F(\lambda) = \varphi'(\pi, \lambda) + \vartheta(\pi, \lambda).$$

This is an entire analytic exponential-type function of order $1/2$, said to be the *Liapunov function*.

Denote the eigenvalues of the problem $\{(10.1) + A\}$ by $\lambda_0 < \lambda_2 \leq \mu_2 < \lambda_4 \leq \mu_4 < \ldots$, those of $\{(10.1) + B\}$ by $\lambda_1 \leq \mu_1 < \lambda_3 \leq \mu_3 < \ldots$ and those of $\{(10.1) + C\}$ by $\nu_1 < \nu_2 < \nu_3 < \ldots$. We know that these numbers can be ordered as

$$\lambda_0 < \lambda_1 \leq \nu_1 \leq \mu_1 < \lambda_2 \leq \nu_2 \leq \mu_2 < \ldots . \tag{10.2}$$

It is not hard to show that the eigenvalues of Problem A satisfy the equation $F(\lambda) = 2$; those of Problem B, $F(\lambda) = -2$; and those of Problem C, $\varphi(\pi, \lambda) = 0$ (see Chapter 1, Section 4). Moreover, the asymptotic expansions

$$\sqrt{\lambda_n} = n + \frac{a_0}{n} + \frac{a_1}{n^3} + \ldots, \qquad \sqrt{\mu_n} = n + \frac{a_0}{n} + \frac{a_1}{n^3} + \ldots \tag{10.3}$$

hold, with the number of exact terms depending on the order of smoothness of $q(x)$. Since the coefficients in the asymptotic series coincide, it follows from (10.2) that the values ν_n have the same asymptotic behaviour (10.3).

6.10.2

We now consider the question of reconstructing equation (10.1), i.e., the function $q(x)$, from the Liapunov function.

Thus, let the Liapunov function be given. It suffices to specify either the sequence of $\lambda_0, \lambda_2, \mu_2, \lambda_4, \mu_4, \ldots$ or $\lambda_1, \mu_1, \lambda_3, \mu_3, \ldots$, for the expansions

$$F(\lambda) - 2 = \pi(\lambda_0 - \lambda) \prod_{n=1}^{\infty} \frac{\lambda_{2n} - \lambda}{(2n)^2} \cdot \frac{\mu_{2n} - \lambda}{(2n)^2}$$

$$F(\lambda) + 2 = \pi \prod_{n=1}^{\infty} \frac{\lambda_{2n+1} - \lambda}{(2n+1)^2} \cdot \frac{\mu_{2n+1} - \lambda}{(2n+1)^2}$$

are valid.

The question arises : How can an equation of the form (10.1) be constructed, whose Liapunov function coincides with a given function $F(\lambda)$? We see that there are infinitely many such equations (or $q(x)$).

The intervals $[\lambda_k, \mu_k]$ are called *gaps*. Select an arbitrary point in each, viz. $\lambda_k \leq \nu_k \leq \mu_k$. The sequence of ν_k has the asymptotic behaviour (10.3), and, therefore, can be the spectrum of the fixed-endpoint problem (if the boundary conditions of condition C are of the form $y(0) = 0$, $y'(\pi) + Hy(\pi) = 0$, $H \neq \infty$, then the principal term of the asymptotic expansion of ν_n is $(n + (1/2))^2)$.

To determine an equation of the form (10.1) uniquely, the norming constants α_n should also be specified. We show that they can be determined (but not uniquely) from a spectrum ν_n and the Liapunov function, and we derive the formula.

Let λ, $\mu \neq \lambda$, be two numbers, and let $\varphi(x, \lambda) = y$, $\varphi(x, \mu) = z$. The equations

$$-y'' + q(x)y = \lambda y, \qquad -z'' + q(x)z = \mu z$$

then hold. Multiplying the former throughout by z, the latter by y, we subtract the first from the second, and obtain $(y'z - z'y)' = (\mu - \lambda)yz$. Integrating, we have

$$\int_0^\pi \varphi(x, \lambda)\varphi(x, \mu)\,dx = \frac{\varphi'(\pi, \lambda)\varphi(\pi, \mu) - \varphi(\pi, \lambda)\varphi'(\pi, \mu)}{\mu - \lambda},$$

and get

$$\int_0^\pi \varphi^2(x, \lambda)\,dx = \varphi'(\pi, \lambda)\dot\varphi(\pi, \lambda) - \varphi(\pi, \lambda)\dot\varphi'(\pi, \lambda)$$

as we pass to the limit when $\mu \to \lambda$, where the prime denotes differentiation with respect to x, and the dot with respect to λ. It follows that

$$\alpha_n = \int_0^\pi \varphi^2(x, \nu_n)\,dx = \varphi'(\pi, \nu_n)\dot\varphi(\pi, \nu_n) \qquad (10.4)$$

for $\lambda = \nu_n$. Since $\varphi(x, \lambda) = \pi \prod_{n=1}^\infty \frac{\nu_n - \lambda}{n^2}$, given the spectrum ν_n, the values of $\dot\varphi(\pi, \nu_n)$ are also known. It therefore follows from (10.4) that, given the spectrum, it suffices to know the quantities $\varphi'(x, \nu_n)$ to determine α_n. We show that $\varphi'(\pi, \nu_n)$ are determined (but not uniquely) from the Liapunov function. By its definition,

$$F(\nu_n) = \varphi'(\pi, \nu_n) + \vartheta(\pi, \nu_n). \qquad (10.5)$$

Since the Wronskian is constant,

$$\varphi'(\pi, \nu_n)\vartheta(\pi, \nu_n) - \varphi(\pi, \nu_n)\vartheta'(\pi, \nu_n) = \varphi'(\pi, \nu_n)\vartheta(\pi, \nu_n) = 1. \qquad (10.6)$$

It follows from the two equations that $\varphi'(\pi, \nu_n)$ and $\vartheta(\pi, \nu_n)$ are the roots of the quadratic equation $z^2 - F(\nu_n)z + 1 = 0$. Therefore,

$$\varphi'(\pi, \nu_n) = \frac{1}{2}F(\nu_n) \pm \frac{1}{2}\sqrt{F^2(\nu_n) - 4}, \qquad (10.7)$$

the sign for the given Liapunov function $F(\lambda)$ (and not the equation) being arbitrary.

Thus, for any choice of ν_n in the gaps and of signs in (10.7), an equation of the form (10.1) holds. It remains to show that the Liapunov function $\tilde F(\lambda)$ of the equation so formed coincides with the original $F(\lambda)$.

Along with (10.7),

$$\varphi'(\pi, \nu_n) = \frac{1}{2}\tilde{F}(\nu_n) \pm \sqrt{\tilde{F}^2(\nu_n) - 4} \qquad (10.7')$$

holds. It follows from (10.7) that

$$\left[\varphi'(\pi, \nu_n) - \frac{1}{2}F(\nu_n)\right]^2 = \frac{1}{4}F^2(\nu_n) - 1,$$

whence $F(\nu_n) = \{\varphi'^2(\pi, \nu_n) + 1\}/\varphi'(\pi, \nu_n)$. Similarly, we derive from (10.7') that $\tilde{F}(\nu_n) = \{\varphi'^2(\pi, \nu_n) + 1\}/\varphi'(\pi, \nu_n)$ and

$$\tilde{F}(\nu_n) = F(\nu_n), \quad n = 1, 2, \ldots . \qquad (10.8)$$

Put $F(\lambda) = 2\cos\sqrt{\lambda}\pi + f(\lambda)$, $\tilde{F}(\lambda) = 2\cos\sqrt{\lambda}\pi + f(\lambda)$, where $f(\lambda)$ and $\tilde{f}(\lambda)$ are two entire analytic functions such that

$$f(\lambda) = O(e^{\pi\sqrt{|\lambda|}}/\sqrt{|\lambda|}), \qquad \tilde{f}(\lambda) = O(e^{\pi\sqrt{|\lambda|}}/\sqrt{|\lambda|}).$$

It follows from (10.8) that $f(\nu_n) = \tilde{f}(\nu_n)$. Therefore, by the familiar interpolation theorems[3], $\tilde{f}(\lambda) \equiv f(\lambda)$ with the consequence that $\tilde{F}(\lambda) \equiv F(\lambda)$.

Note in conclusion that the results of this section cannot be considered as the complete solution to the Sturm-Liouville inverse periodic problem, for which a satisfactory description of the Liapunov function set should be given[4].

6.11   Determination of the regular operator from two spectra

6.11.1   The derivation of the formula for the norming constants

Consider the differential equation

$$-y'' + q(x)y = \lambda y, \qquad (11.1)$$

where $q(x)$, $0 \le x \le \pi$, is a real-valued continuous function. Subjoin the boundary conditions

$$y'(0) - h_1 y(0) = 0, \qquad y'(\pi) + H y(\pi) = 0 \qquad (11.2)$$

$$y'(0) - h_2 y(0) = 0, \qquad y'(\pi) + H y(\pi) = 0 \qquad (11.3)$$

where h_1, h_2 and H are three real numbers, $h_1 \ne h_2$. Let $\lambda_0 < \lambda_1 < \lambda_2 < \ldots$, $\mu_0 < \mu_1 < \mu_2 < \ldots$ be the eigenvalues of the problems (11.1), (11.2) and (11.1),

---

[3]See, e.g., B. Ya. Levin (1959).
[4]This description is given by V. A. Marchenko and I. V. Ostrovsky (1975).

(11.3), respectively, and let $\varphi(x, \lambda)$, $\psi(x, \lambda)$ be the solutions of (11.1), satisfying the initial conditions

$$\varphi(0, \lambda) = 1, \qquad \varphi'(0, \lambda) = h_1, \qquad \psi(0, \lambda) = 1, \qquad \psi'(0, \lambda) = h_2.$$

The eigenvalues of the boundary-value problems (11.1), (11.2) and (11.1), (11.3) coincide with the zeros of

$$\Phi_1(\lambda) = \varphi'(\pi, \lambda) + H\varphi(\pi, \lambda)$$

$$\Phi_2(\lambda) = \psi'(\pi, \lambda) + H\psi(\pi, \lambda) \tag{11.4}$$

respectively. It is obvious that $\varphi(x, \lambda_n) \equiv \varphi_n(x)$ is the eigenfunction of the boundary-value problem (11.1), (11.2). As before, we call $\alpha_n = \int_0^\pi \varphi_n^2(x)\, dx$ a *norming constant*. Here, we derive a formula to determine α_n from the two given spectra $\{\lambda_n\}_{n=0}^\infty$, $\{\mu_n\}_{n=0}^\infty$ of the problems (11.1), (11.2) and (11.1), (11.3). The solution of the inverse problem from two spectra is thereby reduced to the inverse problem from the spectral function, which we already discussed.

Put $f(x, \lambda) = \psi(x, \lambda) + m(\lambda)\varphi(x, \lambda)$, and require that $f'(\pi, \lambda) + Hf(\pi, \lambda) = 0$, with the consequence that

$$m(\lambda) = -\frac{\psi'(\pi, \lambda) + H\psi(\pi, \lambda)}{\varphi'(\pi, \lambda) + H\varphi(\pi, \lambda)} = -\frac{\Phi_2(\lambda)}{\Phi_1(\lambda)}.$$

Hence, it can be seen that $m(\lambda)$ is a meromorphic function with poles and zeros coinciding with the eigenvalues of the problems (11.1), (11.2) and (11.1), (11.3), respectively. On the other hand, it follows from Green's formula that

$$(\lambda - \mu) \int_0^\pi f(x, \lambda) f(x, \mu)\, dx = f'(0, \lambda) f(0, \mu) - f(0, \lambda) f'(0, \mu)$$

$$= (h_1 - h_2)\{m(\lambda) - m(\mu)\}.$$

Putting $\mu = \bar{\lambda}$, we obtain

$$\int_0^\pi |f(x, \lambda)|^2\, dx = (h_2 - h_1)\frac{\operatorname{Im} m(\lambda)}{\operatorname{Im} \lambda}.$$

It follows that if $h_1 > h_2$ (resp. $h_1 < h_2$), the function $m(\lambda)$ maps the upper half-plane onto itself (resp. onto the lower half-plane). Therefore, the zeros and poles of $m(\lambda)$, i.e., the eigenvalues of the problems (11.1), (11.2) and (11.1), (11.3), interlace. Applying Green's formula again, we obtain

$$(\lambda - \lambda_n) \int_0^\pi f(x, \lambda)\varphi(x, \lambda_n)\, dx = f'(0, \lambda)\varphi(0, \lambda_n) - f(0, \lambda)\varphi'(0, \lambda_n) = h_2 - h_1.$$

On the other hand,

$$(\lambda - \lambda_n) \int_0^\pi f(x, \lambda)\varphi(x, \lambda_n)\, dx = (\lambda - \lambda_n) \int_0^\pi \psi(x, \lambda)\varphi(x, \lambda_n)\, dx$$
$$-(\lambda - \lambda_n)\frac{\Phi_2(\lambda)}{\Phi_1(\lambda)} \int_0^\pi \varphi(x, \lambda)\varphi(x, \lambda_n)\, dx = h_2 - h_1.$$

Letting $\lambda \to \lambda_n$, we get

$$\alpha_n = (h_1 - h_2)\Phi_1'(\lambda_n)/\Phi_2(\lambda_n), \tag{11.5}$$

which plays an important role in the sequel. Using the elementary properties of the functions $\Phi_1(\lambda)$ and $\Phi_2(\lambda)$, we prove that the formula

$$\alpha_n = \frac{h_2 - h_1}{\mu_n - \lambda_n} \prod_{k=0}^{\infty}{}' \frac{\lambda_k - \lambda_n}{\mu_k - \mu_n} \tag{11.6}$$

follows from (11.5), where the symbol $\prod'$ (as well as $\sum'$ below) means that the term indexed by n is omitted in the infinite product (resp. sum). It is well known that $\Phi_1(\lambda)$ and $\Phi_2(\lambda)$, determined by (11.4), are two entire analytic functions of order $1/2$ and, therefore, that they can also be determined by their zeros, viz.

$$\Phi_1(\lambda) = C_1 \prod_{k=0}^{\infty} \left(1 - \frac{\lambda}{\lambda_k}\right), \qquad \Phi_2(\lambda) = C_2 \prod_{k=0}^{\infty} \left(1 - \frac{\lambda}{\mu_k}\right),$$

up to constant multipliers, where C_1, C_2 are two constant numbers. It follows, also by (11.5), that

$$\alpha_n = \frac{h_2 - h_1}{\mu_n - \lambda_n} \frac{C_1}{C_2} \prod_{k=0}^{\infty} \frac{\mu_k}{\lambda_k} \prod_{k=0}^{\infty}{}' \frac{\lambda_k - \lambda_n}{\mu_k - \lambda_n}.$$

Therefore, to obtain (11.6), we have to show that

$$\frac{C_1}{C_2} \prod_{k=1}^{\infty} \frac{\mu_k}{\lambda_k} = 1. \tag{11.7}$$

Note that, by the asymptotic formulas for the solution of the equation (11.1) (see Section 2, Chapter 1), $\lim_{\lambda \to -\infty} \Phi_1(\lambda)/\Phi_2(\lambda) = 1$, i.e.,

$$\lim_{\lambda \to -\infty} \frac{C_1}{C_2} \prod_{k=0}^{\infty} \left(1 - \frac{\lambda}{\lambda_k}\right)\left(1 - \frac{\lambda}{\mu_k}\right)^{-1} = \frac{C_1}{C_2} \prod_{k=0}^{\infty} \frac{\mu_k}{\lambda_k} \lim_{\lambda \to -\infty} \prod_{k=0}^{\infty} \frac{\lambda_k - \lambda}{\mu_k - \lambda} = 1. \tag{11.8}$$

We now prove that

$$\lim_{\lambda \to -\infty} \prod_{k=0}^{\infty} \frac{\lambda_k - \lambda}{\mu_k - \lambda} = \lim_{\lambda \to -\infty} \left(1 + \frac{\lambda_k - \mu_k}{\mu_k - \lambda}\right) = 1. \tag{11.9}$$

It follows from the familiar asymptotic formulas for the eigenvalues of the Sturm-Liouville problem (see Section 2, Chapter 1) that $\lambda_k = k^2 + O(1)$, $\mu = k^2 + O(1)$. Therefore, $\lambda_k - \mu_k = O(1)$ and the series

$$\sum_{k=0}^{\infty} \left| \frac{\lambda_k - \mu_k}{\mu_k - \lambda} \right|$$

converges uniformly in the vicinity of $\lambda = -\infty$.

Hence, we can pass to the limit in each term of the infinite product (11.9), or

$$\lim_{\lambda \to -\infty} \prod_{k=0}^{\infty} \left(1 + \frac{\lambda_k - \mu_k}{\mu_k - \lambda} \right) = 1,$$

and (11.7) follows, also by (11.8).

6.11.2 The asymptotic formula for α_k

Suppose two sequences of the eigenvalues $\{\lambda_n\}_{n=0}^{\infty}$, $\{\mu_n\}_{n=0}^{\infty}$ of the problems (11.1), (11.2) and (11.1), (11.3) are given. If $q(x)$ is a sufficiently smooth (twice differentiable) function, then the asymptotic formulas

$$\sqrt{\lambda_n} = n + \frac{a_0}{n} + \frac{a_1}{n^3} + O\left(\frac{1}{n^4}\right),$$

$$\sqrt{\mu_n} = n + \frac{a_0'}{n} + \frac{a_1'}{n^3} + O\left(\frac{1}{n^4}\right)$$

(11.10)

are valid (see Section 2, Chapter 1), where

$$a_0 = \frac{1}{\pi} \left\{ h_1 + H + \frac{1}{2} \int_0^{\pi} q(s)\,ds \right\}, \qquad a_0' = \frac{1}{\pi} \left\{ h_2 + H + \frac{1}{2} \int_0^{\pi} q(s)\,ds \right\}$$

and, therefore,

$$a_0 - a_0' = \frac{1}{\pi} (h_2 - h_1) \neq 0. \tag{1.11}$$

Squaring (11.10), we obtain

$$\lambda_n = n^2 + 2a_0 + \frac{c_0}{n^2} + O\left(\frac{1}{n^3}\right)$$

$$\mu_n = n^2 + 2a_0' + \frac{c_0'}{n^2} + O\left(\frac{1}{n^3}\right)$$

(11.12)

with $c_0 = a_0^2 + 2a_1$, $c_0' = a_0'^2 + 2a_1'$.

To calculate the asymptotic behaviour of α_n, we use (11.6), noticing first of all that the difference $h_2 - h_1$ can be found on the basis of (11.11). To derive the asymptotic behaviour of α_n is not easy, and we break up the procedure into several stages.

(a) Consider the infinite product $\psi_n = \prod_{k=0}^{\infty}{}' \left(1 + \frac{\lambda_k - \mu_k}{\mu_k - \lambda_n}\right)$. Taking logs, we obtain

$$\ln \psi_n = \sum_{k=0}^{\infty}{}' \ln \left(1 + \frac{\lambda_k - \mu_k}{\mu_k - \lambda_n}\right).$$

It follows from the asymptotic formulas (11.12) that, for sufficiently large $n \neq k$, $|(\lambda_k - \mu_k)/(\mu_k - \lambda_n)| < C/n$ (here and onwards, C does not necessarily mean the same constant). Expanding the logs under the summation sign into power series, we get

$$\ln \psi_n = \sum_{k=0}^{\infty}{}' \sum_{p=1}^{\infty} \frac{(-1)^{p-1}}{p} \left(\frac{\lambda_k - \mu_k}{\mu_k - \lambda_n}\right)^p. \tag{11.13}$$

Lemma 6.11.1 *Let $|\mu_k - \lambda_k| \leq a$, $(k = 0, 1, 2, \ldots)$. Then*

$$\sum_{k=0}^{\infty}{}' \left|\frac{\lambda_k - \mu_k}{\mu_k - \lambda_n}\right| \leq \begin{cases} C \ln n/n, & p = 1 \\ \\ Ca^p/n^p, & p \geq 2 \end{cases}. \tag{11.14}$$

Proof. We have

$$\sum_{k=0}^{\infty}{}' \left|\frac{\lambda_k - \mu_k}{\mu_k - \lambda_n}\right|^p \leq a^p \sum_{k=0}^{\infty} \frac{1}{|\mu_k - \lambda_n|^p}.$$

It is not hard to see that the sum on the right is of the same order for large n, as

$$\int_0^{n-1} \frac{dx}{(\mu_n - x^2)^p} + \int_{n+1}^{\infty} \frac{dx}{(x^2 - \mu_n)^p}.$$

These integrals can be estimated as

$$\int_0^{n-1} \frac{dx}{(\mu_n - x^2)^p} \leq \frac{C}{n^p} \int_0^{n-1} \frac{dx}{(\sqrt{\mu_n} - x)^p} = \begin{cases} O(\ln n/n), & p = 1 \\ \\ O(1/n^p), & p > 1 \end{cases}$$

$$\int_{n+1}^{\infty} \frac{dx}{(x^2 - \mu_n)^p} \leq \frac{C}{n^p} \int_{n+1}^{\infty} \frac{dx}{(x - \sqrt{\mu_n})^p} = O\left(\frac{1}{n^p}\right), \quad p > 1$$

$$\int_{n+1}^{\infty} \frac{dx}{x^2 - \mu_n} = \frac{1}{2\sqrt{\mu_n}} \ln \frac{x - \sqrt{\mu_n}}{x + \sqrt{\mu_n}}\Big|_{x=n+1}^{\infty} = O\left(\frac{\ln n}{n}\right),$$

which proves the lemma.   ∎

We now continue studying $\ln \psi_n$. It follows from (11.14) that

$$
\begin{aligned}
\left| \sum_{k=0}^{\infty}{}' \sum_{p=3}^{\infty} \frac{(-1)^{p-1}}{p} \left(\frac{\lambda_k - \mu_k}{\mu_k - \lambda_n} \right)^p \right| &\leq \sum_{p=3}^{\infty} \frac{1}{p} \sum_{k=0}^{\infty}{}' \left| \frac{\lambda_k - \mu_k}{\mu_k - \lambda_n} \right|^p \\
&\leq C \sum_{p=3}^{\infty} \frac{a^p}{n^p} \\
&= C \frac{a^3}{n^3} \sum_{p=0}^{\infty} \frac{a^p}{n^p} \\
&= O \left(\frac{1}{n^3} \right).
\end{aligned}
$$

Therefore,

$$
\ln \psi_n = \sum_{k=0}^{\infty}{}' \frac{\lambda_k - \mu_k}{\mu_k - \lambda_n} - \frac{1}{2} \sum_{k=0}^{\infty}{}' \left(\frac{\lambda_k - \mu_k}{\mu_k - \lambda_n} \right)^2 + O \left(\frac{1}{n^3} \right). \tag{11.15}
$$

(b) So, we should study the asymptotic behaviour of the sums on the right, and represent the former as

$$
\begin{aligned}
\sum_{k=0}^{\infty}{}' \frac{\lambda_k - \mu_k}{\mu_k - \lambda_n} &= \frac{\lambda_0 - \mu_0}{\mu_0 - \lambda_n} + \sum_{k=1}^{\infty}{}' \frac{2(a_0 - a_0')}{\mu_k - \lambda_n} + \sum_{k=1}^{\infty} [(\lambda_k - \mu_k) - 2(a_0 - a_0')] \left[\frac{1}{\mu_k - \lambda_n} + \frac{1}{\lambda_n} \right] \\
&\quad - \frac{1}{\lambda_n} \sum_{k=1}^{\infty}{}' [(\lambda_k - \mu_k) - 2(a_0 - a_0')] = S_0 + 2(a_0 - a_0')S_1 + S_2 + S_3 \tag{11.16}
\end{aligned}
$$

with the obvious notation. It follows from (11.10) that

$$
S_0 = \frac{\lambda_0 - \mu_0}{\mu_0 - \lambda_n} = \frac{\mu_0 - \lambda_0}{\lambda_n(1 - \mu_0/\lambda_n)} = \frac{\mu_0 - \lambda_0}{n^2} + O \left(\frac{1}{n^4} \right)
$$

$$
S_3 = \frac{1}{\lambda_n} \sum_{k=1}^{\infty}{}' [(\mu_k - \lambda_k) - 2(a_0 - a_0')] - \frac{1}{\lambda_n} [(\mu_n - \lambda_n) - 2(a_0 - a_0')] = \frac{A}{n^2} + O \left(\frac{1}{n^4} \right),
$$

where $A = \sum_{k=1}^{\infty}{}' [\mu_k - \lambda_k - 2(a_0 - a_0')]$.

(c) We now consider

$$
S_2 = \sum_{k=1}^{\infty}{}' \frac{\mu_k[\lambda_k - \mu_k - 2(a_0 - a_0')] - (c_0 - c_0')}{\lambda_n(\mu_k - \lambda_n)} + \frac{c_0 - c_0'}{\lambda_n} \sum_{k=1}^{\infty}{}' \frac{1}{\mu_k - \lambda_n}.
$$

It follows from (11.12) that

$$
\mu_k[(\lambda_k - \mu_k) - 2(a_0 - a_0')] - (c_0 - c_0') = O(k^{-1});
$$

therefore,

$$\sum_{k=1}^{\infty}{}' \frac{\mu_k[(\lambda_k - \mu_k) - 2(a_0 - a_0')] - (c_0 - c_0')}{\lambda_n(\mu_k - \lambda_n)} \le \frac{C}{n^2}\sum_{k=1}^{\infty}{}'\frac{1}{k|\mu_k - \lambda_n|} = O\left(\frac{1}{n^3}\right),$$

and $S_2 = \dfrac{c_0 - c_0'}{\lambda_n}S_1 + O\left(\dfrac{1}{n^3}\right)$. It follows from the asymptotic formulas for S_0, S_2, S_3 and (11.16) that

$$\sum_{k=0}^{\infty}{}' \frac{\lambda_k - \mu_k}{\mu_k - \lambda_n} = \frac{A + \mu_0 - \lambda_0}{n^2} + 2(a_0 - a_0') + \frac{c_0 - c_0'}{\lambda_n}S_1 + O\left(\frac{1}{n^3}\right). \qquad (11.17)$$

We now study the behaviour of the sum S_1. It is obvious that

$$\begin{aligned}
S_1 &= \sum_{k=1}^{\infty}\frac{1}{\mu_k - \lambda_n} \\
&= \sum_{k=1}^{\infty}{}'\frac{1}{k^2 + 2a_0' + O(k^{-2}) - \lambda_n} \\
&= \sum_{k=1}^{\infty}{}'\frac{1}{(k^2 - \lambda_n)\{1 + 2a_0'/(k^2 - \lambda_n) + O[k^{-2}(k^2 - \lambda_n)^{-1}]\}} \\
&= \sum_{k=1}^{\infty}{}'\frac{1}{k^2 - \lambda_n} - 2a_0'\sum_{k=1}^{\infty}{}'\frac{1}{(k^2 - \lambda_n)^2} - \sum_{k=1}^{\infty}{}'\frac{1}{(k^2 - \lambda_n)^2 O(k^2)} \\
&\quad + \sum_{k=1}^{\infty}{}'\frac{1}{k^2 - \lambda_n}\frac{\left[\dfrac{2a_0'}{k^2 - \lambda_n} + O\left(\dfrac{1}{k^2(k^2 - \lambda_n)}\right)\right]^2}{1 + \dfrac{2a_0'}{k^2 - \lambda_n} + O\left(\dfrac{1}{k^2(k^2 - \lambda_n)}\right)},
\end{aligned}$$

where we have used the simple identity $\dfrac{1}{1+x} = 1 - x + \dfrac{x^2}{1+x}$.

Invoking the obvious estimate $\displaystyle\sum_{k=1}^{\infty}{}'\frac{1}{k^2(k^2 - \lambda_n)^2} = O(1/n^3)$ and Lemma 6.11.1, we obtain

$$S_1 = \sum_{k=1}^{\infty}{}'\frac{1}{k^2 - \lambda_n} - 2a_0\sum_{k=1}^{\infty}{}'\frac{1}{(k^2 - \lambda_n)^2} + O\left(\frac{1}{n^4}\right). \qquad (11.18)$$

(d) Consider the latter sum in the expansion (11.15). It is not difficult to derive from the asymptotic formulas (11.12) that

$$-\frac{1}{2}\sum_{k=1}^{\infty}{}'\left(\frac{\lambda_k - \mu_k}{\mu_k - \lambda_n}\right)^2 = -2(a_0 - a_0')\sum_{k=1}^{\infty}{}'\frac{1}{(k^2 - \lambda_n)^2} + O\left(\frac{1}{n^3}\right). \qquad (11.19)$$

It follows from (11.17), (11.18) and (11.19) that

$$\ln \Psi_n = \frac{A + \mu_0 - \lambda_n}{n^2} + \left[2(a_0 - a_0') + \frac{c_0 - c_0'}{n^2}\right] \sum_{k=1}^{\infty}{}' \frac{1}{k^2 - \lambda_n}$$

$$- [4a_0'(a_0 - a_0') + 2(a_0 - a_0')^2] \sum_{k=1}^{\infty} \frac{1}{(k^2 - \lambda_n)^2} + O\left(\frac{1}{n^3}\right). \quad (11.20)$$

To study the behaviour of $\sum_{k=1}^{\infty}{}' (k^2 - \lambda_n)^{-p}$, $p = 1, 2$, we consider the familiar formula

$$\sum_{k=1}^{\infty} \frac{1}{k^2 - \lambda} = -\left[\cot \pi\sqrt{\lambda} - \frac{1}{\pi\sqrt{\lambda}}\right] \frac{\lambda}{2\sqrt{\lambda}}, \quad (11.21)$$

and examine the case $p = 1$ in detail. We put $\sqrt{\lambda_n} = n + \epsilon$, where

$$\epsilon = \frac{a_0}{n} + \frac{a_1}{n^3} + O\left(\frac{1}{n^4}\right).$$

It follows from (11.21) that

$$\sum_{k=1}^{\infty}{}' \frac{1}{k^2 - \lambda_n} = \frac{1}{2\lambda_n} - \left[\frac{n}{2\sqrt{\lambda_n}} \cot \pi\sqrt{\lambda_n} + \frac{1}{n^2 - \lambda_n}\right]$$

$$= \frac{1}{2n^2} - \left[\frac{\pi \cot \epsilon}{2(n + \epsilon)} - \frac{1}{2\pi\epsilon + \epsilon^2}\right] + O\left(\frac{1}{n^4}\right)$$

$$= \frac{1}{2n^2} - \frac{\pi\epsilon(2n + \epsilon)\cos \pi\epsilon - 2(n + \epsilon)\sin \pi\epsilon}{2\epsilon(n + \epsilon)(2n + \epsilon)\sin \pi\epsilon} + O\left(\frac{1}{n^4}\right)$$

$$= \frac{3}{4n^2} + \frac{\pi^2 a_0}{6n} + O\left(\frac{1}{n^4}\right). \quad (11.22)$$

It can be proved similarly that

$$\sum_{k=1}^{\infty}{}' \frac{1}{(k^2 - \lambda_n)^2} = \frac{1}{12} \frac{\pi}{n^2} + O\left(\frac{1}{n^4}\right). \quad (11.23)$$

After elementary transformations, it follows from (11.20), (11.22) and (11.23) that

$$\ln \Psi_n = \frac{1}{n^2}\left[A + \mu_0 - \lambda_0 + \frac{\pi^2}{6}(a_0 - a_0')^2 + \frac{3}{2}(a_0 - a_0')\right] + O\left(\frac{1}{n^3}\right).$$

Thus,

$$\Psi_n = 1 + \frac{1}{n^2}\left[A + \mu_0 - \lambda_0 + \frac{\pi^2}{6}(a_0 - a_0')^2 + \frac{3}{2}(a_0 - a_0')\right] + O\left(\frac{1}{n^3}\right).$$

We now consider the factor $(h_2 - h_1)/(\mu_n - \lambda_n)$ in (11.6). By the asymptotic formulas (11.12) and (11.11), we have

$$\frac{h_2 - h_1}{\mu_n - \lambda_n} = \frac{h_2 - h_1}{2(a_0' - a_0) + (c_0' - c_0)/n^2 + O(n^{-3})}$$

$$= \frac{h_2 - h_1}{2(a_0' - a_0)}\left[1 - \frac{c_0' - c_0}{2(a_0' - a_0)n^2}\right] + O\left(\frac{1}{n^3}\right).$$

Since

$$\frac{h_2 - h_1}{2(a_0' - a_0)} = \frac{\pi}{2}, \qquad \frac{c_0' - c_0}{2(a_0' - a_0)} = \frac{a_0' + a_0}{2} + \frac{a_1' - a_1}{a_0' - a_0},$$

relying on (11.10) and (11.11), we finally obtain the asymptotic formula

$$\alpha_n = \frac{\pi}{2} + \frac{\pi}{2}\left[S + \frac{\pi^2}{6}(a_0 - a_0')^2 - a_0 - \frac{a_1' - a_1}{a_0' - a_0}\right]\frac{1}{n^2} + O\left(\frac{1}{n^3}\right) \qquad (11.24)$$

for the norming constants, where

$$S = \mu_0 - \lambda_0 + \sum_{k=1}^{\infty}[(\mu_k - \lambda_k) - 2(a_0' - a_0)].$$

Remark 1 The formula (11.24) can be used, beginning with sufficiently large n. If n is not large, α_n should be calculated by (11.6) directly. The difference $h_2 - h_1$ is determined by the equality (11.11). We also note that it follows from the fact that $\{\lambda_n\}_{n=0}^{\infty}$ and $\{\mu_n\}_{n=0}^{\infty}$ interlace that all α_n are positive.

Remark 2 If the subsequent terms are specified in the asymptotic formulas (11.10), then the subsequent terms can be calculated similarly also in the asymptotic expansion of α_n. However, the procedure becomes much more tedious.

6.11.3 Sufficient conditions for the inverse problem from two spectra to be solvable

The results of the preceding two items enable us to indicate necessary and sufficient conditions for an equation of the form (11.1) to hold, where two increasing sequences of real numbers are the spectra with the conditions (11.2) and (11.3), respectively, and not only yield the procedure of reconstructing the classical Sturm-Liouville operator from the sequences.

Theorem 6.11.1 *Given two sequences of real numbers satisfying the conditions :*

1. The numbers λ_n and μ_n interlace, i.e.,

$$\lambda_0 < \mu_0 < \lambda_1 < \mu_1 < \lambda_2 < \mu_2 < \dots \quad \text{or} \quad \mu_0 < \lambda_0 < \mu_1 < \lambda_1 < \mu_2 < \lambda_2 < \dots.$$

2. The asymptotic formulas (11.10) hold, with $a_0 \neq a_0'$.

Then there exists an absolutely continuous function $q(x)$ and three numbers h_1, h_2 and H such that $\{\lambda_n\}$ is the spectrum of the problem (11.1), (11.2) and $\{\mu_n\}$ is the spectrum of the problem (11.1), (11.3).

Proof. Suppose that two sequences $\{\lambda_n\}$ and $\{\mu_n\}$ with the above properties are given. We then determine the sequence $\{\alpha_n\}$ by (11.6), for which the asymptotic formula (11.24) holds.

Consider the function

$$m(\lambda) = \frac{1}{h_2 - h_1} \prod_{n=0}^{\infty} \frac{\mu_n - \lambda}{\lambda_n - \lambda}.$$

Given that its zeros and poles interlace, and, obviously, as $\lambda \to \infty$, $m(\lambda) \to -1/(h_2 - h_1)$, we have that

$$m(\lambda) = -\frac{1}{h_2 - h_1} + \sum_{n=0}^{\infty} \frac{1}{\alpha_n(\lambda - \lambda_n)}, \tag{11.25}$$

where α_n are all of the same sign. Since $\alpha_n > 0$ for large n (see the formula (11.6)), all $\alpha_n > 0$.

It was shown in Section 8 that an absolutely continuous function $q(x)$ and two numbers h, H can be determined uniquely from $\{\lambda_n\}$ and $\{\alpha_n\}$, $\{\lambda_n\}_{n=0}^{\infty}$ being the sequence of eigenvalues of the problem (11.1), (11.2).

Determine the value h_2 from the equation $h_2 = h_1 + \pi(a_0' - a_0)$. Let $\{\tau_n\}_{n=0}^{\infty}$ be the sequence of eigenvalues of the equation (11.1), reconstructed from $\{\lambda_n\}$ and $\{\alpha_n\}$, with the boundary condition

$$y'(0) - h_2 y(0) = 0, \qquad y'(\pi) + H y(\pi) = 0,$$

where h_2 and H were determined above. If we show that

$$\tau_n = \mu_n, \quad n = 0, 1, 2, \ldots,$$

then the theorem is proved completely.

Denote by $\varphi(x, \lambda)$ and $\psi(x, \lambda)$ the solutions of (1.1) (reconstructed from $\{\lambda_n\}$ and $\{\alpha_n\}$), with the initial conditions

$$\varphi(0, \lambda) = 1, \qquad \varphi'(0, \lambda) = h_1$$

$$\psi(0, \lambda) = 1, \qquad \psi'(0, \lambda) = h_2.$$

Then the poles and zeros of the function

$$\tilde{m}(\lambda) = -\frac{\psi'(\pi, \lambda) + H\psi(\pi, \lambda)}{\varphi'(\pi, \lambda) + H\varphi(\pi, \lambda)}$$

coincide with $\{\lambda_m\}$ and $\{\tau_m\}$, respectively. It was shown above that $\{\lambda_n\}$ and $\{\tau_n\}$ interlace; as $\lambda \to \infty$, $\tilde{m}(\lambda) \to -1$. On the other hand (see the formula (11.6)),

$$\frac{1}{\alpha_n} = \frac{1}{h_2 - h_1} \operatorname{Res} \tilde{m}(\lambda)_{\lambda = \lambda_n}.$$

Hence,

$$\frac{\tilde{m}(\lambda)}{h_2 - h_1} = -\frac{1}{h_2 - h_1} + \sum_{n=0}^{\infty} \frac{1}{\alpha_n(\lambda - \lambda_n)}. \tag{11.26}$$

It follows from the formulas (11.25) and (11.26) that $m(\lambda) = \tilde{m}(\lambda)/(h_2 - h_1)$; therefore, $\tau_n = \mu_n$, which completes the proof. ∎

Remark 3

As we know from Section 2, Chapter 1, for the asymptotic formula (11.10) to exist, we have to require that the second derivative of $q(x)$ exists, and that it is absolutely continuous. Hence, there is a gap between the necessary and the sufficient conditions in the solution to the inverse problem (since the absolute continuity of the first derivative of $q(x)$ was only proved for the asymptotic behaviour (11.10)). To bridge the gap, we should study the asymptotic behaviour of α_n in more detail. We refer the reader to the paper of B. M. Levitan and M. G. Gasymov (1964).

Notes

Section 1

Transformation operators are associated in linear algebra with the concept of similarity (linear equivalence) of matrices.

The concept for differential operators arose from the theory of general displacement operators, developed by J. Delsarte (1938a,b), B. M. Levitan (1964b, 1968), and J. Delsarte and J. Lions (1957). The term is due to V. A. Marchenko (1950). In the French literature, the term 'transmutation' is common (see J. Delsarte and J. Lions (1957)). Its general definition was borrowed from the latter paper.

Section 2

We already noted that the idea of a transformation operator is due to J. Delsarte, who considered a particular case of differential operators (J. Delsarte (1938b)). In the general case, the formula (2.2) was first derived by A. Ya. Povzner (1948).

The derivation of the integral equation (2.2) was borrowed from the monograph of V. A. Marchenko (1977).

Section 3

The fundamental integral equation of the inverse problem from the spectral function was first derived in the paper of I. M. Gelfand and B. M. Levitan (1951), published after the important paper of V. A. Marchenko (1950), who had proved that the Sturm-Liouville operator is uniquely determined by the spectral function.

Sections 4 to 7

The major results were first published by I. M. Gelfand and B. M. Levitan (1951). Our proofs are somewhat different. Another approach to the inverse problems can be found in the papers of M. G. Krein. In the paper of I. M. Gelfand and B. M. Levitan (1951), there was a gap in one derivative between the necessary and the sufficient conditions for the solvability of the inverse problem.

Theorem 6.6.1 without a gap between the necessary and sufficient conditions was proved differently by M. G. Krein (1953a).

Sections 8 and 9

The results were first published in the paper of I. M. Gelfand and B. M. Levitan (1951).

Section 10

The first to study the periodic inverse Sturm-Liouville problem was I. V. Stankevich (1970). The discussed results and methods are due to him. Another approach can be found in the monograph of V. A. Marchenko (1977).

Section 11

The uniqueness theorem for the classical Sturm-Liouville operator from two spectra is due to G. Borg (1946); see also the papers of N. Levinson (1949), and B. M. Levitan and M. G. Gasymov (1964).

The solvability of the classical inverse Sturm-Liouville problem from two spectra, indicating sufficient conditions, was first studied by B. M. Levitan (1964). Another approach which we mostly used can be found in the paper of B. M. Levitan and M. G. Gasymov (1964), where sufficient conditions for the solvability of the inverse Sturm-Liouville problem from two spectra on the half-line are also considered.

M. G. Krein (1951a,b, 1952) found an original method for solving inverse problems; in particular, from two spectra.

Part two

One-dimensional Dirac operators

Chapter 7

SPECTRAL THEORY IN THE REGULAR CASE

7.1 Definition of the operator—basic properties

7.1.1

Consider the matrix equation

$$B\frac{dy}{dx} + P(x)y = \lambda y, \quad y(x) = \begin{pmatrix} y_1(x) \\ y_2(x) \end{pmatrix}, \tag{1.1}$$

where $B = \begin{pmatrix} 0 & 1 \\ -1 & 0 \end{pmatrix}$, $P(x) = \begin{pmatrix} p_{11}(x) & p_{12}(x) \\ p_{21}(x) & p_{22}(x) \end{pmatrix}$, $p_{12}(x) \equiv p_{21}(x)$, $p_{ik}(x)$, $i, k = 1, 2$, are real-valued functions defined and continuous on the interval $[0, \pi]$, and λ is a parameter.

The equation (1.1) is equivalent to the system of two consistent ordinary first-order differential equations

$$\begin{aligned} y_2' + p_{11}(x)y_1 + p_{12}(x)y_2 &= \lambda y_1, \\ -y_1' + p_{21}(x)y_1 + p_{22}(x)y_2 &= \lambda y_2. \end{aligned} \tag{1.2}$$

In the case where $p_{12}(x) = p_{21}(x) \equiv 0$, $p_{11}(x) = V(x) + m$, $p_{22}(x) = V(x) - m$, $V(x)$ is a potential function and m the mass of a particle, (1.2) is called a *one-dimensional stationary Dirac system* in relativistic quantum theory.

Let $H = H(x)$ be a smooth and orthogonal transformation of a two-dimensional space. Any orthogonal transformation of a two-dimensional space is known to have a matrix of the form

$$H(x) = \begin{pmatrix} \cos \varphi(x) & -\sin \varphi(x) \\ \sin \varphi(x) & \cos \varphi(x) \end{pmatrix}$$

with respect to a fixed, orthogonal and normalized basis. It is easy to see that the matrices B and H are commuting, i.e., $BH = HB$. Put $y = H(x)z$. Substituting y in (1.1) and pre-multiplying throughout by H^{-1}, we obtain

$$H^{-1}B\frac{d}{dx}(Hz) + H^{-1}PHz = H^{-1}\lambda Hz,$$

or

$$B\frac{dz}{dx} + \left(H^{-1}B\frac{d}{dx}H + H^{-1}PH\right)z = \lambda z. \qquad (1.3)$$

To calculate the matrix $Q(x) \equiv H^{-1}B(d/dx)H + H^{-1}PH$, we have

$$H^{-1}B\frac{d}{dx}H = \begin{pmatrix} \varphi'(x) & 0 \\ 0 & \varphi'(x) \end{pmatrix},$$

$$H^{-1}PH = \begin{pmatrix} p_{11}\cos^2\varphi + p_{12}\sin 2\varphi + p_{22}\sin^2\varphi & p_{12}\cos 2\varphi + \frac{1}{2}(p_{22} - p_{11})\sin 2\varphi \\ p_{12}\cos 2\varphi + \frac{1}{2}(p_{22} - p_{11})\sin 2\varphi & p_{11}\sin^2\varphi - p_{12}\sin 2\varphi + p_{22}\cos^2\varphi \end{pmatrix}.$$

Therefore, the matrix Q is of the form

$$Q(x) = \begin{pmatrix} q_{11}(x) & q_{12}(x) \\ q_{12}(x) & q_{22}(x) \end{pmatrix}$$

$$= \begin{pmatrix} \varphi' + p_{11}\cos^2\varphi + p_{12}\sin 2\varphi + p_{22}\sin^2\varphi & p_{12}\cos 2\varphi + \frac{1}{2}(p_{22} - p_{11})\sin 2\varphi \\ p_{12}\cos 2\varphi + \frac{1}{2}(p_{22} - p_{11})\sin 2\varphi & \varphi' + p_{11}\sin^2\varphi - p_{12}\sin 2\varphi + p_{22}\cos^2\varphi \end{pmatrix}.$$

We select the function $\varphi(x)$ so that $q_{12}(x) \equiv 0$. Then

$$p_{12}(x)\cos 2\varphi(x) + \frac{1}{2}\{p_{22}(x) - p_{11}(x)\}\sin 2\varphi(x) = 0.$$

Hence, if[1] $p_{11}(x) \neq p_{22}(x)$, then

$$\varphi(x) = \frac{1}{2}\tan^{-1}\frac{2p_{12}(x)}{p_{11}(x) - p_{22}(x)},$$

and the matrix $Q(x)$ takes the form

$$Q(x) = \begin{pmatrix} q_{11}(x) & 0 \\ 0 & q_{22}(x) \end{pmatrix} \equiv \begin{pmatrix} p(x) & 0 \\ 0 & r(x) \end{pmatrix}.$$

Therefore, the equation (1.3) can be rewritten as

$$\begin{pmatrix} 0 & 1 \\ -1 & 0 \end{pmatrix}\frac{dz}{dx} + \begin{pmatrix} p(x) & 0 \\ 0 & r(x) \end{pmatrix}z = \lambda z. \qquad (1.4)$$

We now select $\varphi(x)$ so that trace $Q(x) = q_{11}(x) + q_{22}(x) = 0$, i.e., $2\varphi'(x) + p_{11}(x) + p_{22}(x) = 0$. Hence,

$$\varphi(x) = -\frac{1}{2}\int_0^x \{p_{11}(s) + p_{22}(s)\}\,ds.$$

---

[1]If $p_{11}(x) \equiv p_{22}(x)$, then $\varphi(x) = \pi/4$. Otherwise, reducing the system (1.1) to (1.4) requires additional study.

The equation (1.3) then takes the form

$$\begin{pmatrix} 0 & 1 \\ -1 & 0 \end{pmatrix} \frac{dz}{dx} + \begin{pmatrix} p(x) & q(x) \\ q(x) & -p(x) \end{pmatrix} z = \lambda z. \tag{1.5}$$

In the sequel, we shall call the equations (1.4) and (1.5) the *canonical forms* of the equation (1.1) (or of the system (1.2)). In solving various problems in the spectral theory of (1.1), it is convenient to use some or other canonical form; e.g., in studying the asymptotic behaviour of eigenvalues and vector-valued eigenfunctions, and also questions of expansion of arbitrary vector-valued functions in terms of vector-valued eigenfunctions of (1.1) (with some homogeneous boundary conditions at 0 and π), it is convenient to use the canonical equation (1.4). In studying such questions as the asymptotic distribution of the eigenvalues of (1.1), given on an infinite interval, and also the inverse problem, it is convenient to apply the canonical equation (1.5).

7.1.2

Thus, we consider the boundary-value problem for the equation (1.1), a priori reducing the latter to the canonical form (1.4) as in

$$y_2' - \{\lambda + p(x)\}y_1 = 0, \qquad y_1' + \{\lambda + r(x)\}y_2 = 0 \tag{1.6}$$

$$y_1(0)\sin\alpha + y_2(0)\cos\alpha = 0 \tag{1.7}$$

$$y_1(\pi)\sin\beta + y_2(\pi)\cos\beta = 0. \tag{1.8}$$

We assume that the functions $p(x)$ and $r(x)$ are continuous on the interval $[0, \pi]$.

Let the boundary-value problem under consideration have a nontrivial solution $y(x, \lambda_0) = \begin{pmatrix} y_1(x, \lambda_0) \\ y_2(x, \lambda_0) \end{pmatrix}$ for certain λ_0, called an *eigenvalue*; the corresponding solution $y(x, \lambda_0)$ is called a *vector-valued eigenfunction.*

Lemma 7.1.1 *The vector-valued eigenfunctions $y(x, \lambda_1)$ and $z(x, \lambda_2)$ corresponding to different eigenvalues $\lambda_1 \neq \lambda_2$ are orthogonal, i.e., $(y^T = (y_1 \ y_2))$,*

$$\int_0^\pi y^T z \, dz \equiv \int_0^\pi \{y_1(x, \lambda_1)z_1(x, \lambda_2) + y_2(x, \lambda_1)z_2(x, \lambda_2)\} \, dx = 0.$$

Proof. Since $y(x, \lambda_1)$ and $z(x, \lambda_2)$ are solutions of the system (1.6),

$$y_2' - \{\lambda_1 + p(x)\}y_1 = 0, \quad z_2' - \{\lambda_2 + p(x)\}z_1 = 0$$

$$y_1' + \{\lambda_1 + r(x)\}y_2 = 0, \quad z_1' + \{\lambda_2 + r(x)\}z_2 = 0.$$

Multiplying by $z_1, -z_2, -y_1$ and y_2, respectively, and adding together, we obtain

$$\frac{d}{dx}\{y_1(x, \lambda_1)z_2(x, \lambda_2) - y_2(x, \lambda_1)z_1(x, \lambda_2)\}$$

$$= (\lambda_1 - \lambda_2)\{y_1(x, \lambda_1)z_1(x, \lambda_2) + y_2(x, \lambda_1)z_2(x, \lambda_2)\}.$$

Now, integrating from 0 to π, we find

$$(\lambda_1 - \lambda_2) \int_0^\pi y^T(x, \lambda_1) z(x, \lambda_2) \, dx = \{y_1(x, \lambda_1) z_2(x, \lambda_2) - y_2(x, \lambda_1) z_1(x, \lambda_2)\}_0^\pi,$$

where the right-hand side vanishes because of the boundary conditions (1.7) and (1.8).   ∎

Lemma 7.1.2 *The eigenvalues of the boundary-value problem (1.6), (1.7), (1.8) are real.*

The proof is left to the reader (see Lemma 1.1.2).

7.2   Asymptotic formulas for the eigenvalues and for the vector-valued eigenfunctions

7.2.1

We now derive the asymptotic formulas for the eigenvalues and for the vector-valued eigenfunctions of the boundary-value problem (1.6), (1.7), (1.8). In particular, similarly to the Sturm-Liouville problem, it follows that there exist infinitely many eigenvalues.

In deriving the asymptotic formulas for the eigenvalues and eigenfunctions of the Sturm-Liouville problem in Section 2, Chapter 1, we essentially used the integral equations (2.4) and (2.5) of Chapter 1 for the fundamental system of solutions of the Sturm-Liouville equation. In the case of the Dirac system, formulas similar to (2.4) and (2.5) can also be written out, and the further reasoning can be done along the same lines. However, we offer another method. In studying the asymptotic behaviour of the eigenvalues and vector-valued eigenfunctions of a Dirac system, it is very convenient to introduce so-called *transformation matrix operators* whose derivation and some applications are in Section 3 of Chapter 10. Here, we only give their definition and final form.

Let A and B be two linear differential operators, and let E_1, E_2 be two linear function spaces.

Definition 7.2.1 *A continuous linear operator X mapping the space E_1 into E_2 is called a* transformation operator *if it satisfies the conditions*

   1. $AX = XB$, and

   2. there exists a continuous inverse operator X^{-1}.

Let A and B be of the form

$$A \equiv \begin{pmatrix} p_1(x) & \dfrac{d}{dx} \\ -\dfrac{d}{dx} & r_1(x) \end{pmatrix}, \qquad B \equiv \begin{pmatrix} p_2(x) & \dfrac{d}{dx} \\ -\dfrac{d}{dx} & r_2(x) \end{pmatrix}, \qquad (2.1)$$

where $p_k(x)$, $r_k(x)$, $k = 1, 2$ are real-valued and continuous functions on the interval $[0, \pi]$.

Let E_1 and E_2 be the sets of continuously differentiable vector-valued functions $f(x)$ and $g(x)$ satisfying the boundary conditions

$$f_1(0) \sin \gamma + f_2(0) \cos \gamma = 0$$

$$g_1(0) \sin \delta + g_2(0) \cos \delta = 0 \tag{2.2}$$

where γ and δ are two arbitrary real numbers.

It was shown in Section 3, Chapter 10 (see the Remark) that the transformation matrix operator X can then be specified in the form

$$X\{f(x)\} = R(x)f(x) + \int_0^x K(x, s)f(s)\, ds, \tag{2.3}$$

where $R(x)$ and $K(x, s)$ are two continuously differentiable matrices of order two,

$$R(x) = \begin{pmatrix} \alpha(x) & \beta(x) \\ -\beta(x) & \alpha(x) \end{pmatrix},$$

and the functions $\alpha(x)$, $\beta(x)$ can be calculated explicitly :

$$\alpha(x) = \frac{1}{\kappa} \sin \left\{ \frac{1}{2} \int_0^x \mathrm{trace}[A - B]\, d\tau + \sin^{-1} \frac{1}{\kappa} \right\},$$

$$\beta(x) = \frac{1}{\kappa} \cos \left\{ \frac{1}{2} \int_0^x \mathrm{trace}[A - B]\, d\tau + \sin^{-1} \frac{1}{\kappa} \right\}, \tag{2.4}$$

$\kappa = \sec(\delta - \gamma)$.

7.2.2

Denote by $\varphi(x, \lambda)$ the solution of the system (1.6), satisfying the initial conditions

$$\varphi_1(0, \lambda) = \cos \alpha, \qquad \varphi_2(0, \lambda) = -\sin \alpha. \tag{2.5}$$

It is obvious that $\varphi(x, \lambda)$ satisfies the boundary condition (1.7). Consider the problem (1.6), (2.5) for $p(x) = r(x) \equiv 0$. It can easily be seen that

$$\psi_1(x, \lambda) = \cos(\lambda x - \alpha)$$

$$\psi_2(x, \lambda) = \sin(\lambda x - \alpha) \tag{2.6}$$

denoting the solution of the problem by $\psi(x, \lambda)$.

We now apply the transformation matrix operator to the solution of the problem (1.6), (1.7). The system (1.6) is of the form

$$A_1 y \equiv \begin{pmatrix} -p(x) & \dfrac{d}{dx} \\ -\dfrac{d}{dx} & -r(x) \end{pmatrix} y = \lambda y. \tag{2.7}$$

On the other hand, the vector-valued function $\psi(x, \lambda)$ determined by the equalities (2.6), is the solution of

$$B_1 y = \begin{pmatrix} 0 & \dfrac{d}{dx} \\ -\dfrac{d}{dx} & 0 \end{pmatrix} y = \lambda y. \tag{2.8}$$

Since $\psi(x, \lambda)$ is the solution of $B_1 \psi = \lambda \psi$, by the definition of X, we have

$$A_1 X\{\psi\} = X B_1\{\psi\} = X\{\lambda\psi\} = \lambda X\{\psi\},$$

i.e., $\varphi = X\{\psi\}$ is the solution of $A_1\{\psi\} = \lambda \psi$. Thus, if $\psi(x, \lambda)$ is the solution of (2.8), then $\varphi(x, \lambda) = X\{\psi(x, \lambda)\}$ is the solution of (2.7). In the case under consideration,

$$p_1(x) = -p(x), \qquad r_1(x) = -r(x), \qquad p_2(x) = r_2(x) = 0.$$

Therefore, $\text{trace}[A_1 - B_1] = -[p(x) + r(x)]$. Also, since (2.2) are now replaced by the boundary condition (1.7), $\gamma = \delta$, so that $\kappa = 1$. By (2.4),

$$\alpha(x) = \cos\left\{\frac{1}{2}\int_0^x [p(\tau) + r(\tau)]\, d\tau\right\},$$

$$\beta(x) = \sin\left\{\frac{1}{2}\int_0^x [p(\tau) + r(\tau)]\, d\tau\right\}. \tag{2.9}$$

We derive from (2.3) for the solution $\varphi(x, \lambda)$ of the problem (1.6), (2.5) that

$$\varphi(x, \lambda) = R(x)\psi(x, \lambda) + \int_0^x K(x, s)\psi(s, \lambda)\, ds.$$

Hence, taking into account the expression of the matrix $R(x)$ in terms of the functions $\alpha(x)$, $\beta(x)$, their values from (2.9) and the explicit form of the vector-valued function $\psi(x, \lambda)$ from (2.6), we obtain the formulas

$$\varphi_1(x, \lambda) = \cos\{\xi(x, \lambda) - \alpha\} + \int_0^x \{K_{11}(x, s)\cos(\lambda s - \alpha) + K_{12}(x, s)\sin(\lambda s - \alpha)\}\, ds \tag{2.10}$$

$$\varphi_2(x, \lambda) = \sin\{\xi(x, \lambda) - \alpha\} + \int_0^x \{K_{21}(x, s)\cos(\lambda s - \alpha) - K_{22}(x, s)\sin(\lambda s - \alpha)\}\, ds \tag{2.11}$$

for the components of the vector-valued function $\varphi(x, \lambda)$, where

$$\xi(x, \lambda) = \lambda x - \frac{1}{2}\int_0^x [p(\tau) - r(\tau)]\, d\tau, \tag{2.12}$$

and $K_{ij}(x, s)$, $i, j = 1, 2$ are the entries of the kernel matrix $K(x, s)$ from (2.3).

Lemma 7.2.1 *As $|\lambda| \to \infty$ uniformly in x (for $0 \le x \le \pi$), the estimates*

$$\varphi_1(x, \lambda) = \cos\{\xi(x, \lambda) - \alpha\} + O(\lambda^{-1})$$

$$\varphi_2(x, \lambda) = \sin\{\xi(x, \lambda) - \alpha\} + O(\lambda^{-1})$$

(2.13)

$$\frac{\partial}{\partial \lambda} \varphi_1(x, \lambda) = -x \sin\{\xi(x, \lambda) - \alpha\} + O(1)$$

$$\frac{\partial \varphi_2(x, \lambda)}{\partial \lambda} = x \cos\{\xi(x, \lambda) - \alpha\} + O(1)$$

(2.14)

are valid.

Proof. To obtain (2.13), it suffices to integrate by parts the integrals involved in (2.10), (2.11), which is possible because of the differentiability of the functions $K_{ij}(x, s)$. However, (2.14) are proved similarly if we first differentiate (2.10), (2.11) with respect to λ, and the lemma is thus proved. ∎

Lemma 7.2.2 *The eigenvalues of the boundary-value problem (1.6), (1.7), (1.8) are simple.*

Proof. Since $\varphi(x, \lambda)$ satisfies the boundary condition (1.7), to determine the eigenvalues of the problem under consideration, the functions $\varphi_1(x, \lambda)$, $\varphi_2(x, \lambda)$ should be substituted in the boundary condition (1.8), and its roots should be found. Put

$$D(\lambda) = \varphi_1(\pi, \lambda) \sin \beta + \varphi_2(\pi, \lambda) \cos \beta.$$

Then

$$\frac{dD(\lambda)}{d\lambda} = \frac{\partial \varphi_1}{\partial \lambda} \sin \beta + \frac{\partial \varphi_2}{\partial \lambda} \cos \beta.$$

Let λ_0 be a double eigenvalue, and $\varphi^0(x, \lambda_0)$ one of the corresponding vector-valued eigenfunctions. Then the conditions $D(\lambda_0) = 0$, $\dfrac{dD(\lambda_0)}{d\lambda} = 0$ should be fulfilled simultaneously, i.e.,

$$\varphi_1^0(\pi, \lambda_0) \sin \beta + \varphi_2^0(\pi, \lambda_0) \cos \beta = 0$$

$$\frac{\partial}{\partial \lambda} \varphi_1^0(\pi, \lambda_0) \sin \beta + \frac{\partial}{\partial \lambda} \varphi_2^0(\pi, \lambda_0) \cos \beta = 0.$$

Since $\sin \beta$ and $\cos \beta$ cannot vanish simultaneously, it follows from the last two equalities that

$$\varphi_2^0(\pi, \lambda) \frac{\partial \varphi_1^0(\pi, \lambda_0)}{\partial \lambda} - \varphi_1^0(\pi, \lambda) \frac{\partial \varphi_2^0(\pi, \lambda_0)}{\partial \lambda} = 0.$$

(2.15)

Now, differentiating the system (1.6) with respect to λ, we have

$$\left(\frac{\partial y_2}{\partial \lambda}\right)'_x - \{\lambda + p(x)\}\frac{\partial y_1}{\partial \lambda} = y_1$$

$$\left(\frac{\partial y_1}{\partial \lambda}\right)'_x + \{\lambda + r(x)\}\frac{\partial y_2}{\partial \lambda} = -y_2 \qquad (2.16)$$

Multiplying the equations of (1.6) and (2.16) by $\dfrac{\partial y_1}{\partial \lambda}$, $-\dfrac{\partial y_2}{\partial \lambda}$, $-y_1$ and y_2, respectively, adding them together and integrating with respect to x from 0 to π, we obtain

$$\left\{ y_1(x,\lambda)\frac{\partial y_2}{\partial \lambda} - y_2(x,\lambda)\frac{\partial y_1}{\partial \lambda} \right\}_0^\pi = \int_0^\pi [y_1^2(x,\lambda) + y_2^2(x,\lambda)]\, dx.$$

Putting $\lambda = \lambda_0$, taking into account that

$$\left.\frac{\partial \varphi_1^0(x,\lambda_0)}{\partial \lambda}\right|_{x=0} = \left.\frac{\partial \varphi_2^0(x,\lambda_0)}{\partial \lambda}\right|_{x=0} = 0$$

by (2.10) and (2.11), and using the equality (2.15), we obtain the relation

$$\int_0^\pi [\{\varphi_1^0(x,\lambda_0)\}^2 + \{\varphi_2^0(x,\lambda_0)\}^2]\, dx = \varphi_1^0(\pi,\lambda_0)\frac{\partial \varphi_2^0(\pi,\lambda_0)}{\partial \lambda} - \varphi_2^0(\pi,\lambda)\frac{\partial \varphi_1^0(\pi,\lambda)}{\partial \lambda} = 0.$$

Hence, $\varphi_1^0(x,\lambda_0) = \varphi_2^0(x,\lambda_0) \equiv 0$, or $\varphi^0(x,\lambda_0) \equiv 0$, which is impossible. The lemma is thus proved. ∎

7.2.3

We have already noted above that the eigenvalues of the boundary-value problem (1.6), (1.7), (1.8) coincide with the roots of

$$\varphi_1(\pi,\lambda)\sin\beta + \varphi_2(\pi,\lambda)\cos\beta = 0.$$

Substituting $\varphi_1(\pi,\lambda)$ and $\varphi_2(\pi,\lambda)$ from the estimates (2.13), we obtain

$$\sin(\lambda\pi - \vartheta) + O(\lambda^{-1}) = 0, \qquad (2.17)$$

where

$$\vartheta = \beta - \alpha - \frac{1}{2}\int_0^\pi \{p(\tau) + r(\tau)\}\, d\tau \qquad (2.18)$$

by (2.12).

It is obvious that the equation (2.17), for large $|\lambda|$, has solutions of the form

$$\pi\lambda_n - \vartheta = n\pi + \delta_n,$$

which can be proved as for the Sturm-Liouville operator in Section 2, Chapter 1. Substituting the values in (2.17), we see that $\sin \delta_n = O(n^{-1})$, i.e., $\delta_n = O(n^{-1})$. Therefore, we obtain the asymptotic formula

$$\lambda_{\pm n} = \frac{\vartheta}{\pi} \pm n + O\left(\frac{1}{n}\right) \tag{2.19}$$

for the eigenvalues, where ϑ is determined by the formula (2.18).

If we assume that the functions $p(x)$ and $r(x)$ from the system (1.6) are differentiable, the formulas (2.19) can be sharpened considerably, viz. the expansion terms

$$\lambda_n = \frac{\vartheta}{\pi} + n + \frac{C_1}{n} + \frac{C_2}{n^2} + O\left(\frac{1}{n^3}\right) \tag{2.20}$$

can be written out.

Using (2.19), we obtain the asymptotic formulas for the vector-valued eigenfunctions

$$\varphi_1(x, \lambda_n) \equiv u_n(x), \qquad \varphi_2(x, \lambda_n) \equiv v_n(x),$$

viz.

$$u_n(x) = \cos(\xi_n - \alpha) + O(n^{-1}),$$
$$\tag{2.21}$$
$$v_n(x) = \sin(\xi_n - \alpha) + O(n^{-1}),$$

where $\xi_n = \xi(x, \lambda_n) = \lambda_n x - \frac{1}{2} \int_0^x \{p(\tau) + r(\tau)\} \, d\tau$.

To obtain the asymptotic expansion of the normalized vector-valued eigenfunctions, consider the integral

$$\alpha_n^2 = \int_0^\pi \{u_n^2(x) + v_n^2(x)\} \, dx = \pi + O\left(\frac{1}{n}\right).$$

Therefore, the normalized vector-valued eigenfunctions are of the form

$$\frac{1}{\alpha_n} \varphi(x, \lambda_n) = \begin{pmatrix} \frac{1}{\sqrt{\pi}} \cos(\xi_n - \alpha) + O\left(\frac{1}{n}\right) \\ \frac{1}{\sqrt{\pi}} \sin(\xi_n - \alpha) + O\left(\frac{1}{n}\right) \end{pmatrix}.$$

7.3   Proof of the expansion theorem by the method of integral equations

7.3.1

Here, we show that, as in the case of the Sturm-Liouville operator, we can prove the completeness of the vector-valued eigenfunctions of a Dirac system on a finite interval by the method of integral equations.

We shall look for the solution of the system

$$y_2' - \{\lambda + p(x)\} y_1 = f_1(x) \tag{3.1}$$

$$y_1' + \{\lambda + r(x)\}y_2 = -f_2(x) \qquad (3.2)$$

($f(x) \not\equiv 0$ being a continuous vector-valued function), which fulfils the boundary conditions

$$y_1(a)\sin\alpha + y_2(a)\cos\alpha = 0 \qquad (3.3)$$

$$y_1(b)\sin\beta + y_2(b)\cos\beta = 0. \qquad (3.4)$$

Let λ be a fixed complex number, let $\varphi(x,\lambda)$ be the solution of the system (1.6), fulfilling the initial conditions

$$\varphi_1(a,\lambda) = \cos\alpha, \qquad \varphi_2(a,\lambda) = -\sin\alpha, \qquad (3.5)$$

and $\psi(x,\lambda)$ the one satisfying

$$\psi_1(b,\lambda) = \cos\beta, \qquad \psi_2(b,\lambda) = -\sin\beta. \qquad (3.6)$$

If $\varphi(x,\lambda)$ and $\psi(x,\lambda)$ are linearly independent, i.e., $\varphi(x,\lambda)$ is not a vector-valued eigenfunction (for if $\varphi(x,\lambda) = C\psi(x,\lambda)$, $\varphi(x,\lambda)$ also fulfils the boundary condition (3.4) and, therefore, is a vector-valued eigenfunction), then the Wronskian

$$W\{\varphi,\psi\} = \begin{vmatrix} \varphi_1(x,\lambda) & \varphi_2(x,\lambda) \\ \psi_1(x,\lambda) & \psi_2(x,\lambda) \end{vmatrix} \neq 0.$$

Conversely, if for certain λ, the Wronskian $W\{\varphi,\psi\} = 0$, then $\varphi(x,\lambda) = C\psi(x,\lambda)$ and, therefore, $\varphi(x,\lambda)$ is a vector-valued eigenfunction. Thus, the eigenvalues of the boundary-value problem (1.6), (3.3), (3.4) coincide with the zeros of $W\{\varphi,\psi\}$ if we consider the Wronskian as a function of λ.

We prove that $W\{\varphi,\psi\}$ does not depend on x. We have

$$\varphi_2' - \{\lambda + p(x)\}\varphi_1 = 0, \quad \psi_2' - \{\lambda + p(x)\}\psi_1 = 0,$$
$$\varphi_1' + \{\lambda + r(x)\}\varphi_2 = 0, \quad \psi_1' + \{\lambda + r(x)\}\psi_2 = 0. \qquad (3.7)$$

Multiplying the equations by ψ_1, $-\psi_2$, $-\varphi_1$, φ_2 and adding them together, we obtain

$$(\varphi_2'\psi_1 - \varphi_1'\psi_2 - \psi_2'\varphi_1 + \psi_1'\varphi_2)(x,\lambda) = 0,$$

i.e.,

$$\frac{d}{dx}(\varphi_1\psi_2 - \varphi_2\psi_1)(x,\lambda) = \frac{d}{dx}W\{\varphi,\psi\} = 0,$$

which proves the statement. Therefore, $W\{\varphi,\psi\}$ only depends on λ, and $W\{\varphi,\psi\} \equiv \omega(\lambda)$.

We now introduce the notation $u^T = (u_1\ u_2)$ for the transpose of the matrix $u = \begin{pmatrix} u_1 \\ u_2 \end{pmatrix}$. Therefore, according to the rule for matrix multiplication, we have

$$u^T v = (u_1\ u_2)\begin{pmatrix} v_1 \\ v_2 \end{pmatrix} = u_1 v_1 + u_2 v_2,$$

$$uv^T = \begin{pmatrix} u_1 \\ u_2 \end{pmatrix} (v_1 \ v_2) = \begin{pmatrix} u_1v_1 & u_1v_2 \\ u_2v_1 & u_2v_2 \end{pmatrix}.$$

Let the vector-valued functions $\varphi(x, \lambda)$ and $\psi(x, \lambda)$ have the above values, and let $\omega(\lambda) = W\{\varphi, \psi\}$. Put

$$G(x, y; \lambda) = \begin{cases} \dfrac{1}{\omega(\lambda)} \psi(x, \lambda)\varphi^T(y, \lambda), & y < x \\[3mm] \dfrac{1}{\omega(\lambda)} \varphi(x, \lambda)\psi^T(y, \lambda), & x < y \end{cases} \qquad (3.8)$$

which is called *Green's matrix*.

We show that the vector-valued function

$$y(x, \lambda) = \int_a^b G(x, y; \lambda) f(y)\, dy \qquad (3.9)$$

(where $f(x)$ is the same vector-valued function as in the system (3.1), (3.2)) is the solution of (3.1), (3.2), satisfying the boundary conditions (3.5), (3.4).

In fact, by definition,

$$G(x, y; \lambda) = \begin{cases} \dfrac{1}{\omega(\lambda)} \begin{pmatrix} \varphi_1(x, \lambda)\psi_1(y, \lambda) & \varphi_1(x, \lambda)\psi_2(y, \lambda) \\ \varphi_2(x, \lambda)\psi_1(y, \lambda) & \varphi_2(x, \lambda)\psi_2(y, \lambda) \end{pmatrix}, & x < y \\[5mm] \dfrac{1}{\omega(\lambda)} \begin{pmatrix} \varphi_1(y, \lambda)\psi_1(x, \lambda) & \varphi_1(y, \lambda)\psi_2(x, \lambda) \\ \varphi_2(y, \lambda)\psi_1(x, \lambda) & \varphi_2(y, \lambda)\psi_2(x, \lambda) \end{pmatrix}, & y < x \end{cases}.$$

Therefore,

$$G(x, y; \lambda) f(y) = \begin{cases} \dfrac{1}{\omega(\lambda)} \begin{pmatrix} \varphi_1(x, \lambda)\psi_1(y, \lambda)f_1(y) + \varphi_1(x, \lambda)\psi_2(y, \lambda)f_2(y) \\ \varphi_2(x, \lambda)\psi_1(y, \lambda)f_1(y) + \varphi_2(x, \lambda)\psi_2(y, \lambda)f_2(y) \end{pmatrix}, & x < y \\[5mm] \dfrac{1}{\omega(\lambda)} \begin{pmatrix} \varphi_1(y, \lambda)\psi_1(x, \lambda)f_1(y) + \varphi_1(y, \lambda)\psi_2(x, \lambda)f_2(y) \\ \varphi_2(y, \lambda)\psi_1(x, \lambda)f_1(y) + \varphi_2(y, \lambda)\psi_2(x, \lambda)f_2(y) \end{pmatrix}, & y < x \end{cases}$$

and, putting $[u^T v](y) = u_1(y)v_1(y) + u_2(y)v_2(y)$, we derive from (3.9) that

$$y_1(x, \lambda) = \frac{1}{\omega(\lambda)} \left\{ \psi_1(x, \lambda) \int_a^x [\varphi^T f](y)\, dy + \varphi_1(x, \lambda) \int_x^b [\psi^T f](y)\, dy \right\}, \qquad (3.10)$$

$$y_2(x, \lambda) = \frac{1}{\omega(\lambda)} \left\{ \psi_2(x, \lambda) \int_a^x [\varphi^T f](y)\, dy + \varphi_2(x, \lambda) \int_x^b [\psi^T f](y)\, dy \right\}. \qquad (3.11)$$

Differentiating the equality (3.10) with respect to x, we find

$$y_1'(x, \lambda) = \frac{1}{\omega(\lambda)} \left\{ \psi_1'(x, \lambda) \int_a^x [\varphi^T f](y)\, dy + \varphi_1'(x, \lambda) \int_x^b [\psi^T f](y)\, dy \right.$$

$$\left. -\varphi_1(x, \lambda)[\psi_1(x, \lambda)f_1(x) + \psi_2(x, \lambda)f_2(x)] + \psi_1(x, \lambda)[\varphi_1(x, \lambda)f_1(x) + \varphi_2(x, \lambda)f_2(x)] \right\}.$$

Substituting $\varphi_1'(x, \lambda)$ and $\psi_1'(x, \lambda)$ from the system (3.7), we have

$$y_1'(x, \lambda) = -\frac{\lambda + r(x)}{\omega(\lambda)} \left\{ \psi_2(x, \lambda) \int_a^x [\varphi^T f](y)\, dy + \varphi_2(x, \lambda) \int_x^b [\psi^T f](y)\, dy \right\}$$
$$- \frac{1}{\omega(\lambda)} \{\varphi_1(x, \lambda)\psi_2(x, \lambda) - \varphi_2(x, \lambda)\psi_1(x, \lambda)\} f_2(x). \quad (3.12)$$

Taking into account the equality (3.11) and that

$$\varphi_1(x, \lambda)\psi_2(x, \lambda) - \varphi_2(x, \lambda)\psi_1(x, \lambda) = W\{\varphi, \psi\} = \omega(\lambda),$$

(3.12) can be rewritten in the form

$$y_1'(x, \lambda) = -\{\lambda + r(x)\}y_2(x, \lambda) - f_2(x),$$

which coincides with (3.2). The validity of (3.1) is proved similarly. Therefore, the vector-valued function (3.9) is the solution of the system (3.1), (3.2). It can be checked directly that (3.9) also satisfies the boundary conditions (3.3), (3.4).

We have thus proved the following.

Theorem 7.3.1 *If λ is not an eigenvalue of the homogeneous boundary-value problem (1.6), (3.3), (3.4), then the nonhomogeneous system (3.1), (3.2), (3.3), (3.4) is solvable for any vector-valued function $f(x)$, and the solution is given by the formula (3.9) (the so-called* resolvent*). Conversely, if λ is an eigenvalue of the homogeneous boundary-value problem (1.6), (3.3), (3.4), then the nonhomogeneous system (3.1), (3.2), (3.3), (3.4) is, generally speaking, unsolvable.*

If λ is not an eigenvalue of the homogeneous boundary-value problem, then the nonhomogeneous system has a unique solution. In fact, it is obvious that the difference of two solutions of the nonhomogeneous problem is a vector-valued eigenfunction of the homogeneous problem; by assumption, this difference is identically zero.

We can assume that the number $\lambda = 0$ is not an eigenvalue. Otherwise, we select a fixed number η, and consider the boundary-value problem

$$y_2' - \{(\lambda - \eta) + p(x)\}y_1 = 0, \qquad y_1(a) \sin \alpha + y_2(a) \cos \alpha = 0,$$

$$y_1' + \{(\lambda - \eta) + r(x)\}y_2 = 0, \qquad y_1(b) \sin \beta + y_2(b) \cos \beta = 0.$$

Its vector-valued eigenfunctions are the same as in the problem without η, while the eigenvalues are shifted through η. It is obvious that η can be chosen so that zero is not an eigenvalue for the new problem.

Put $G(x, y; 0) = G(x, y)$. Then the vector-valued function

$$y(x) = \int_a^b G(x, t)f(t)\, dt$$

is the solution of the system

$$y_2' - p(x)y_1 = f_1(x), \qquad y_1' + r(x)y_1 = -f_2(x),$$

fulfilling the boundary conditions (3.3), (3.4) (proved). We rewrite the system (3.1), (3.2) in the form

$$By \equiv \begin{pmatrix} -p(x) & \frac{d}{dx} \\ -\frac{d}{dx} & -r(x) \end{pmatrix} y = f(x) + \lambda y. \tag{3.9'}$$

It follows from the above that the system (3.1), (3.2), (3.3), (3.4) is equivalent to the system of integral equations

$$y(x) = \int_a^b G(x,t)\{f(t) + \lambda y(t)\}\, dt,$$

or

$$y(x) - \lambda \int_a^b G(x,t)y(t)\, dt = \int_a^b G(x,t)f(t)\, dt \equiv g(x).$$

In particular, the homogeneous boundary-value problem (1.6), (3.3), (3.4) is equivalent to the system of integral equations

$$y(x) - \lambda \int_a^b G(x,t)y(t)\, dt = 0. \tag{3.13}$$

7.3.2

Denote by $\lambda_0, \lambda_{\pm 1}, \lambda_{\pm 2}, \ldots, \lambda_{\pm n}, \ldots$ the eigenvalues of the boundary-value problem (1.6), (3.3), (3.4), and by $v_0(x), v_{\pm 1}(x), v_{\pm 2}(x), \ldots, v_{\pm n}(x), \ldots$ the corresponding normalized vector-valued eigenfunctions. Consider the matrix kernel

$$H(x,\xi) = \sum_{n=-\infty}^{\infty} \frac{v_n(x)v_n^T(\xi)}{\lambda_n}. \tag{3.14}$$

By the asymptotic formulas of Section 2, the series for $H(x,\xi)$ converges uniformly outside each neighborhood of the diagonal $x = \xi$. Furthermore, the boundedness of the partial sums of the series (3.14) follows from that of $\sum_{n=-\infty}^{\infty} \sin nx/n$ which, therefore, can be integrated termwise.

We now consider the matrix kernel

$$Q(x,\xi) = H(x,\xi) - G(x,\xi) = -G(x,\xi) + \sum_{n=-\infty}^{\infty} \frac{v_n(x)v_n^T(\xi)}{\lambda_n}$$

easily seen to be continuous, since $H(x,\xi)$ and $G(x,\xi)$ are continuous except for $\xi = x$, and have the same jump

$$G(x, x+0) - G(x, x-0) = H(x, x+0) - H(x, x-0) = \begin{pmatrix} 0 & 1 \\ -1 & 0 \end{pmatrix}$$

on it. Furthermore, $Q(x, \xi)$ satisfies $Q(x, \xi) = Q^T(\xi, x)$, providing the self-adjointness of the integral operator with $Q(x, \xi)$. Any matrix satisfying this condition is called a *self-adjoint kernel* in the sequel.

By the familiar theorem in the theory of integral equations, any self-adjoint kernel $Q(x, \xi)$ not identically zero has at least one eigenfunction, i.e., there exists a number λ_0 and a vector-valued function $u(x) \not\equiv 0$ such that

$$u(x) + \lambda_0 \int_a^b Q(x, \xi)u(\xi) \, d\xi = 0. \tag{3.15}$$

Thus, if we show that $Q(x, \xi)$ has no vector-valued eigenfunctions, then $Q(x, \xi) \equiv 0$, i.e.,

$$G(x, \xi) = \sum_{n=-\infty}^{\infty} \frac{v_n(x)v_n^T(\xi)}{\lambda_n}. \tag{3.16}$$

It follows from (3.13) that

$$\int_a^b G(x, \xi)v_k(\xi) \, d\xi = \frac{1}{\lambda_k} v_k(x). \tag{3.17}$$

We now calculate the integral

$$\int_a^b Q(x, \xi)v_k(\xi) \, d\xi \equiv q(x) = \begin{pmatrix} q_1(x) \\ q_2(x) \end{pmatrix}.$$

By the definition of the matrix $Q(x, \xi)$, we have

$$q(x) = \int_a^b H(x, \xi)v_k(x) \, d\xi - \int_a^b G(x, \xi)v_k(\xi) \, d\xi. \tag{3.18}$$

On the other hand, each term of the series (3.14) is of the form

$$\frac{1}{\lambda_n} v_n(x)v_n^T(\xi) = \frac{1}{\lambda_n} \begin{pmatrix} v_{n1}(x)v_{n1}(\xi) & v_{n1}(x)v_{n2}(\xi) \\ v_{n2}(x)v_{n1}(x) & v_{n2}(x)v_{n2}(\xi) \end{pmatrix}.$$

Therefore, letting

$$\int_a^b H(x, \xi)v_k(\xi) \, d\xi \equiv h(x) = \begin{pmatrix} h_1(x) \\ h_2(x) \end{pmatrix},$$

relying on the orthonormality of the vector-valued eigenfunctions $v_n(x)$, we obtain

$$h_1(x) = \frac{1}{\lambda_k} \int_a^b \{v_{k1}(x)v_{k1}^2(\xi) + v_{k1}(x)v_{k2}^2(\xi)\} \, d\xi = \frac{1}{\lambda_k} v_{k1}(x),$$

$$h_2(x) = \frac{1}{\lambda_k} \int_a^b \{v_{k2}(x)v_{k1}^2(\xi) + v_{k2}(x)v_{k2}^2(\xi)\} \, d\xi = \frac{1}{\lambda_k} v_{k2}(x),$$

i.e., $h(x) = v_k(x)/\lambda_k$. By (3.17) and (3.18), we have

$$q(x) = \int_a^b Q(x,\xi)v_k(\xi)\,d\xi = \frac{1}{\lambda_k}v_k(x) - \frac{1}{\lambda_k}v_k(x) = 0, \qquad (3.18')$$

i.e., the kernel $Q(x,\xi)$ is orthogonal to all vector-valued eigenfunctions of the boundary-value problem (1.6), (3.3), (3.4).

Let $u(x)$ be the solution of the equation (3.15). We show that it is orthogonal to all $v_n(x)$. In fact, it follows from (3.15) that

$$u_1(x) + \lambda_0 \int_a^b \{q_{11}(x,\xi)u_1(\xi) + q_{12}(x,\xi)u_2(\xi)\}\,d\xi = 0$$

$$u_2(x) + \lambda_0 \int_a^b \{q_{21}(x,\xi)u_1(\xi) + q_{22}(x,\xi)u_2(\xi)\}\,d\xi = 0$$

$(Q \equiv (q_{ij}))$. Multiplying the first equation throughout by $v_{n1}(x)$, the second by $v_{n2}(x)$, integrating with respect to x from a to b, and adding together, we obtain

$$\int_a^b u^T(x)v_n(x)\,dx + \lambda_0 \int_a^b u^T(\xi)\left\{\int_a^b Q^T(x,\xi)v_n(x)\,dx\right\}\,d\xi = 0. \qquad (3.19)$$

Since $Q(x,\xi) = Q^T(\xi,x)$,

$$\int_a^b Q^T(x,\xi)v_n(x)\,dx = \int_a^b Q(\xi,x)v_n(x)\,dx = 0,$$

also by (3.18'). Therefore, it follows from (3.19) that

$$\int_a^b u^T(x)v_n(x)\,dx = 0.$$

By the definition of $Q(x,\xi)$, we have

$$u(x) + \lambda_0 \int_a^b Q(x,\xi)u(\xi)\,d\xi = u(x) - \lambda_0 \int_a^b G(x,\xi)\,d\xi = 0,$$

i.e., $u(x)$ is the vector-valued eigenfunction of the boundary-value problem (1.6), (3.3), (3.4). Since $u(x)$ is orthogonal to all $v_n(x)$, $u_n(x) \equiv 0$; therefore, $Q(x,\xi) \equiv 0$.

Theorem 7.3.2 *(Expansion Theorem.) If $f(x)$ has a continuous derivative and fulfils the boundary conditions (3.3), (3.4), then $f(x)$ can be expanded into an absolutely and uniformly convergent Fourier series of the vector-valued eigenfunctions of the boundary-value problem (1.6), (3.3), (3.4), viz.*

$$f(x) = \sum_{n=-\infty}^{\infty} a_n v_n(x), \quad \text{where} \quad a_n = \int_a^b f^T(x)v_n(x)\,dx. \qquad (3.20)$$

Proof. Put

$$f_2'(x) - p(x)f_1(x) \equiv h_1(x), \qquad f_1'(x) + r(x)f_2(x) \equiv h_2(x).$$

By the formula (3.9′), we then have

$$f(x) = \int_a^b G(x,\xi)h(\xi)\,d\xi, \qquad h(x) = \begin{pmatrix} h_1(x) \\ h_2(x) \end{pmatrix},$$

or, using the expression for $G(x,\xi)$ from (3.16),

$$f(x) = \sum_{n=-\infty}^{\infty} v_n(x)\frac{1}{\lambda_n} \int_a^b h^T(\xi)v_n(\xi)\,d\xi \equiv \sum_{n=-\infty}^{\infty} a_n v_n(x).$$

Furthermore, it follows that

$$f^T(x) = \sum_{n=-\infty}^{\infty} a_n v_n^T(x),$$

which, multiplied on the right by $v_n(x)$ and integrated from a to b, yields

$$a_n = \int_a^b f^T(x)v_n(x)\,dx, \tag{3.20′}$$

since $v_n(x)$ are orthonormalized, and the theorem is thus proved. ∎

Theorem 7.3.3 *For any square-integrable vector-valued function $f(x)$ in an interval $[a, b]$, the Parseval equation*

$$\int_a^b f^2(x)\,dx = \sum_{n=-\infty}^{\infty} a_n^2, \quad f^2(x) = f_1^2(x) + f_2^2(x) \tag{3.21}$$

holds.

Proof. If $f(x)$ fulfils the conditions of the preceding theorem, then the equality (3.21) immediately follows from the uniform convergence of the series (3.20). Indeed, multiplying the expansion (3.20) on the left by $f^T(x)$ and integrating from a to b, we obtain, also relying on (3.20′), that

$$\int_a^b f^T(x)f(x)\,dx \equiv \int_a^b f^2(x)\,dx = \sum_{n=-\infty}^{\infty} a_n \int_a^b f^T(x)v_n(x)\,dx = \sum_{n=-\infty}^{\infty} a_n^2,$$

i.e., the equality (3.21).

The extending the Parseval equation to arbitrary vector-valued functions of class $\mathcal{L}^2(a, b)$ is done by the usual methods. ∎

7.3.3

We now retrace our steps to the formula (3.9). It has already been noticed that its right-hand side is called the *resolvent*. We showed that a resolvent exists for all λ which are not eigenvalues of the boundary-value problem (1.6), (3.3), (3.4). We now show how to expand the resolvent into a Fourier series if the expansion of $f(x)$ is given.

The vector-valued function $y(x, \lambda)$ determined by (3.9) satisfies the boundary conditions (3.3), (3.4). Therefore, integrating by parts, we get

$$
\begin{aligned}
\int_a^b (By(x, \lambda))^T v_n(x)\, dx &= \int_a^b \{[y_2' - p(x)y_1]v_{n1} - [y_1' + r(x)y_2]v_{n2}\}\, dx \\
&= \{y_2(x, \lambda)v_{n1}(x) - y_1(x, \lambda)v_{n2}(x)\}_a^b \\
&\quad - \int_a^b \{[v_{n1}' + r(x)v_{n2}]y_2 - [v_{n2}' - p(x)v_{n1}]y_1\}\, dx \\
&= \lambda_n \int_a^b v_n^T(x)y(x, \lambda)\, dx \\
&\equiv \lambda_n d_n(\lambda) \qquad (3.22)
\end{aligned}
$$

(see (3.9')). Let $y(x, \lambda) = \sum_{n=-\infty}^{\infty} d_n(\lambda)v_n(x)$, $a_n = \int_a^b f^T(x)v_n(x)\, dx$. Then, by (3.9') and (3.22),

$$
\begin{aligned}
a_n = \int_a^b f^T(x)v_n(x)\, dx &= \int_a^b (By(x, \lambda))^T v_n(x)\, dx - \lambda \int_a^b y^T(x, \lambda)v_n(x)\, dx \\
&= (\lambda_n - \lambda)d_n(\lambda).
\end{aligned}
$$

Hence, $d_n(\lambda) = a_n/(\lambda_n - \lambda)$, and the resolvent expansion is of the form

$$
y(x, \lambda) = \sum_{n=-\infty}^{\infty} \frac{a_n}{\lambda_n - \lambda} v_n(x). \qquad (3.23)
$$

7.4 Periodic and semi-periodic problems

Consider the eigenvalue problem for the system

$$
\begin{pmatrix} 0 & 1 \\ -1 & 0 \end{pmatrix} \frac{dy}{dx} + \begin{pmatrix} p(x) & 0 \\ 0 & r(x) \end{pmatrix} y = \lambda y, \quad y(x) = \begin{pmatrix} y_1(x) \\ y_2(x) \end{pmatrix},
$$

or, in expanded form,

$$
y_2' + p(x)y_1 = \lambda y_1, \qquad -y_1' + r(x)y_2 = \lambda y_2. \qquad (4.1)
$$

Assume that the coefficients of the system, i.e., $p(x)$ and $r(x)$, are two real-valued smooth periodic functions with period a, or $p(x + a) = p(x)$, $r(x + a) = r(x)$ for any real x.

In connection with (4.1), consider the boundary conditions

$$y_1(0) = y_1(a), \qquad y_2(0) = y_2(a) \tag{4.2}$$

$$y_1(0) = -y_1(a), \qquad y_2(0) = -y_2(a). \tag{4.3}$$

The boundary-value problem (4.1), (4.2) is said to be *periodic*, while (4.1), (4.3) is said to be *semi-periodic*.

Along with the problems (4.1), (4.2) and (4.1), (4.3), those for (4.1) with boundary conditions

$$y_1(0) = y_1(a) \tag{4.4}$$

$$y_2(0) = y_2(a) \tag{4.4'}$$

are also useful.

Let

$$\varphi(x, \lambda) = \begin{pmatrix} \varphi_1(x, \lambda) \\ \varphi_2(x, \lambda) \end{pmatrix} \quad \text{and} \quad \vartheta(x, \lambda) = \begin{pmatrix} \vartheta_1(x, \lambda) \\ \vartheta_2(x, \lambda) \end{pmatrix}$$

be the solutions of (4.1), satisfying the initial conditions

$$\varphi_1(0, \lambda) = \vartheta_2(0, \lambda) = 0, \qquad \varphi_2(0, \lambda) = \vartheta_1(0, \lambda) = 1. \tag{4.5}$$

It follows from the system (4.1) that

$$\frac{d\varphi_2}{dx} + p(x)\varphi_1 = \lambda\varphi_1, \qquad \frac{d\vartheta_2}{dx} + p(x)\vartheta_1 = \lambda\vartheta_1,$$

$$-\frac{d\varphi_1}{dx} + r(x)\varphi_2 = \lambda\varphi_2, \qquad -\frac{d\vartheta_1}{dx} + r(x)\vartheta_2 = \lambda\vartheta_2. \tag{4.6}$$

Multiplying these by $-\vartheta_1$, $-\vartheta_2$, φ_1, φ_2, respectively, and then adding, we obtain

$$\frac{d}{dx}\{\varphi_1(x, \lambda)\vartheta_2(x, \lambda) - \varphi_2(x, \lambda)\vartheta_1(x, \lambda)\} = 0.$$

Therefore,

$$\varphi_1(x, \lambda)\vartheta_2(x, \lambda) - \varphi_2(x, \lambda)\vartheta_1(x, \lambda) = -1. \tag{4.7}$$

Thus, the solutions $\varphi(x, \lambda)$ and $\vartheta(x, \lambda)$ are linearly independent, and form a basis for the solutions of (4.1).

Assume that λ_0 is an eigenvalue of the problem (4.1), (4.2), and that $y(x, \lambda_0)$ is the corresponding vector-valued eigenfunction. Then there are two constants C_1 and C_2 such that

$$y(x, \lambda) = C_1\varphi(x, \lambda_0) + C_2\vartheta(x, \lambda_0),$$

or, in coordinates,

$$y_1(x, \lambda_0) = C_1\varphi_1(x, \lambda_0) + C_2\vartheta_1(x, \lambda_0),$$

$$y_2(x, \lambda_0) = C_1 \varphi_2(x, \lambda_0) + C_2 \vartheta_2(x, \lambda_0).$$

Relying on (4.2) and (4.5), we obtain

$$C_2 = C_1 \varphi_1(a, \lambda_0) + C_2 \vartheta_1(a, \lambda_0),$$

$$C_1 = C_1 \varphi_2(a, \lambda_0) + C_2 \vartheta_2(a, \lambda_0). \tag{4.8}$$

For the system (4.8) of algebraic equations to have nontrivial solutions, it is necessary and sufficient that its determinant vanishes, i.e.,

$$\begin{vmatrix} \varphi_1(a, \lambda_0) & \vartheta_1(a, \lambda_0) - 1 \\ \varphi_2(a, \lambda_0) - 1 & \vartheta_2(a, \lambda_0) \end{vmatrix} =$$

$$\varphi_1(a, \lambda_0)\vartheta_2(a, \lambda_0) - \varphi_2(a, \lambda_0)\vartheta_1(a, \lambda_0) - 1 + \varphi_2(a, \lambda_0) + \vartheta_1(a, \lambda_0)$$

$$= \varphi_2(a, \lambda_0) + \vartheta_1(a, \lambda_0) - 2 = 0$$

(taking the equality (4.7) into account).

Thus, the eigenvalues of the periodic problem (4.1), (4.2) are the roots of the equation

$$\vartheta_1(a, \lambda) + \varphi_2(a, \lambda) = 2. \tag{4.9}$$

It can be proved similarly that the eigenvalues of the semi-periodic problem (4.1), (4.3) are the roots of the equation

$$\vartheta_1(a, \lambda) + \varphi_2(a, \lambda) = -2. \tag{4.10}$$

In the case of the periodic (or semi-periodic) problem, the eigenvalues can repeat (with multiplicity not exceeding two). In contrast to the Sturm-Liouville problem, the functions $\vartheta_1(x, \lambda)$ and $\varphi_2(x, \lambda)$ both oscillate towards positive and negative λ. Therefore, the spectrum extends from $-\infty$ to $+\infty$ for the system (4.1), as for separated boundary conditions.

The following is proved similarly as in the case of the Sturm-Liouville problem.

Lemma 7.4.1 *For an eigenvalue $\bar{\lambda}$ of the periodic (resp. semi-periodic) problem to be repeated, it is necessary and sufficient that*

$$\vartheta_1(a, \bar{\lambda}) = \varphi_2(a, \bar{\lambda}) = 1, \qquad \vartheta_2(a, \bar{\lambda}) = \varphi_1(a, \bar{\lambda}) = 0.$$

$$(\vartheta_1(a, \bar{\lambda}) = \varphi_2(a, \bar{\lambda}) = -1, \quad \vartheta_2(a, \bar{\lambda}) = \varphi_1(a, \bar{\lambda}) = 0.) \tag{4.11}$$

Remark *It suffices to verify only the second part.* In fact, if $\vartheta_2(a, \lambda) = \varphi_1(a, \lambda) = 0$, then it follows from the Wronskian being constant (the equality (4.7)) that $\varphi_2(a, \bar{\lambda})\vartheta_1(a, \bar{\lambda}) = 1$. Furthermore, since $\bar{\lambda}$ is an eigenvalue, $\vartheta_1(a, \bar{\lambda}) + \varphi_2(a, \bar{\lambda}) = \pm 2$; therefore, $\vartheta_1(a, \bar{\lambda})$, $\varphi_2(a, \bar{\lambda})$ are the roots of the quadratic equation $X^2 \pm 2X + 1 = 0$, and hence $\vartheta_1(a, \bar{\lambda}) = \varphi_2(a, \bar{\lambda}) = \pm 1$. In a similar way to the Sturm-Liouville problem, the multiplicity of an eigenvalue is not so simply related to the zeros of the function

$$F(\lambda) \mp 2 \equiv \vartheta_1(a, \lambda) + \varphi_1(a, \lambda) \mp 2.$$

Theorem 7.4.1 *A point $\bar{\lambda}$ is a multiple root of the equations $F(\lambda) \mp 2 = 0$ if and only if*

$$\vartheta_2(a, \bar{\lambda}) = \varphi_1(a, \bar{\lambda}) = 0, \qquad (4.12)$$

i.e., $\bar{\lambda}$ is a repeated eigenvalue.

Proof. If the conditions (4.12) are fulfilled, then, by the Remark after Lemma 7.4.1, $\vartheta_1(a, \bar{\lambda}) = \varphi_2(a, \bar{\lambda}) = \pm 1$. Furthermore, the equality (4.7) can be written in the form

$$-\varphi_1 \vartheta_2 = 1 - \varphi_2 \vartheta_1 = 1 - \frac{1}{4}\{(\vartheta_1 + \varphi_2)^2 - (\vartheta_1 - \varphi_2)^2\}.$$

If (4.12) holds, then the right-hand side of the last equality has a multiple root at the point $\lambda = \bar{\lambda}$, since the left-hand side also possesses this property. However, the right-hand side can be written in the form

$$\frac{1}{4}\{[2 - (\vartheta_1 + \varphi_2)][2 + (\vartheta_1 + \varphi_2)]\} + \frac{1}{4}(\vartheta_1 - \varphi_2)^2.$$

Since $\vartheta_1(a, \bar{\lambda}) = \varphi_2(a, \bar{\lambda})$, $(\vartheta_1 - \vartheta_2)^2$ has a multiple root at the point $\lambda = \bar{\lambda}$. Therefore, the same holds for $[2 - (\vartheta_1 + \varphi_2)][2 + (\vartheta_1 + \varphi_2)]$. But, if $\bar{\lambda}$ is an eigenvalue, say, of the periodic problem, then the second factor $2 + (\vartheta_1 + \varphi_2) = 4$, and, therefore, the function $2 - (\vartheta_1 + \varphi_2) \equiv 2 - F(\lambda)$ has a multiple root for $\lambda = \bar{\lambda}$.

Conversely, let $F'(\bar{\lambda}) = 0$, i.e., let $\bar{\lambda}$ be a multiple root of the equation $F(\lambda) = \pm 2$. Differentiating the equations (4.6) with respect to λ, we have

$$\frac{\partial^2 \varphi_2}{\partial x \partial \lambda} + \{p(x) - \lambda\}\frac{\partial \varphi_1}{\partial \lambda} = \varphi_1, \qquad \frac{\partial^2 \vartheta_2}{\partial x \partial \lambda} + \{p(x) - \lambda\}\frac{\partial \vartheta_1}{\partial \lambda} = \vartheta_1,$$

$$-\frac{\partial^2 \varphi_1}{\partial x \partial \lambda} + \{r(x) - \lambda\}\frac{\partial \varphi_2}{\partial \lambda} = \varphi_2, \qquad -\frac{\partial^2 \vartheta_1}{\partial x \partial \lambda} + \{r(x) - \lambda\}\frac{\partial \vartheta_2}{\partial \lambda} = \vartheta_2,$$

$$\frac{\partial \varphi_1(0, \lambda)}{\partial \lambda} = \frac{\partial \varphi_2(0, \lambda)}{\partial \lambda} = 0, \qquad \frac{\partial \vartheta_1(0, \lambda)}{\partial \lambda} = \frac{\partial \vartheta_2(0, \lambda)}{\partial \lambda} = 0.$$

Therefore, solving these problems for the vector-valued functions $\partial \varphi / \partial \lambda$ and $\partial \vartheta / \partial \lambda$ by the method of variation of the constants, we have

$$\frac{\partial \varphi_1}{\partial \lambda} = \varphi_1(x, \lambda)\int_0^x \varphi^T(\xi, \lambda)\vartheta(\xi, \lambda)\, d\xi - \vartheta_1(x, \lambda)\int_0^x \varphi^2(\xi, \lambda)\, d\xi$$

$$\frac{\partial \varphi_2}{\partial \lambda} = \varphi_2(x, \lambda)\int_0^x \varphi^T(\xi, \lambda)\vartheta(\xi, \lambda)\, d\xi - \vartheta_2(x, \lambda)\int_0^x \varphi^2(\xi, \lambda)\, d\xi$$

$$\frac{\partial \vartheta_1}{\partial \lambda} = \vartheta_1(x, \lambda)\int_0^x \varphi^T(\xi, \lambda)\vartheta(\xi, \lambda)\, d\xi - \varphi_1(x, \lambda)\int_0^x \vartheta^2(\xi, \lambda)\, d\xi$$

$$\frac{\partial \vartheta_2}{\partial \lambda} = \vartheta_2(x, \lambda)\int_0^x \varphi^T(\xi, \lambda)\vartheta(\xi, \lambda)\, d\xi - \varphi_2(x, \lambda)\int_0^x \vartheta^2(\xi, \lambda)\, d\xi$$

$$(4.13)$$

where $\varphi^T \vartheta = \varphi_1 \vartheta_1 + \varphi_2 \vartheta_2$, $\varphi^2 = \varphi_1^2 + \varphi_2^2$. It follows from (4.13) that $(F(\lambda) = \vartheta_1(a, \lambda) + \varphi_2(a, \lambda))$,

$$\frac{dF(\lambda)}{d\lambda} = (\varphi_2 - \vartheta_1) \int_0^a \varphi^T(\xi, \lambda) \vartheta(\xi, \lambda)\, d\xi + \varphi_1 \int_0^a \vartheta^2(\xi, \lambda)\, d\xi - \vartheta_2 \int_0^a \varphi^2(\xi, \lambda)\, d\xi \quad (4.14)$$

for $x = a$ (here and onwards, $\varphi(a, \lambda)$, $\vartheta(a, \lambda)$ are denoted by φ and ϑ).

Now, let $-2 < F(\lambda) < 2$. Then

$$(\vartheta_1 + \varphi_2)^2 = \vartheta_1^2 + \varphi_2^2 + 2\vartheta_1 \varphi_2 < 4 = 4(\varphi_2 \vartheta_1 - \varphi_1 \vartheta_2)$$

$$(\vartheta_1 - \varphi_2)^2 = (\vartheta_1 + \varphi_2)^2 - 4\vartheta_1 \varphi_2 = (\vartheta_1 + \varphi_2)^2 - 4 - 4\varphi_1 \vartheta_2 < -4\varphi_1 \vartheta_2.$$

Therefore, φ_1 and ϑ_2 are nonzero, and of opposite signs.
The identity (4.14) is transformed as

$$\frac{dF(\lambda)}{d\lambda} = \varphi_1 \int_0^a \left\{ \vartheta_1^2(\xi, \lambda) + \frac{\varphi_2 - \vartheta_1}{\varphi_1} \vartheta_1(\xi, \lambda) \varphi_1(\xi, \lambda) - \frac{\vartheta_2}{\varphi_1} \varphi_1^2(\xi, \lambda) \right\} d\xi$$

$$+ \varphi_1 \int_0^a \left\{ \vartheta_2^2(\xi, \lambda) + \frac{\varphi_2 - \vartheta_1}{\varphi_1} \vartheta_2(\xi, \lambda) \varphi_2(\xi, \lambda) - \frac{\vartheta_2}{\varphi_1} \varphi_2^2(\xi, \lambda) \right\} d\xi$$

$$= \varphi_1 \int_0^a \left\{ \vartheta_1(\xi, \lambda) + \frac{\varphi_2 - \vartheta_1}{2\varphi_1} \varphi_1(\xi, \lambda) \right\}^2 d\xi + \frac{4 - F^2(\lambda)}{4\varphi_1} \int_0^a \varphi_1^2(\xi, \lambda)\, d\xi$$

$$+ \varphi_2 \int_0^a \left\{ \vartheta_2(\xi, \lambda) + \frac{\varphi_2 - \vartheta_1}{2\varphi_1} \varphi_2(\xi, \lambda) \right\}^2 d\xi + \frac{4 - F^2(\lambda)}{4\varphi_1} \int_0^a \varphi_2^2(\xi, \lambda)\, d\xi, \quad (4.15)$$

where the right-hand side does not vanish, and its sign coincides with that of φ_1. Hence, $F(\lambda)$ cannot have either a maximum or a minimum at the point with $F^2(\lambda) < 4$, and if $F(\bar{\lambda}) = 2$, $F'(\bar{\lambda}) < 0$ at a certain point $\bar{\lambda}$, then $F(\lambda)$ decreases until it attains the value -2. Since $F(\lambda)$ oscillates both in the directions of increasing and decreasing λ, which follows from the asymptotic formulas for $\vartheta_1(x, \lambda)$, $\varphi_2(x, \lambda)$ (similarly to (2.13)), in contrast to the Sturm-Liouville problem, the eigenvalues of the periodic and semi-periodic problems for (4.1) interlacing, extend to infinity in both directions, i.e.,

$$\ldots < \lambda_{-2} \leq \lambda_{-1} < \mu_{-2} \leq \mu_{-1} < \lambda_0 \leq \lambda_1 < \mu_0 \leq \mu_1 < \lambda_2 \leq \lambda_3 < \mu_2 \leq \mu_3 < \ldots,$$

where $\lambda_0, \lambda_{\pm 1}, \lambda_{\pm 2}, \ldots$ are the eigenvalues of the periodic problem, and $\mu_0, \mu_{\pm 1}, \mu_{\pm 2}, \ldots$ those of the semi-periodic problem.

We now assume that $\lambda = \bar{\lambda}$ is a zero of the function $F(\lambda) - 2$, of order two or higher. Then, if we assume that $\varphi_1(a, \bar{\lambda}) \neq 0$, we run up against a contradiction with $F'(\bar{\lambda}) = 0$, due to the formula (4.15). Therefore, $\varphi_1(a, \bar{\lambda}) = 0$. It is proved similarly that $\vartheta_2(a, \bar{\lambda}) = 0$. Hence, $\vartheta_1(a, \bar{\lambda}) \varphi_2(a, \bar{\lambda}) = 1$, $\vartheta_1(a, \bar{\lambda}) - 2 = -\varphi_2(a, \bar{\lambda}) = -1/\vartheta_1(a, \bar{\lambda})$, $[\vartheta_1(a, \lambda) - 1]^2 = 0$, i.e., $\vartheta_1(a, \bar{\lambda}) = 1$, and $\varphi_2(a, \bar{\lambda}) = 1$.

Thus, the conditions

$$\vartheta_1(a, \lambda) = \varphi_2(a, \lambda) = 1, \qquad \vartheta_2(a, \lambda) = \varphi_1(a, \lambda) = 0 \tag{4.16}$$

are necessary for $F(\lambda) - 2$ to have a zero of order two. It follows from the formula (4.14) that they are also sufficient.

It can be proved similarly that the necessary and sufficient conditions for a multiple root of the function $F(\lambda) + 2$ to appear are of the form

$$\vartheta_1(a, \lambda) = \varphi_2(\lambda) = -1, \qquad \vartheta_2(0, \lambda) = \varphi_1(a, \lambda) = 0. \tag{4.17}$$

We show that the function $F(\lambda) - 2$ cannot have zeros of order higher than two. Differentiating the equations (4.6) with respect to λ twice, we obtain the problems

$$\frac{\partial^3 \vartheta_2}{\partial x \partial \lambda^2} + \{p(x) - \lambda\} \frac{\partial^2 \vartheta_1}{\partial \lambda^2} = 2\frac{\partial \vartheta_1}{\partial \lambda}$$

$$-\frac{\partial^3 \vartheta_1}{\partial x \partial \lambda^2} + \{r(x) - \lambda\} \frac{\partial^2 \vartheta_2}{\partial \lambda^2} = 2\frac{\partial \vartheta_2}{\partial \lambda}$$

$$\frac{\partial^2 \vartheta_1(0, \lambda)}{\partial \lambda^2} = \frac{\partial^2 \vartheta_2(0, \lambda)}{\partial \lambda^2} = 0$$

$$\frac{\partial^3 \varphi_2}{\partial x \partial \lambda^2} + \{p(x) - \lambda\} \frac{\partial^2 \varphi_1}{\partial \lambda^2} = 2\frac{\partial \varphi_1}{\partial \lambda}$$

$$-\frac{\partial^3 \varphi_1}{\partial x \partial \lambda^2} + \{r(x) - \lambda\} \frac{\partial^2 \varphi_2}{\partial \lambda^2} = 2\frac{\partial \varphi_2}{\partial \lambda}$$

$$\frac{\partial^2 \varphi_1(0, \lambda)}{\partial \lambda^2} = \frac{\partial^2 \varphi_2(0, \lambda)}{\partial \lambda^2} = 0$$

for the vector-valued functions $\partial^2 \vartheta / \partial \lambda^2$ and $\partial^2 \varphi / \partial \lambda^2$, which can be solved by the method of variation of the constants. For the purpose, it suffices to have expressions for

$$\frac{\partial^2 \vartheta_1(x, \lambda)}{\partial \lambda^2} = 2\varphi_1(x, \lambda) \int_0^x \left\{ \frac{\partial \vartheta_1(\xi, \lambda)}{\partial \lambda} \vartheta_1(\xi, \lambda) + \frac{\partial \vartheta_2(\xi, \lambda)}{\partial \lambda} \vartheta_2(\xi, \lambda) \right\} d\xi$$

$$-2\vartheta_1(x, \lambda) \int_0^x \left\{ \frac{\partial \vartheta_1(\xi, \lambda)}{\partial \lambda} \varphi_1(x, \lambda) + \frac{\partial \vartheta_2(\xi, \lambda)}{\partial \lambda} \varphi_2(\xi, \lambda) \right\} d\xi$$

and

$$\frac{\partial^2 \varphi_2(x, \lambda)}{\partial \lambda^2} = 2\varphi_2(x, \lambda) \int_0^x \left\{ \frac{\partial \varphi_1(\xi, \lambda)}{\partial \lambda} \vartheta_1(\xi, \lambda) + \frac{\partial \varphi_2(\xi, \lambda)}{\partial \lambda} \vartheta_2(\xi, \lambda) \right\} d\xi$$

$$-2\vartheta_2(x, \lambda) \int_0^x \left\{ \frac{\partial \varphi_1(\xi, \lambda)}{\partial \lambda} \varphi_1(\xi, \lambda) + \frac{\partial \varphi_2(\xi, \lambda)}{\partial \lambda} \varphi_2(x, \lambda) \right\} d\xi.$$

Putting $x = a$, taking into account (4.16) and adding together, we obtain

$$\frac{d^2 F(\lambda)}{d\lambda^2} = 2 \int_0^a \left\{ \frac{\partial \varphi_1(\xi, \lambda)}{\partial \lambda} \vartheta_1(\xi, \lambda) + \frac{\partial \varphi_2(\xi, \lambda)}{\partial \lambda} \vartheta_2(\xi, \lambda) \right\} d\xi$$
$$- 2 \int_0^a \left\{ \frac{\partial \vartheta_1(\xi, \lambda)}{\partial \lambda} \vartheta_1(\xi, \lambda) + \frac{\partial \vartheta_2(\xi, \lambda)}{\partial \lambda} \vartheta_2(\xi, \lambda) \right\} d\xi.$$

If we substitute the expressions from the equalities (4.13) for $\partial\varphi/\partial\lambda$ and $\partial\vartheta/\partial\lambda$, then, after simple calculations, we obtain

$$\frac{d^2 F(\lambda)}{d\lambda^2} = -\int_0^a d\xi \int_0^\xi \{\vartheta_1(\xi, \lambda)\varphi_1(t, \lambda) - \vartheta_1(t, \lambda)\varphi_1(\xi, \lambda)\}^2 dt$$
$$- \int_0^a d\xi \int_0^\xi \{\vartheta_1(\xi, \lambda)\varphi_2(t, \lambda) - \vartheta_2(t, \lambda)\varphi_1(\xi, \lambda)\}^2 dt,$$

where the right-hand side is negative; therefore, $d^2 F(\lambda)/d\lambda^2|_{\lambda=\tilde{\lambda}} \neq 0$, as required.

Proceeding from (4.17), the proof for the semi-periodic problem is obtained similarly. The theorem is thus proved. ∎

7.5   Trace calculation

7.5.1

Consider the boundary-value problem

$$\begin{aligned} u'(x) - \{p(x) + \lambda\}v(x) &= 0, \\ v'(x) + \{q(x) + \lambda\}u(x) &= 0, \end{aligned} \quad 0 \leq x \leq \pi \tag{5.1}$$

$$u(0)\cos\alpha + v(0)\sin\alpha = 0 \tag{5.2}$$

$$u(\pi)\cos\beta + v(\pi)\sin\beta = 0 \tag{5.3}$$

where $p(x), q(x) \in C^1(0, \pi)$ and α, β are two real numbers.

Let $\{\lambda_n\}_{n=-\infty}^{\infty}$ be the eigenvalues of (5.1–3). As was indicated in Section 2, the asymptotic formula

$$\lambda_n = c_0 + n + \frac{c_1}{n} + O\left(\frac{1}{n^2}\right), \tag{5.4}$$

where

$$c_0 = \beta - \alpha + \frac{1}{2} \int_0^\pi \{p(x) + q(x)\} \, dx, \tag{5.5}$$

holds for sufficiently large n and under the above assumptions; c_1 will be calculated below.

It follows from (5.4) that the series

$$S_\lambda = \lambda_0 + \sum_{n=1}^{\infty} (\lambda_n - \lambda_{-n} - c_0)$$

converges. This is called the *regularized trace of a one-dimensional Dirac operator.*
If we assume that the functions $p(x)$ and $q(x)$ are such that

$$c_0 = \frac{1}{2} \int_0^\pi \{p(x) + q(x)\}\, dx + \beta - \alpha = 0, \tag{5.6}$$

then we obtain the series

$$S_\lambda^0 = \sum_{-\infty}^\infty \lambda_n \tag{5.7}$$

called the *Dirac operator trace*, and not S_λ.

Our goal is to find S_λ^0. We prove that

$$S_\lambda^0 = \frac{1}{4}\{[q(0) - p(0)]\cos 2\alpha + [q(\pi) - p(\pi)]\cos 2\beta\}. \tag{5.8}$$

7.5.2

It is easy to see that the solution of the system (5.1), fulfilling the condition $u(0, \lambda) = -\sin \alpha$, $v(0, \lambda) = \cos \alpha$, satisfies the system of integral equations

$$u(x, \lambda) = \sin(\lambda x - \alpha) - \int_0^x u(s, \lambda)q(s)\sin\lambda(x - s)\, ds + \int_0^x v(s, \lambda)p(s)\cos\lambda(x - s)\, ds, \tag{5.9}$$

$$v(x, \lambda) = \cos(\lambda x - \alpha) - \int_0^x v(s, \lambda)p(s)\sin\lambda(x - s)\, ds - \int_0^x u(s, \lambda)q(s)\cos\lambda(x - s)\, ds. \tag{5.10}$$

It is obvious that the functions $u(x, \lambda)$ and $v(x, \lambda)$ also satisfy the boundary conditions (5.2).

Solving the system (5.9), (5.10) by the method of successive approximations, we obtain

$$u(x, \lambda) = \sin\{\eta(x) + \lambda x - \alpha\} + \frac{1}{\lambda}U(x, \lambda) + O\left(\frac{1}{\lambda^2}\right) \tag{5.11}$$

$$v(x, \lambda) = \cos\{\eta(x) + \lambda x - \alpha\} + \frac{1}{\lambda}V(x, \lambda) + O\left(\frac{1}{\lambda^2}\right) \tag{5.12}$$

for large $|\lambda|$, where

$$U(x, \lambda) = \frac{1}{4}\{p(x) - q(x)\}\sin\{\eta(x) + \lambda x - \alpha\} + \frac{1}{4}\{p(0) - q(0)\}\sin\{\eta(x) + \lambda x + \alpha\}$$
$$- \frac{1}{8}\int_0^x \{p(s) - q(s)\}^2\, ds \cdot \cos\{\eta(x) + \lambda x - \alpha\}$$

$$V(x, \lambda) = \frac{1}{4}\{p(0) - q(0)\}\cos\{\eta(x) + \lambda x + \alpha\} - \frac{1}{4}\{p(x) - q(x)\}\cos\{\eta(x) + \lambda x - \alpha\}$$
$$+ \frac{1}{8}\int_0^x \{p(s) - q(s)\}^2\, ds \cdot \sin\{\eta(x) + \lambda x - \alpha\},$$

$$\eta(x) = \frac{1}{2} \int_0^x \{p(s) + q(s)\}\, ds.$$

The eigenvalues of the boundary-value problem (5.1–3) satisfy the equation

$$u(\pi, \lambda) \cos \beta + v(\pi, \lambda) \sin \beta = 0. \tag{5.13}$$

Therefore, using the expressions (5.11) and (5.12), we derive from (5.13) that

$$\tan \lambda \pi = -\tan\{\eta(\pi) - \alpha + \beta\} - \frac{1}{\lambda}\frac{1}{4\cos^2\{\eta(\pi) - \alpha + \beta\}} \{[p(0) - q(0)]\sin 2\alpha$$
$$-[p(\pi) - q(\pi)]\sin 2\beta - \frac{1}{2}\int_0^\pi [p(s) - q(s)]^2\, ds\} + O\left(\frac{1}{\lambda^2}\right).$$

Substituting $\lambda_n = n + \delta_n$ for λ, we have

$$\lambda_n = -\frac{1}{\pi}\{\eta(\pi) - \alpha + \beta\} + n + \frac{c_1}{n} + O\left(\frac{1}{n^2}\right), \tag{5.14}$$

where

$$c_1 = \frac{1}{4\pi}\frac{1}{\cos^2\{\eta(\pi) - \alpha + \beta\}} \{[p(\pi) - q(\pi)]\sin 2\beta - [p(0) - q(0)]\sin 2\alpha$$
$$+ \frac{1}{2}\int_0^\pi [p(s) - q(s)]^2\, ds\}. \tag{5.15}$$

By the condition (5.6), it follows from (5.14) that

$$\lambda_n = n + \frac{a_1}{n} + O\left(\frac{1}{n^2}\right), \tag{5.16}$$

with

$$a_1 = \frac{1}{4\pi}\{[p(\pi) - q(\pi)]\sin 2\beta - [p(0) - q(0)]\sin 2\alpha + \frac{1}{2}\int_0^\pi [p(s) - q(s)]^2\, ds\}. \tag{5.17}$$

7.5.3

Let the left-hand side of the equality (5.13) be

$$\Phi(\lambda) = u(\pi, \lambda) \cos \beta + v(\pi, \lambda) \sin \beta. \tag{5.18}$$

The eigenvalues of the problem (5.1–3) are zeros of $\Phi(\lambda)$. Moreover, it is an entire analytic function; therefore, it can be expanded into the infinite product

$$\Phi(\lambda) = A\left(\frac{\lambda}{\lambda_0} - 1\right)\prod_{n=1}^\infty \left(1 - \frac{\lambda}{\lambda_{-n}}\right)\left(1 - \frac{\lambda}{\lambda_n}\right), \tag{5.19}$$

where A is a constant quantity to be determined in the sequel. The convergence of (5.19) easily follows from the asymptotic formula (5.16).

After studying the asymptotic behaviour of both sides of (5.18) for large imaginary $\lambda = i\mu$, comparing the coefficients of the same powers of λ on the left and right, in particular, we obtain the value of the sum S_λ^0.

It follows from the equalities (5.2), (5.6), (5.11) and (5.12) that

$$u(\pi, i\mu)\cos\beta + v(\pi, i\mu)\sin\beta$$
$$= \frac{1}{2}e^{\pi\mu}\left\{ i + \frac{1}{\mu}\left[\frac{p(\pi) - q(\pi)}{4}\cos 2\beta + \frac{p(0) - q(0)}{4}\cos 2\alpha \right] + \frac{i}{\mu}\left[\frac{p(\pi) - q(\pi)}{4}\sin 2\beta \right.\right.$$
$$\left.\left. - \frac{p(0) - q(0)}{4}\sin 2\alpha + \frac{1}{8}\int_0^\pi \{p(s) - q(s)\}^2\, ds \right]\right\} + O\left(\frac{1}{\mu^2}\right). \quad (5.20)$$

On the other hand, relying on the familiar equality

$$\sin z = z\prod_{n=1}^\infty \left(1 - \frac{z^2}{n^2\pi^2} \right),$$

we have

$$\Phi(i\mu) = A\left(\frac{i\mu}{\lambda_0} - 1 \right)\prod_{n=1}^\infty \left(1 - \frac{i\mu}{\lambda_{-n}} \right)\left(1 - \frac{i\mu}{\lambda_n} \right)\frac{\sinh \pi\mu}{\prod_{n=1}^\infty \left(1 + \frac{\mu^2}{n^2} \right)\pi\mu}$$
$$= AB(i\mu - \lambda_0)\frac{\sinh \pi\mu}{\pi\mu}\psi(\mu) \quad (5.21)$$

on the basis of (5.19), where

$$B = \frac{1}{\lambda_0}\prod_{n=1}^\infty \left(-\frac{n^2}{\lambda_n\lambda_{-n}} \right)$$

$$\psi(\mu) = \prod_{n=1}^\infty \left(1 - \frac{\lambda_n\lambda_{-n} + n^2 - i\mu(\lambda_n + \lambda_{-n})}{n^2 + \mu^2} \right).$$

To study the asymptotic behaviour of the function $\psi(\mu)$ for large μ, we consider

$$\ln\psi(\mu) = -\sum_{n=1}^\infty\sum_{k=1}^\infty \frac{1}{k}\left(\frac{n^2 + \lambda_n\lambda_{-n} - i\mu(\lambda_n + \lambda_{-n})}{n^2 + \mu^2} \right)^k$$
$$= -\sum_{k=1}^\infty \frac{1}{k}\sum_{n=1}^\infty \left(\frac{n^2 + \lambda_n\lambda_{-n} - i\mu(\lambda_n + \lambda_{-n})}{n^2 + \mu^2} \right)^k.$$

It is easy to establish that

$$-\sum_{k=2}^\infty \frac{1}{k}\sum_{n=1}^\infty \left(\frac{n^2 + \lambda_n\lambda_{-n} - i\mu(\lambda_n + \lambda_{-n})}{n^2 + \mu^2} \right)^k = O\left(\frac{1}{\mu^2} \right).$$

Furthermore, by the familiar formula

$$\sum_{n=1}^{\infty} \frac{1}{n^2 + \mu^2} = \frac{\pi \coth \pi\mu}{2\mu} - \frac{1}{2\mu^2} = \frac{\pi}{2\mu} - \frac{1}{2\mu^2} + O(e^{-2\pi\mu}),$$

we obtain

$$-\sum_{n=1}^{\infty} \frac{n^2 + \lambda_n \lambda_{-n}}{n^2 + \mu^2} = \frac{a_1 \pi}{\mu} + O\left(\frac{1}{\mu^2}\right)$$

$$i \sum_{n=1}^{\infty} \frac{\mu(\lambda_n + \lambda_{-n})}{n^2 + \mu^2} = \frac{i}{\mu} \sum_{n=1}^{\infty} (\lambda_n + \lambda_{-n}) + O\left(\frac{1}{\mu^2}\right)$$

(using (5.16)). Therefore,

$$\ln \psi(\mu) = \frac{a_1 \pi}{\mu} + \frac{i}{\mu} \sum_{n=1}^{\infty} (\lambda_n + \lambda_{-n}) + O\left(\frac{1}{\mu^2}\right)$$

and

$$\psi(\mu) = 1 + \frac{a_1 \pi}{\mu} + \frac{i}{\mu} \sum_{n=1}^{\infty} (\lambda_n + \lambda_{-n}) + O\left(\frac{1}{\mu^2}\right).$$

Substituting the latter expression in (5.21), we get

$$\Phi(i\mu) = \frac{AB}{2\pi} e^{\pi\mu} \left\{ i + \frac{i}{\mu} a_1 \pi - \frac{1}{\mu} \sum_{n=-\infty}^{\infty} \lambda_n \right\} + O\left(\frac{1}{\mu^2}\right). \tag{5.22}$$

Now, comparing (5.20) with (5.22), we obtain (5.8) and $AB = \pi$, i.e., $A = \pi/B$.

Notes

Section 1

The canonical form (1.4) of the system of two equations of order one is due to I. S. Sargsjan (1966a), while (1.5) is due to M. G. Gasymov (1968).

Section 2

Here, the results are mostly due to W. Hurwitz (1921). The idea to apply transformation operators is due to B. M. Levitan and I. S. Sargsjan (1975).

Section 3

The completeness of a system of vector-valued eigenfunctions for a system of equations of order one was proved by W. Hurwitz (1921). The use of the method of integral equations is due to B. M. Levitan and I. S. Sargsjan (1975).

Section 4

The results of this section are similar to those for the Sturm-Liouville operator; see E. Titchmarsh (1962).

Section 5

The trace was calculated by E. Abdukadyrov (1967).

Chapter 8

SPECTRAL THEORY IN THE SINGULAR CASE

8.1 Proof of the Parseval equation on the half-line

8.1.1

In a similar way to the case of the Sturm-Liouville operator, considered in Chapter 2, we prove the expansion theorem and the Parseval equation for the singular Dirac operator, considering the latter as the limit of regular problems.

Consider the case of $[0, \infty)$.

Thus, we study the problem $(0 \leq x < \infty)$

$$y_2' - \{\lambda + q_1(x)\}y_1 = 0, \tag{1.1}$$

$$y_1' + \{\lambda + q_2(x)\}y_2 = 0, \tag{1.2}$$

where the coefficients $q_1(x)$ and $q_2(x)$ are continuous in each finite interval $[0, b]$.

Let $\varphi(x, \lambda)$ be the solution of the system (1.1), (1.2), satisfying the initial conditions

$$\varphi_1(0, \lambda) = \cos \alpha, \qquad \varphi_2(0, \lambda) = -\sin \alpha, \tag{1.3}$$

where α is an arbitrary real number. It is obvious that $\varphi(x, \lambda)$ satisfies the boundary condition

$$y_1(0, \lambda) \sin \alpha + y_2(0, \lambda) \cos \alpha = 0. \tag{1.4}$$

Furthermore, let b be an arbitrary positive number (letting it increase indefinitely in the sequel), and let β be an arbitrary real number; subjoin the boundary condition

$$y_1(b, \lambda) \sin \beta + y_2(b, \lambda) \cos \beta = 0 \tag{1.5}$$

to the problem (1.1), (1.2), (1.4).

The problem (1.1), (1.2), (1.4), (1.5) is regular. Let $\lambda_{n,b}$ be the eigenvalues, and $\varphi_{n,b}(x) = \varphi(x, \lambda_{n,b})$ the corresponding vector-valued eigenfunctions of the problem, satisfying the initial conditions (1.3). Then, if $f(x) \in \mathcal{L}^2(0, \infty)$ and

$$\alpha_{n,b}^2 = \int_0^b \{\varphi_1^2(x, \lambda_{n,b}) + \varphi_2^2(x, \lambda_{n,b})\} \, dx,$$

we have

$$\int_0^b \{f_1^2(x) + f_2^2(x)\}\, dx = \sum_{n=-\infty}^{\infty} \frac{1}{\alpha_{n,b}^2} \left\{ \int_0^b f^T(x)\varphi_{n,b}(x)\, dx \right\}^2 \qquad (1.6)$$

owing to the Parseval equation (see the formula (3.20), Chapter 7).

If we introduce the monotonically increasing step function

$$\rho_b(\lambda) = \begin{cases} \displaystyle\sum_{0 \leq \lambda_{n,b} < \lambda} \frac{1}{\alpha_{n,b}^2}, & \lambda \geq 0 \\[3mm] -\rho_b(-\lambda), & \lambda < 0 \end{cases} \qquad (1.7)$$

then the Parseval equation (1.6) can be written as

$$\int_0^b \{f_1^2(x) + f_2^2(x)\}\, dx = \int_{-\infty}^{\infty} F^2(\lambda)\, d\rho_b(\lambda), \qquad (1.8)$$

where $F(\lambda) = \int_0^b f^T(x)\varphi(x,\lambda)\, dx$.

We show in the sequel that the Parseval equation for the problem (1.1), (1.2), (1.4) is obtained from (1.8) as $b \to \infty$. The following is valid.

Lemma 8.1.1 *For any positive number N, there is a positive constant number $A = A(N)$, not depending on b, such that*

$$\bigvee_{-N}^{N} \{\rho_b(\lambda)\} = \sum_{-N \leq \lambda_{n,b} \leq N} \frac{1}{\alpha_{n,b}^2} = \rho_b(N) - \rho_b(-N) < A.$$

The proof is exactly the same as that of the corresponding lemma for the Sturm-Liouville operator (see Lemma 2.1.1).

We now show that, by Lemma 8.1.1 and by the familiar theorems on taking the limit in the integrand of the Stieltjes integral, we can derive the Parseval equation for (1.1), (1.2), (1.4).

First, let the vector-valued function $f_n(x)$ vanish outside the interval $[0, n]$, $n < b$, have continuous first derivative, and satisfy the boundary condition (1.4). Applying the Parseval equation (1.8) to $f_n(x)$, we obtain

$$\int_0^b \{f_{1n}^2(x) + f_{2n}^2(x)\}\, dx = \int_{-\infty}^{\infty} F_n^2(\lambda)\, d\rho_b(\lambda), \qquad (1.9)$$

where

$$F_n(\lambda) = \int_0^b f_n^T(x)\varphi(x,\lambda)\, dx. \qquad (1.10)$$

Since $\varphi(x,\lambda)$ satisfies the system (1.1), (1.2),

$$\varphi_1(x,\lambda) = \frac{1}{\lambda}[\varphi_2' - q_1(x)\varphi_1],$$

$$\varphi_2(x, \lambda) = \frac{1}{\lambda}[-\varphi_1' - q_2(x)\varphi_2],$$

and it follows from (1.10) that

$$F_n(\lambda) = \frac{1}{\lambda} \int_0^b f_{1n}(x)[\varphi_2'(x, \lambda) - q_1(x)\varphi_1(x, \lambda)] \, dx$$

$$+ \frac{1}{\lambda} \int_0^b f_{2n}(x)[-\varphi_1'(x, \lambda) - q_2(x)\varphi_2(x, \lambda)] \, dx.$$

Furthermore, since both $f_n(x)$ and $\varphi(x, \lambda)$ satisfy the boundary condition (1.4), while $f(x)$ is identically zero in the vicinity of the point b, integrating by parts on the right-hand side, we obtain

$$F_n(\lambda) = \frac{1}{\lambda} \int_0^b \{\varphi_1(x, \lambda)[f_{2n}'(x) - q_1(x)f_{1n}(x)] + \varphi_2(x, \lambda)[-f_{1n}'(x) - q_2(x)f_{2n}(x)]\} \, dx.$$

Hence, for arbitrary finite $N > 0$, by the Parseval equation (1.8),

$$\int_{|\lambda| > N} F_n^2(\lambda) \, d\rho_b(\lambda) \leq \frac{1}{N^2} \int_{|\lambda| > N} \left\{ \int_0^b [\varphi_1(x, \lambda)\{f_{2n}' - q_1(x)f_{1n}\} \right.$$

$$\left. + \varphi_2(x, \lambda)\{-f_{1n}' - q_2(x)f_{2n}\}] \, dx \right\}^2 \, d\rho_b(\lambda)$$

$$\leq \frac{1}{N^2} \int_{-\infty}^{\infty} \left\{ \int_0^b [\varphi_1(x, \lambda)\{f_{2n}' - q_1(x)f_{1n}\} \right.$$

$$\left. + \varphi_2(x, \lambda)\{-f_{1n}' - q_2(x)f_{2n}\}] \, dx \right\}^2 \, d\rho_b(\lambda)$$

$$= \frac{1}{N^2} \int_0^n \{[f_{2n}' - q_1(x)f_{1n}]^2 + [f_{1n}' + q_2(x)f_{2n}]^2\} \, dx.$$

It follows that

$$\left| \int_0^n \{f_{1n}^2(x) + f_{2n}^2(x)\} \, dx - \int_{-N}^N F_n^2(\lambda) \, d\rho_b(\lambda) \right|$$

$$< \frac{1}{N^2} \int_0^n \{[f_{2n}' - q_1(x)f_{1n}]^2 + [f_{1n}' + q_2(x)f_{2n}]^2\} \, dx, \quad (1.11)$$

also relying on (1.9).

By Lemma 8.1.1, the set of monotone functions $\{\rho_b(\lambda)\}$, $-N \leq \lambda \leq N$, is bounded. Therefore, we can pick out a sequence b_k such that the functions $\rho_{b_k}(\lambda)$ converge to a monotone function $\rho(\lambda)$. Passing to the limit with respect to b_k in (1.11), we obtain

$$\left| \int_0^n \{f_{1n}^2(x) + f_{2n}^2(x)\} \, dx - \int_{-N}^N F_n^2(\lambda) \, d\rho(\lambda) \right|$$

$$\leq \frac{1}{N^2} \int_0^n \{[f_{2n}' - q_1(x)f_{1n}]^2 + [f_{1n}' + q_2(x)f_{2n}]^2\} \, dx.$$

Finally, letting $N \to \infty$, we have

$$\int_0^n \{f_{1n}^2(x) + f_{2n}^2(x)\} \, dx = \int_{-\infty}^\infty F_n^2(\lambda) \, d\rho(\lambda),$$

i.e., the Parseval equation for the vector-valued functions $f_n(x)$ vanishing outside a finite interval, with continuous first derivative and boundary condition (1.4). The extending the equation to arbitrary vector-valued functions of class $\mathcal{L}^2(0, \infty)$ is done exactly as for the Sturm-Liouville operator (Chapter 2, Section 1).

We have thus proved the following.

Theorem 8.1.1 *Let a vector-valued function $f(x) \in \mathcal{L}^2(0, \infty)$. There exist a monotonically increasing function $\rho(\lambda)$, $-\infty < \lambda < \infty$, not depending on $f(x)$, and a function $F(\lambda)$ (the generalized Fourier transform of $f(x)$ with respect to the vector-valued eigenfunctions of the Dirac operator, i.e., of the problem (1.1), (1.2), (1.4)), such that*

$$\int_0^\infty \{f_1^2(x) + f_2^2(x)\} \, dx = \int_{-\infty}^\infty F^2(\lambda) \, d\rho(\lambda). \tag{1.12}$$

$F(\lambda)$ *is the mean-square limit of the sequence of continuous functions*

$$F_n(\lambda) = \int_0^n \{f_1(x)\varphi_1(x, \lambda) + f_2(x)\varphi_2(x, \lambda)\} \, dx,$$

i.e., $\lim_{n \to \infty} \int_{-\infty}^\infty \{F(\lambda) - F_n(\lambda)\}^2 \, d\rho(\lambda) = 0.$

8.1.2

Let two vector-valued functions $f(x)$ and $g(x)$ be of class $\mathcal{L}^2(0, \infty)$, and let $F(\lambda)$, $G(\lambda)$ be their Fourier transforms. It is obvious that $f(x) \pm g(x)$ have $F(\lambda) \pm G(\lambda)$ as Fourier transforms. Therefore,

$$\int_0^\infty \{[f_1 + g_1]^2(x) + [f_2 + g_2]^2(x)\} \, dx = \int_{-\infty}^\infty \{F + G\}^2(\lambda) \, d\rho(\lambda),$$

$$\int_0^\infty \{[f_1 - g_1]^2(x) + [f_2 - g_2]^2(x)\} \, dx = \int_{-\infty}^\infty \{F - G\}^2(\lambda) \, d\rho(\lambda),$$

relying on (1.12). Subtracting the lower equality from the upper, we obtain

$$\int_0^\infty \{f_1 \cdot g_1 + f_2 \cdot g_2\}(x) \, dx = \int_{-\infty}^\infty F(\lambda)G(\lambda) d\rho(\lambda), \tag{1.13}$$

which is called the *generalized Parseval equation.*

Theorem 8.1.2 *(Expansion Theorem.) Let $f(x) \in \mathcal{L}^2(0, \infty)$ be a continuous vector-valued function, $0 \le x < \infty$, let $\varphi(x, \lambda)$ and $F(\lambda)$ be as before, and let the integrals*

$$\int_{-\infty}^\infty F(\lambda)\varphi_1(x, \lambda) \, d\rho(\lambda), \qquad \int_{-\infty}^\infty F(\lambda)\varphi_2(x, \lambda) \, d\rho(\lambda) \tag{1.14}$$

converge absolutely and uniformly in x, in each finite interval. Then

$$f_1(x) = \int_{-\infty}^{\infty} F(\lambda)\varphi_1(x, \lambda) \, d\rho(\lambda),$$

$$f_2(x) = \int_{-\infty}^{\infty} F(\lambda)\varphi_2(x, \lambda) \, d\rho(\lambda).$$

Proof. Let $g(x)$ be a continuous vector-valued function vanishing outside a finite interval $[0, n]$. Then (1.13) is written as

$$\int_0^n \{f_1 g_1 + f_2 g_2\}(x) \, dx = \int_{-\infty}^{\infty} F(\lambda) \left\{ \int_0^n [g_1 \varphi_1 + g_2 \varphi_2](x) \, dx \right\} d\rho(\lambda).$$

The absolute convergence allows us to change the order of integration in the last integral, and we rewrite the whole equality as

$$\int_0^n \left\{ \left[f_1(x) - \int_{-\infty}^{\infty} F(\lambda)\varphi_1(x, \lambda) \, d\rho(\lambda) \right] g_1(x) \right.$$
$$\left. + \left[f_2(x) - \int_{-\infty}^{\infty} F(\lambda)\varphi_2(x, \lambda) \, d\rho(\lambda) \right] g_2(x) \right\} dx = 0.$$

Since $g(x)$ is arbitrary, and $f(x)$ and the functions in (1.14) are continuous (the continuity of the latter following from the assumed uniform convergence of the integrals (1.14)), the coefficients of $g_1(x)$ and $g_2(x)$ in the last integrand are zero. The proof is complete. ∎

8.2 The limit-circle and the limit-point cases

Here, we still consider the interval $[0, \infty)$, and assume that the functions $q_1(x)$ and $q_2(x)$ are continuous in each finite interval.

If each solution $\varphi(x, \lambda_0)$ of the equation

$$Ly \equiv \begin{pmatrix} 0 & 1 \\ -1 & 0 \end{pmatrix} \frac{dy}{dx} + \begin{pmatrix} q_1(x) & 0 \\ 0 & q_2(x) \end{pmatrix} y = \lambda_0 y$$

satisfies the condition

$$\int_0^{\infty} |\varphi(x, \lambda_0)|^2 \, dx \equiv \int_0^{\infty} \{|\varphi_1(x, \lambda)|^2 + |\varphi_2(x, \lambda_0)|^2\} \, dx < \infty$$

for a certain complex value λ_0, i.e., $\varphi(x, \lambda_0) \in \mathcal{L}^2(0, \infty)$, then we will say that we are in the *limit-circle case at infinity*; otherwise, we are in the *limit-point case at infinity* for the operator L. To justify the definition, it is necessary to show that this classification only depends on L, and does not depend on the particular value λ_0 selected. This fact is proved exactly as for the Sturm-Liouville operator (Theorem 2.2.2). We are not, however, going to do this, since we are in the limit-point case only for the Dirac operator with the above assumptions about $q_1(x)$ and $q_2(x)$, as is proved in Section 6 of the present chapter.

Let $\varphi(x, \lambda)$ be as above, and let $\vartheta(x, \lambda)$ be the solution of the system (1.1), (1.2), satisfying the conditions

$$\vartheta_1(0, x) = \sin \alpha, \qquad \vartheta_2(0, \lambda) = \cos \alpha. \tag{2.1}$$

Now, if $F(x)$ and $G(x)$ are two solutions of the system (1.1), (1.2), corresponding to λ and λ', then

$$
\begin{aligned}
(\lambda' - \lambda) \int_{x_1}^{x_2} F^T(x) G(x) \, dx &= \int_{x_1}^{x_2} \{ F_1(x)[G_2' + q_1(x)G_1] + F_2(x)[-G_1' + q_2(x)G_2] \\
&\quad + G_1(x)[-F_2' - q_1(x)F_1] + G_2[F_1' - q_2(x)G_2] \} \, dx \\
&= \int_{x_1}^{x_2} \frac{d}{dx} W\{F, G\}(x) \, dx \\
&= W_{x=x_2}\{F, G\} - W_{x=x_1}\{F, G\}.
\end{aligned}
\tag{2.2}
$$

In particular, if $\lambda' = \lambda$, it follows from (2.2) that $W_x\{F, G\} \equiv F_1(x)G_2(x) - F_2(x)G_1(x) = \text{const.}$ Thus, $W\{\varphi, \vartheta\} \equiv 1$, where $\varphi(x, \lambda)$ and $\vartheta(x, \lambda)$ are two linearly independent solutions of (1.1), (1.2), and the general solution of the system can be represented as $\vartheta(x, \lambda) + l\varphi(x, \lambda)$.

Consider those solutions which satisfy the boundary condition

$$\{\vartheta_1(b, \lambda) + l\varphi_1(b, \lambda)\} \cos \beta + \{\vartheta_2(b, \lambda) + l\varphi_2(b, \lambda)\} \sin \beta = 0 \tag{2.3}$$

at the point $x = b$, and determine

$$l = -\frac{\vartheta_1(b, \lambda) \cot \beta + \vartheta_2(b, \lambda)}{\varphi_1(b, \lambda) \cot \beta + \varphi_2(b, \lambda)}.$$

For complex λ, the denominator on the right-hand side is not zero, since the eigenvalues of the regular problem are real. Therefore, if $\cot \beta$ is replaced by a complex variable z, then $l = l(\lambda, z)$ is a meromorphic function of λ; it is regular in the upper and lower half-planes. Therefore,

$$l = l(\lambda, z) = -\frac{\vartheta_1(b, \lambda)z + \vartheta_2(b, \lambda)}{\varphi_1(b, \lambda)z + \varphi_2(b, \lambda)}. \tag{2.4}$$

The real axis in the z-plane is associated with a circle C_b in the l-plane, $l = \infty$ with the point $z = -\varphi_2(b, \lambda)/\varphi_1(b, \lambda)$, while the circle's center is associated with

$$\bar{z} = -\bar{\varphi}_2(b, \lambda)/\bar{\varphi}_1(b, \lambda).$$

Therefore, the center is at the point

$$-\frac{\vartheta_1(b, \lambda)\varphi_2(b, \bar{\lambda}) - \vartheta_2(b, \lambda)\varphi_1(b, \bar{\lambda})}{\varphi_1(b, \lambda)\varphi_2(b, \bar{\lambda}) - \varphi_2(b, \lambda)\varphi_1(b, \bar{\lambda})} \tag{2.5}$$

(noting that $\bar{\varphi}(b, \lambda) = \varphi(b, \bar{\lambda})$).

Furthermore,

$$\text{Im}\left\{-\frac{\varphi_2(b,\lambda)}{\varphi_1(b,\lambda)}\right\} = \frac{1}{2}i\left\{\frac{\varphi_2(b,\lambda)}{\varphi_1(b,\lambda)} - \frac{\varphi_2(b,\bar{\lambda})}{\varphi_1(b,\bar{\lambda})}\right\}$$
$$= \frac{1}{2}i\frac{\varphi_2(b,\lambda)\varphi_1(b,\bar{\lambda}) - \varphi_2(b,\bar{\lambda})\varphi_1(b,\lambda)}{|\varphi_1(b,\lambda)|^2}.$$

If $\lambda' = \bar{\lambda}$, $\lambda = u + iv$, by (2.2) (with the consequence that $G(x) = \bar{F}(x)$), it follows from the last equality that

$$\text{Im}\left\{\frac{\varphi_2(b,\lambda)}{\varphi_1(b,\lambda)}\right\} = v\frac{\int_0^b \{\varphi_1(x,\lambda)\varphi_1(x,\bar{\lambda}) + \varphi_2(x,\lambda)\varphi_2(x,\bar{\lambda})\}\,dx}{|\varphi_1(b,\lambda)|^2},$$

which shows that if $v > 0$, the upper half-plane is associated with the exterior of C_b.

We now calculate the radius. On the one hand, the point $-\vartheta_2(b,\lambda)/\varphi_2(b,\lambda)$ (for $z = 0$) is on C_b; on the other hand, the center is at (2.5) as has already been indicated. Therefore, the radius is

$$r_b = \left|\frac{\vartheta_1(b,\lambda)\varphi_2(b,\bar{\lambda}) - \vartheta_2(b,\lambda)\varphi_1(b,\bar{\lambda})}{\varphi_1(b,\lambda)\varphi_2(b,\bar{\lambda}) - \varphi_2(b,\lambda)\varphi_2(b,\bar{\lambda})} - \frac{\vartheta_2(b,\lambda)}{\varphi_2(b,\lambda)}\right|.$$

Hence, using (2.2) and $W\{\varphi, \vartheta\} \equiv 1$, we find

$$r_b = \frac{1}{2|v|}\left(\int_0^b |\varphi(x,\lambda)|^2\,dx\right)^{-1}. \tag{2.6}$$

Furthermore, l is inside C_b for $v > 0$ if $\text{Im}\,z < 0$, i.e., if $i(z - \bar{z}) > 0$. Therefore, solving (2.4) for z, we obtain

$$i\left\{-\frac{l\varphi_2(b,\lambda) + \vartheta_2(b,\lambda)}{l\varphi_1(b,\lambda) + \vartheta_1(b,\lambda)} + \frac{\bar{l}\bar{\varphi}_2(b,\lambda) + \bar{\vartheta}_2(b,\lambda)}{\bar{l}\bar{\varphi}_1(b,\lambda) + \bar{\vartheta}_1(b,\lambda)}\right\} > 0.$$

Hence,

$$i\{[l\varphi_1(b,\lambda)+\vartheta_1(b,\lambda)][\bar{l}\bar{\varphi}_2(b,\lambda)+\bar{\vartheta}_2(b,\lambda)]-[\bar{l}\bar{\varphi}_1(b,\lambda)+\bar{\vartheta}_1(b,\lambda)][l\varphi_2(b,\lambda)+\vartheta_2(b,\lambda)]\} > 0. \tag{2.7}$$

Now, putting in (2.2) $\lambda = u + iv$, $\lambda' = u - iv$, $G(x) = \bar{F}(x)$, we find

$$2v\int_0^b |F(x)|^2\,dx = i\{F_1(0)\bar{F}_2(0) - F_2(0)\bar{F}_1(0)\} - i\{F_1(b)\bar{F}_2(b) - F_2(b)\bar{F}_1(b)\}. \tag{2.8}$$

Therefore, if $F(x) = \vartheta(x,\lambda) + l\varphi(x,\lambda)$, we derive from (2.8) that

$$2v\int_0^b |\vartheta(x,\lambda) + l\varphi(x,\lambda)|^2\,dx = iW\{\vartheta + l\varphi, \bar{\vartheta} + \bar{l}\bar{\varphi}\}|_b^0, \tag{2.9}$$

and, relying on (2.7), that

$$2v \int_0^b |\vartheta(x,\lambda) + l\varphi(x,\lambda)|^2 \, dx < iW\{\vartheta + l\varphi, \bar{\vartheta} + \bar{l}\bar{\varphi}\}_{x=0}.$$

On the other hand, by (1.2) and (2.1),

$$W\{\vartheta, \bar{\vartheta}\}_{x=0} = W\{\varphi, \bar{\varphi}\}_{x=0} = 0, \qquad W\{\varphi, \bar{\vartheta}\}_{x=0} = W\{\vartheta, \bar{\varphi}\}_{x=0} = 1.$$

Hence,

$$W\{\vartheta + l\varphi, \bar{\vartheta} + \bar{l}\bar{\varphi}\} = l - \bar{l} = 2i\,\text{Im}\,l. \tag{2.10}$$

It then follows from (2.9) that

$$\int_0^b |\vartheta(x,\lambda) + l\varphi(x,\lambda)|^2 \, dx < -\frac{\text{Im}\,l}{v}. \tag{2.11}$$

The same result is obtained in the case $v < 0$ (using the inequality $\text{Im}\,z > 0$). In both cases, the signs of $\text{Im}\,z$ and v are opposite.

Reasoning similarly, it can easily be shown that l is on C_b if and only if

$$\int_0^b |\vartheta(x,\lambda) + l\varphi(x,\lambda)|^2 \, dx = -\frac{\text{Im}\,l}{v}. \tag{2.12}$$

Furthermore, if l is inside C_b and $0 < b' < b$, then

$$\int_0^{b'} |\vartheta + l\varphi|^2 \, dx < \int_0^b |\vartheta + l\varphi|^2 \, dx < -\frac{\text{Im}\,l}{v}.$$

Therefore, l is also inside $C_{b'}$; if $b' < b$, then $C_{b'}$ contains C_b. Consequently, as $b \to \infty$, the circles C_b either converge to the limit circle C_∞ or to the limit point m_∞. If C_b converges to the circle, then its radius $r_\infty = \lim r_b$ is positive, and, relying on (2.6), $\varphi(x,\lambda) \in \mathcal{L}^2(0,\infty)$. If $m = m(\lambda)$ is any point on C_∞, then m is inside any C_b for $b > 0$. Therefore, by (2.11),

$$\int_0^b |\vartheta(x,\lambda) + m\varphi(x,\lambda)|^2 \, dx \leq -\frac{\text{Im}\,m}{v}.$$

Letting $b \to \infty$, we obtain $\vartheta + l\varphi \in \mathcal{L}^2(0,\infty)$. The same argument is valid if C_b converges to m_∞.

Therefore, if $\text{Im}\,\lambda \neq 0$, there always exists a solution of the system (1.1), (1.2), of class $\mathcal{L}^2(0,\infty)$. In the case where $C_b \to C_\infty$, all solutions belong to $\mathcal{L}^2(0,\infty)$ when $\text{Im}\,\lambda \neq 0$, since the solutions φ and $\vartheta + m\varphi$ are such, which identifies the case of the limit-circle with the existence of C_∞. Accordingly, the limit-point case is identified with the existence of m_∞.

In the case where $C_b \to m_\infty$, we have $\lim r_b = 0$, and it follows from (2.6) that $\varphi(x,\lambda) \notin \mathcal{L}^2(0,\infty)$. Hence, there is only one solution of class $\mathcal{L}^2(0,\infty)$. In the case of the limit-circle, l is on C_b if and only if (2.12) holds. Therefore, l is on C_∞ if and only if $v \int_0^\infty |\vartheta + l\varphi|^2 \, dx = -\text{Im}\, l$, and, relying on (2.9), (2.10), (2.11), l is on C_∞ if and only if

$$\lim_{x \to \infty} W\{\vartheta(x, \lambda) + l\varphi(x, \lambda), \bar{\vartheta}(x, \lambda) + \bar{l}\bar{\varphi}(x, \lambda)\} = 0.$$

We have thus proved the following.

Theorem 8.2.1 *If* $\text{Im}\, \lambda \neq 0$ *and* $\varphi(x, \lambda)$, $\vartheta(x, \lambda)$ *are two linearly independent solutions of the system*

$$y_2' - \{\lambda + q_1(x)\}y_1 = 0, \qquad y_1' + \{\lambda + q_2(x)\}y_2 = 0,$$

fulfilling the initial conditions

$$\varphi_1(0, \lambda) = \cos \alpha, \qquad \varphi_2(0, \lambda) = -\sin \alpha,$$

$$\vartheta_1(0, \lambda) = \sin \alpha, \qquad \vartheta_2(0, \lambda) = \cos \alpha$$

then the solution $\psi(x, \lambda) = \vartheta(x, \lambda) + l\varphi(x, \lambda)$ *satisfies the boundary condition*

$$\{\vartheta_1(b, \lambda) + l\varphi_1(b, \lambda)\} \cos \beta + \{\vartheta_2(b, \lambda) + l\varphi_2(b, \lambda)\} \sin \beta = 0$$

if and only if l *is on* C_b *in the complex* l-*plane, with the equation* $W\{\psi, \bar{\psi}\}_{x=b} = 0$. *If* $b \to \infty$, *then* C_b *tends either to the limit circle* C_∞ *or to the limit point* m_∞. *In the former case, all solutions of the system are in the class* $\mathcal{L}^2(0, \infty)$. *In the latter, precisely one linearly independent solution is in* $\mathcal{L}^2(0, \infty)$ *if* $\text{Im}\, \lambda \neq 0$. *Furthermore, in the case of the limit circle, a point is on* C_∞ *if and only if* $\lim_{x \to \infty} W_x\{\psi, \bar{\psi}\} = 0$.

Remark In the case of a limit point, if l is any point on C_b, then l tends to the limit point m_∞ irrespective of the choice of β in (2.3). In particular, this occurs when $\beta = 0$; thus, the limit point is determined by the equality

$$m_\infty(\lambda) = -\lim_{b \to \infty} \frac{\vartheta_1(b, \lambda)}{\varphi_1(b, \lambda)}. \tag{2.13}$$

Put $\psi_b(x, \lambda) = \vartheta(x, \lambda) + l_b\varphi(x, \lambda)$. The following is valid.

Lemma 8.2.1 *For* $\text{Im}\, \lambda \neq 0$ *and as* $b \to \infty$,

$$\psi_b(x, \lambda) \to \psi(x, \lambda), \qquad \int_0^b |\psi_b(x, \lambda)|^2 \, dx \to \int_0^\infty |\psi(x, \lambda)|^2 \, dx.$$

Proof. It is obvious that

$$\psi_b(x, \lambda) = \psi(x, \lambda) + \{l(\lambda, b) - m(\lambda)\}\varphi(x, \lambda),$$

where $\psi(x,\lambda) \in \mathcal{L}^2(0,\infty)$ and $m(\lambda)$ is a point on the limit circle (as is noted above, we prove in Section 6 that we are in the limit-point case only for the Dirac system).

By the formula (2.6), we have

$$|l(x,b) - m(\lambda)| \le 2r_b = \left(|v| \int_0^b |\varphi(x,\lambda)|^2 \, dx\right)^{-1}.$$

Since $r_b \to 0$ as $b \to \infty$,

$$\psi_b(x,\lambda) \to \psi(x,\lambda).$$

Furthermore,

$$\int_0^b |\{l(\lambda,b) - m(\lambda)\}\varphi(x,\lambda)|^2 \, dx$$

$$= |l(\lambda,b) - m(\lambda)|^2 \int_0^b |\varphi(x,\lambda)|^2 \, dx \le \left(v^2 \int_0^b |\varphi(x,\lambda)|^2 \, dx\right)^{-1}.$$

Hence, as $b \to \infty$ for $v = \operatorname{Im}\lambda \ne 0$,

$$\int_0^b |\psi_b(x,\lambda)|^2 \, dx \to \int_0^\infty |\psi(x,\lambda)|^2 \, dx.$$

∎

8.3 Integral representation of the resolvent. The formulas for the functions $\rho(\lambda)$ and $m(z)$.

Let two vector-valued functions $\psi(x,\lambda)$ and $\varphi(x,\lambda)$ have the above values. Put

$$G(x,y;\lambda) = \begin{cases} \psi(x,\lambda)\varphi^T(y,\lambda), & y < x \\ \varphi(x,\lambda)\psi^T(y,\lambda), & y > x \end{cases}, \tag{3.1}$$

or, in more detail,

$$G(x,y;\lambda) = \begin{cases} \begin{pmatrix} \psi_1(x,\lambda)\varphi_1(y,\lambda) & \psi_1(x,\lambda)\varphi_2(y,\lambda) \\ \psi_2(x,\lambda)\varphi_1(y,\lambda) & \psi_2(x,\lambda)\varphi_2(y,\lambda) \end{pmatrix}, & (y < x) \\ \begin{pmatrix} \varphi_1(x,\lambda)\psi_1(y,\lambda) & \varphi_1(x,\lambda)\psi_2(y,\lambda) \\ \varphi_2(x,\lambda)\psi_1(y,\lambda) & \varphi_2(x,\lambda)\psi_2(y,\lambda) \end{pmatrix}, & (y > x) \end{cases} \tag{3.1'}$$

Let $f(x) \in \mathcal{L}^2(0,\infty)$. We show that the vector-valued function determined by

$$\Phi(x,\lambda) = \int_0^\infty G(x,y;\lambda)f(y) \, dy \tag{3.2}$$

satisfies the system

$$y_2' - \{\lambda + q_1(x)\}y_1 = +f_1(x)$$

$$y_1' + \{\lambda + q_2(x)\}y_2 = -f_2(x)$$

(3.3)

and the boundary condition (1.4). By the definition of the matrix $G(x, y; \lambda)$, we have

$$\Phi_1(x, \lambda) = \psi_1(x, \lambda) \int_0^x \varphi^T(y, \lambda) f(y) \, dy + \varphi_1(x, \lambda) \int_x^\infty \psi^T(y, \lambda) f(y) \, dy \qquad (3.4)$$

$$\Phi_2(x, \lambda) = \psi_2(x, \lambda) \int_0^x \varphi^T(y, \lambda) f(y) \, dy + \varphi_2(x, \lambda) \int_x^\infty \psi^T(y, \lambda) f(y) \, dy. \qquad (3.5)$$

Differentiating (3.4) with respect to x, we obtain

$$\Phi_1'(x, \lambda) = \psi_1'(x, \lambda) \int_0^x \varphi^T(y, \lambda) f(y) \, dy + \varphi_1'(x, \lambda) \int_x^\infty \psi^T(y, \lambda) f(y) \, dy$$
$$+ \psi_1(x, \lambda)[\varphi^T(x, \lambda) f(x)] - \varphi_1(x, \lambda)[\psi^T(x, \lambda) f(x)].$$

Since $\psi(x, \lambda)$, $\varphi(x, \lambda)$ are solutions of the system (1.1), (1.2), substituting the values of $\psi_1'(x, \lambda)$, $\varphi_1'(x, \lambda)$ from (1.1), (1.2) in the last equality and taking into account (3.5), we have

$$\Phi_1'(x, \lambda) = -\{\lambda + q_2(x)\} \left\{ \psi_2(x, \lambda) \int_0^x \varphi^T(y, \lambda) f(y) \, dy + \varphi_2(x, \lambda) \int_x^\infty \psi^T(y, \lambda) f(y) \, dy \right\}$$
$$+ \{\psi_1(x, \lambda)\varphi_2(x, \lambda) - \psi_2(x, \lambda)\varphi_1(x, \lambda)\} f_2(x) = -\{\lambda + q_2(x)\}\Phi_2(x, \lambda) - f_2(x),$$

i.e., the second equation of (3.3) is satisfied (we have relied on $W\{\varphi, \psi\} \equiv 1$). Similarly, it can be verified that $\Phi(x, \lambda)$ also satisfies the first equation in (3.3).

Furthermore, it follows from (3.4) and (3.5) that

$$\Phi_1(0, \lambda) = \varphi_1(0, \lambda) \int_0^\infty \psi^T(y, \lambda) f(y) \, dy$$

$$\Phi_2(0, \lambda) = \varphi_2(0, \lambda) \int_0^\infty \psi^T(y, \lambda) f(y) \, dy$$

and then $\Phi(x, y)$ satisfies the boundary condition (1.4).

Let $\lambda_0, \lambda_{\pm 1}, \lambda_{\pm 2}, \ldots, \lambda_{\pm n}, \ldots$ be the eigenvalues of the boundary-value problem (1.1), (1.2), (1.4), (1.5) and let $\varphi_0(x), \varphi_{\pm 1}(x), \varphi_{\pm 2}(x), \ldots, \varphi_{\pm n}(x), \ldots$ be the corresponding vector-valued eigenfunctions. Let $l(\lambda, b)$ and $\psi_b(x, \lambda)$ be as in Section 2. Put

$$G_b(x, y; \lambda) = \begin{cases} \psi_b(x, \lambda)\varphi_b^T(y, \lambda), & y < x \\ \\ \varphi_b(x, \lambda)\psi_b^T(y, \lambda), & y > x \end{cases}$$

$$R_{z,b} f = \int_0^b G_b(x, y; z) f(y) \, dy. \qquad (3.6)$$

Then, also putting $\alpha_n^2 = \int_0^b |\varphi_n(x)|^2 \, dx$ (where $|\varphi_n|^2 = |\varphi_{n1}|^2 + |\varphi_{n2}|^2$), by the expansion (3.23), Chapter 7, we obtain

$$
\begin{aligned}
R_{z,b}f &= -\sum_{n=-\infty}^{\infty} \frac{\varphi_n(x)}{\alpha_n^2(z-\lambda_n)} \int_0^b f^T(y)\varphi_n(y)\,dy \\
&= -\int_{-\infty}^{\infty} \frac{\varphi(x,\lambda)}{z-\lambda} \left\{ \int_0^b f^T(y)\varphi(y,\lambda)\,dy \right\} d\rho_b(\lambda),
\end{aligned} \tag{3.7}
$$

where $\rho_b(\lambda)$ is the same as in Section 1.

Lemma 8.3.1 *For each nonreal z and each fixed x, the estimate*

$$
\int_{-\infty}^{\infty} \left| \frac{\varphi(x,\lambda)}{z-\lambda} \right|^2 d\rho_b(\lambda) < C \tag{3.8}
$$

is valid, where the constant C can be chosen to be independent of b.

Proof. Putting $f(y) = \varphi_n(y)$ in the formula (3.23), Chapter 7, we obtain

$$
\frac{1}{\alpha_n} \int_0^b G_b(x,y;z)\varphi_n(y)\,dy = -\frac{\varphi_n(x)}{\alpha_n(z-\lambda_n)}.
$$

Therefore, by the Parseval equation,

$$
\int_0^b |G_b(x,y;z)|^2\,dx = \sum_{n=-\infty}^{\infty} \frac{|\varphi_n(x)|^2}{\alpha_n^2|z-\lambda_n|^2} = \int_{-\infty}^{\infty} \frac{|\varphi(x,\lambda)|^2}{|z-\lambda|^2}\,d\rho_b(\lambda),
$$

which is equivalent to the equality

$$
|\psi_b(x,z)|^2 \int_0^x |\varphi(y,z)|^2\,dy + |\varphi(x,z)|^2 \int_x^b |\psi_b(y,z)|^2\,dy = \int_{-\infty}^{\infty} \frac{|\varphi(x,\lambda)|^2}{|z-\lambda|^2}\,d\rho_b(\lambda)
$$

by the definition of the matrix $G_b(x,y;z)$; hence, by Lemma 8.2.1, (3.8) holds. ∎

Corollary 8.3.1 *Let $\rho(\lambda)$ be the same as in Section 1. The estimate*

$$
\int_{-\infty}^{\infty} \frac{|\varphi(x,\lambda)|^2}{|z-\lambda|^2}\,d\rho(\lambda) \leq C \tag{3.9}
$$

is then valid with the same constant C as in (3.8).

Proof. In fact, for arbitrary a, it follows from (3.8) that

$$
\int_{-a}^{a} \frac{|\varphi(x,\lambda)|^2}{|z-\lambda|^2}\,d\rho_b(\lambda) < C.
$$

Here, letting $b \to \infty$ and then $a \to \infty$, we obtain (3.9). ∎

Corollary 8.3.2 *For any* $a > 0$,

$$\int_{-\infty}^{a} \frac{d\rho(\lambda)}{\lambda^2} < +\infty, \qquad \int_{a}^{\infty} \frac{d\rho(\lambda)}{\lambda^2} < +\infty.$$

Proof. Indeed, since $\varphi_{n1}(0)$ and $\varphi_{n2}(0)$ cannot vanish simultaneously, i.e., $|\varphi(0, \lambda)|^2 \neq 0$, putting $x = 0$ in (3.9), we obtain $\int_{-\infty}^{\infty} \frac{d\rho(\lambda)}{|z - \lambda|^2} < +\infty$, and the statement follows. ∎

Lemma 8.3.2 *Let* $f(x) \in \mathcal{L}^2(0, \infty)$, *and let* $z = u + iv$. *The estimate*

$$\int_{0}^{\infty} |R_z f|^2 \, dx \leq \frac{1}{v^2} \int_{0}^{\infty} |f(y)|^2 \, dy$$

then holds, where

$$R_z f = \int_{0}^{\infty} G(x, y; z) f(y) \, dy$$

and $G(x, y; z)$ *is determined by (3.1).*

Proof. For arbitrary $b > 0$, it follows from (3.7) and the Parseval equation that

$$
\begin{aligned}
\int_{0}^{b} |R_{z,b} f|^2 \, dx &= \sum_{n=-\infty}^{\infty} \frac{1}{\alpha_n^2 |z - \lambda_n|^2} \left\{ \int_{0}^{b} f^T(y) \varphi_n(y) \, dy \right\}^2 \\
&\leq \frac{1}{v^2} \sum_{n=-\infty}^{\infty} \frac{1}{\alpha_n^2} \left\{ \int_{0}^{b} f^T(y) \varphi_n(y) \, dy \right\}^2 \\
&= \frac{1}{v^2} \int_{0}^{b} |f(y)|^2 \, dy.
\end{aligned}
$$

If a is a fixed positive number and $a < b$, then

$$\int_{0}^{a} |R_{z,b} f|^2 \, dx \leq \int_{0}^{b} |R_{z,b} f|^2 \, dx \leq \frac{1}{v^2} \int_{0}^{b} |f(y)|^2 \, dy.$$

Letting $b \to \infty$ and taking into account the arbitrariness of a, we obtain the statement. ∎

Theorem 8.3.1 *(Integral Representation of the Resolvent.) For each vector-valued function* $f(x) \in \mathcal{L}^2(0, \infty)$ *and for any nonreal* z,

$$R_z f = -\int_{-\infty}^{\infty} \frac{\varphi(x, \lambda) F(\lambda)}{z - \lambda} \, d\rho(\lambda), \tag{3.10}$$

where $F(\lambda) = \underset{n \to \infty}{\text{l.i.m.}} \int_{0}^{n} f^T(x) \varphi(x, \lambda) \, dx$.

Proof. First, let $f(x) = f_n(x)$ vanish outside a finite interval $[0, n]$, satisfy the boundary condition (1.4) and have continuous first derivative. Suppose that $b > n$, and let μ be an arbitrary positive number. Put

$$F_n(\lambda) = \int_0^b f_n^T(x)\varphi(x, \lambda)\, dx = \int_0^n f_n^T(x)\varphi(x, \lambda)\, dx.$$

The equality (3.7) can then be rewritten in the form

$$R_{z,b}f_n = -\int_{-\infty}^{\infty} \frac{\varphi(x, \lambda)F_n(\lambda)}{z - \lambda}\, d\rho_b(\lambda)$$

$$= -\left\{\int_{-\infty}^{-\mu} + \int_{-\mu}^{\mu} + \int_{\mu}^{\infty}\right\} \frac{\varphi(x, \lambda)F_n(\lambda)}{z - \lambda}\, d\rho_b(\lambda) \equiv I_1 + I_2 + I_3. \quad (3.11)$$

We now estimate I_1. By (3.7), we have

$$|I_1| = \left|\int_{-\infty}^{-\mu} \frac{\varphi(x, \lambda)F_n(\lambda)}{z - \lambda}\, d\rho_b(\lambda)\right| = \left|\sum_{\lambda_k < -\mu} \frac{\varphi_k(x)}{\alpha_k^2(z - \lambda_k)} \int_0^n f_n^T(x)\varphi_k(x)\, dx\right|$$

$$\leq \left(\sum_{\lambda_k < -\mu} \frac{|\varphi_k(x)|^2}{\alpha_k^2|z - \lambda_k|^2}\right)^{1/2} \left(\sum_{\lambda_k < -\mu} \frac{1}{\alpha_k^2}\left\{\int_0^n f_n^T(x)\varphi_k(x)\, dx\right\}^2\right)^{1/2}. \quad (3.12)$$

Since the vector-valued function $\varphi_n(x)$ is a solution of the system (1.1), (1.2),

$$\int_0^n f_n^T(x)\varphi_k(x)\, dx$$

$$= \frac{1}{\lambda_k} \int_0^n \{f_{n1}(x)[\varphi_{k2}'(x) - q_1(x)\varphi_{k1}(x)] + f_{n2}(x)[-\varphi_{k1}'(x) - q_2(x)\varphi_{k2}(x)]\}\, dx.$$

Now, integrating the last integral by parts and taking into account that $f_n(x)$ and $\varphi_n(x)$ satisfy (1.4), while $f_n(x)$ vanishes outside $[0, n]$, we get

$$\int_0^n f_n^T(x)\varphi_k(x)\, dx$$

$$= \frac{1}{\lambda_k} \int_0^n \{\varphi_{k1}(x)[f_{n2}'(x) - q_1(x)f_{n1}(x)] + \varphi_{k2}(x)[-f_{n1}'(x) - q_2(x)f_{n2}(x)]\}\, dx.$$

Since, by Lemma 8.3.1, the first sum on the right-hand side of (3.12) is bounded, taking into account the last equality, we derive from (3.12) that

$$|I_1| \leq \frac{C^{1/2}}{\mu^2} \left(\sum_{\lambda_k < -\mu} \frac{1}{\alpha_k^2}\left\{\int_0^n h_n^T(x)\varphi_k(x)\, dx\right\}^2\right)^{1/2}, \quad (3.13)$$

where $h_n(x) = \begin{pmatrix} f_{n2}'(x) - q_1(x)f_{n1}(x) \\ -f_{n1}'(x) - q_2(x)f_{n2}(x) \end{pmatrix}$. By the Bessel inequality,

$$|I_1| \leq \frac{C^{1/2}}{\mu} \left(\int_0^n |h_n(x)|^2\, dx\right)^{1/2} = \frac{C_1}{\mu}.$$

That $|I_3| \leq C_2/\mu$ can be proved similarly. It follows from these estimates that I_1 and I_2 tend to zero uniformly in b as $\mu \to \infty$. Therefore, we can use the generalization of the Helly selection theorem, and derive the formula

$$R_z f_n = -\int_{-\infty}^{\infty} \frac{\varphi(x, \lambda) F_n(\lambda)}{z - \lambda} \, d\rho(\lambda) \tag{3.14}$$

from (3.11).

The formula (3.10) is obtained from (3.14) for an arbitrary vector-valued function $f(x)$ of class $\mathcal{L}^2(0, \infty)$ precisely in the same way as in proving Theorem 2.3.1. The proof is thus complete. ∎

The following is an easy consequence of (3.10).

Theorem 8.3.2 *For each nonreal z,*

$$m(z) - m(z_0) = \int_{-\infty}^{\infty} \left(\frac{1}{\lambda - z} - \frac{1}{\lambda - z_0} \right) d\rho(\lambda) \tag{3.15}$$

(Im $z_0 \neq 0$); if λ and μ are two points of continuity of the function $\rho(\lambda)$, then

$$\rho(\lambda) - \rho(\mu) = \frac{1}{\pi} \lim_{\tau \to 0} \int_{\mu}^{\lambda} \{- \mathrm{Im}[m(\sigma + i\tau)]\} \, d\sigma \tag{3.16}$$

($\tau > 0$).

Proof. Since $f(x)$ is arbitrary, it follows from (3.10) that

$$G(x, y; z) = \int_{-\infty}^{\infty} \frac{\varphi(x, \lambda) \varphi^T(y, \lambda)}{\lambda - z} \, d\rho(\lambda). \tag{3.17}$$

Therefore,

$$G(x, y; z) - G(x, y; z_0) = \int_{-\infty}^{\infty} \varphi_x \varphi_y^T \left(\frac{1}{\lambda - z} - \frac{1}{\lambda - z_0} \right) d\rho(\lambda),$$

where there are matrices on both sides. Hence, their corresponding elements are equal. Equalizing those placed where the first row and the first column meet, by (3.1'), the definition of the product $\psi(x, \lambda) \varphi^T(y, \lambda)$, putting $x = y = 0$ and taking into account the initial conditions (1.3), (2.1), we obtain

$$[\cos \alpha + m(z) \cos \alpha] \cos \alpha - [\cos \alpha + m(z_0) \cos \alpha] \cos \alpha$$
$$= \int_{-\infty}^{\infty} \cos^2 \alpha \cdot \left(\frac{1}{\lambda - z} - \frac{1}{\lambda - z_0} \right) d\rho(\lambda),$$

which exactly proves the formula (3.15).

We now prove (3.16). Let $z = \sigma + i\tau$. It follows from (3.15) that

$$\chi(\sigma,\tau) = \frac{\operatorname{sgn}\tau}{\pi} \frac{m(z) - m(\bar{z})}{2i} = -\frac{1}{\pi} \int_{-\infty}^{\infty} \frac{|\tau|\,d\rho(\lambda)}{(\lambda - \sigma)^2 + \tau^2}.$$

It then follows, also relying on the Stieltjes inversion formula, that if λ and μ are two points of continuity of $\rho(\lambda)$, we have

$$\rho(\lambda) - \rho(\mu) = -\lim_{\tau \to 0} \int_{\mu}^{\lambda} \chi(\sigma,\tau)\,d\sigma. \tag{3.18}$$

On the other hand, by (2.13), $m(\bar{z}) = \bar{m}(z)$. Hence,

$$\chi(\sigma,\tau) = \frac{\operatorname{sgn}\tau}{\pi} \cdot \frac{m(z) - \bar{m}(z)}{2i} = \frac{\operatorname{sgn}\tau}{\pi} \operatorname{Im}\{m(z)\}. \tag{3.19}$$

For $\tau > 0$, the formula (3.16) follows from (3.18) and (3.19). The theorem is thus proved. ∎

8.4 Proof of the expansion theorem in the case of the whole line

We again consider the system

$$y_2' - \{\lambda + q_1(x)\}y_1 = 0$$
$$y_1' + \{\lambda + q_2(x)\}y_2 = 0 \tag{4.1}$$

however assuming that x varies in the interval $(-\infty, \infty)$ and that $q_1(x)$, $q_2(x)$ are two continuous functions in each finite interval. Let $\varphi_1(x,\lambda)$ be the solution of (4.1), satisfying the initial conditions

$$\varphi_{11}(0,\lambda) = 1, \qquad \varphi_{12}(0,\lambda) = 0, \tag{4.2}$$

and let $\varphi_2(x,\lambda)$ be the solution of the system with the initial conditions

$$\varphi_{21}(0,\lambda) = 0, \qquad \varphi_{22}(0,\lambda) = 1. \tag{4.2'}$$

Let $[a,b]$ be an arbitrary finite interval. Consider the boundary-value problem determined by (4.1) and the conditions

$$y_1(a)\sin\alpha + y_2(a)\cos\alpha = 0$$
$$y_1(b)\sin\beta + y_2(b)\cos\beta = 0 \tag{4.3}$$

where α and β are two arbitrary real numbers.

Let $\lambda_0, \lambda_{\pm 1}, \lambda_{\pm 2}, \ldots, \lambda_{\pm n}, \ldots$ be the eigenvalues of the problem, and let the corresponding vector-valued eigenfunctions be $y_0(x), y_{\pm 1}(x), y_{\pm 2}(x), \ldots, y_{\pm n}(x), \ldots$. Since the solutions $\varphi_1(x, \lambda), \varphi_2(x, \lambda)$ of (4.1) are linearly independent,

$$y_n(x) = \beta_n \varphi_1(x, \lambda_n) + \gamma_n \varphi_2(x, \lambda_n).$$

Furthermore, since this problem is homogeneous, we can assume, without loss of generality, that $|\beta_n|, |\gamma_n| \le 1$. Put $\alpha_n^2 = \int_a^b |y_n(x)|^2 \, dx$. If $f(x) \in \mathcal{L}^2[a, b]$, then, applying the Parseval equation, we obtain

$$
\begin{aligned}
\int_a^b |y_n(x)|^2 \, dx &= \sum_{n=-\infty}^{\infty} \frac{1}{\alpha_n^2} \left\{ \int_a^b f^T(x) y_n(x) \, dx \right\}^2 \\
&= \sum_{n=-\infty}^{\infty} \frac{1}{\alpha_n^2} \left\{ \int_a^b f^T [\beta_n \varphi_1(x, \lambda_n) + \gamma_n \varphi_2(x, \lambda_n)] \, dx \right\}^2 \\
&= \sum_{n=-\infty}^{\infty} \frac{\beta_n^2}{\alpha_n^2} \left\{ \int_a^b f^T(x) \varphi_1(x, \lambda_n) \, dx \right\}^2 \\
&\quad + 2 \sum_{n=-\infty}^{\infty} \frac{\beta_n \gamma_n}{\alpha_n^2} \left\{ \int_a^b f^T(x) \varphi_1(x, \lambda_n) \, dx \right\} \left\{ \int_a^b f^T(x) \varphi_2(x, \lambda_n) \, dx \right\} \\
&\quad + \sum_{n=-\infty}^{\infty} \frac{\gamma_n^2}{\alpha_n^2} \left\{ \int_a^b f^T(x) \varphi_2(x, \lambda_n) \, dx \right\}^2.
\end{aligned}
\tag{4.4}
$$

We introduce the notation

$$
\begin{aligned}
\rho_{11,[a,b]}(\lambda) &= \sum_{0 < \lambda_n \le \lambda} \frac{\beta_n^2}{\alpha_n^2} \quad (\lambda > 0) \\
\rho_{11,[a,b]}(\lambda) &= -\rho_{11,[a,b]}(-\lambda) \quad (\lambda \le 0) \\
\rho_{12,[a,b]}(\lambda) &= \sum_{0 < \lambda_n \le \lambda} \frac{\beta_n \gamma_n}{\alpha_n^2} \quad (\lambda > 0) \\
\rho_{12,[a,b]}(\lambda) &= -\rho_{12,[a,b]}(-\lambda) \quad (\lambda \le 0) \\
\rho_{21,[a,b]}(\lambda) &= \rho_{12,[a,b]}(\lambda) \\
\rho_{22,[a,b]}(\lambda) &= \sum_{0 < \lambda_n \le \lambda} \frac{\gamma_n^2}{\alpha_n^2} \quad (\lambda > 0) \\
\rho_{22,[a,b]}(\lambda) &= -\rho_{22,[a,b]}(-\lambda) \quad (\lambda \le 0).
\end{aligned}
$$

Then the Parseval equation (4.4) can be written in the form

$$\int_a^b |f(x)|^2 \, dx = \int_{-\infty}^{\infty} \sum_{i,j=1}^{2} F_i(\lambda) F_j(\lambda) d\rho_{ij,[a,b]}(\lambda), \tag{4.5}$$

where

$$F_i(\lambda) = \int_a^b f^T(x) \varphi_i(x, \lambda) \, dx. \qquad i = 1, 2. \tag{4.6}$$

Lemma 8.4.1 *For any positive number N, there exists a positive constant number $A = A(N)$ not depending on either a or b, so that*

$$\bigvee_{-N}^{N} \{\rho_{ij,[a,b]}(\lambda)\} < A, \quad i,j = 1,2.$$

Proof. Since the functions $\varphi_{ij}(x,\lambda)$, $i,j = 1,2$, are continuous as functions of the variables and λ, by the initial conditions (4.2), (4.2'),

$$\varphi_{ij}(0,\lambda) = \delta_{ij},$$

for any positive ϵ and the given N, there is $h > 0$ such that

$$|\varphi_{ij}(x,\lambda) - \delta_{ij}| < \epsilon \tag{4.7}$$

for $0 \leq x \leq h$, $|\lambda| \leq N$.

Let $f_h(x) = \begin{pmatrix} f_{h1}(x) \\ f_{h2}(x) \end{pmatrix}$ be a nonnegative vector-valued function, with $f_{h2}(x) \equiv 0$, and $f_{h1}(x)$ vanishing outside the interval $[0,h]$, and let f_{h1} be normed so that

$$\int_0^h f_{h1}(x)\,dx = 1. \tag{4.8}$$

Put

$$F_{hi}(\lambda) = \int_0^h f_h^T(x)\varphi_i(x,\lambda)\,dx = \int_0^h f_{h1}(x)\varphi_{i1}(x,\lambda)\,dx, \quad i,j = 1,2.$$

By (4.7), (4.8) we get

$$1 - \epsilon < F_{h1}(x) < 1 + \epsilon, \qquad |F_{h2}(\lambda)| < \epsilon, \qquad |\lambda| < N. \tag{4.9}$$

Furthermore, applying the Parseval equation (4.5) to the vector-valued function $f_h(x)$, we obtain

$$\int_0^h f_{h1}^2(x)\,dx \;\geq\; \int_{-N}^{N} F_{h1}^2(\lambda)\,d\rho_{11}(\lambda) + 2\int_{-N}^{N} F_{h1}(\lambda)F_{h2}(\lambda)\,d\rho_{12}(\lambda)$$

$$+ \int_{-N}^{N} F_{h2}^2(\lambda)\,d\rho_{22}(\lambda)$$

$$\geq \int_{-N}^{N} F_{h1}^2(\lambda)\,d\rho_{11}(\lambda) - 2\int_{-N}^{N} |F_{h1}(\lambda)|\,|F_{h2}(\lambda)|\,|d\rho_{12}(\lambda)|$$

(omitting the subscript $[a,b]$ of the functions $\rho_{ij,[a,b]}(\lambda)$ in the whole proof). It follows from (4.9) that

$$\int_0^h f_{h1}^2(x)\,dx > \int_{-N}^{N} (1-\epsilon)^2\,d\rho_{11}(\lambda) - 2\int_{-N}^{N} \epsilon(1+\epsilon)\,|d\rho_{12}(\lambda)| = I_1 + I_2. \tag{4.10}$$

We now calculate the integrals I_1 and I_2. It is obvious that

$$I_1 = (1 - \epsilon)^2 \{\rho_{11}(N) - \rho_{11}(-N)\}. \tag{4.11}$$

Furthermore,

$$I_2 = -2\epsilon(1 + \epsilon) \bigvee_{-N}^{N} \{\rho_{12}(\lambda)\}. \tag{4.12}$$

Since

$$2 \bigvee_{-N}^{N} \{\rho_{12}(\lambda)\} \le \rho_{11}(N) - \rho_{11}(-N) + \rho_{22}(N) - \rho_{22}(-N), \tag{4.13}$$

it follows from the relations (4.10–13) that

$$\int_0^h f_{h1}^2(x)\, dx > (1 - 3\epsilon)\{\rho_{11}(N) - \rho_{11}(-N)\} - \epsilon(1 + \epsilon)\{\rho_{22}(N) - \rho_{22}(-N)\}. \tag{4.14}$$

Let the vector-valued function $g_h(x) = \begin{pmatrix} g_{h1}(x) \\ g_{h2}(x) \end{pmatrix}$ satisfy the same conditions as $f_h(x)$, however, replacing $g_{h1}(x)$ by $g_{h2}(x)$, i.e., $g_{h1}(x) \equiv 0$, $\int_0^h g_{h2}(x) = 1$.

Proceeding as above, we then obtain the inequality

$$\int_0^h g_{h2}^2(x)\, dx > (1 - 3\epsilon)\{\rho_{22}(N) - \rho_{22}(-N)\} - \epsilon(1 + \epsilon)\{\rho_{11}(N) - \rho_{11}(-N)\}. \tag{4.14'}$$

Now, adding (4.14) and (4.14′) together, we have

$$\int_0^b \{f_{h1}^2(x) + g_{h2}^2(x)\}\, dx > (1 - 4\epsilon - \epsilon^2)\{\rho_{11}(N) - \rho_{11}(-N) + \rho_{22}(N) - \rho_{22}(-N)\},$$

hence (if the number $\epsilon > 0$ is selected so that $1 - 4\epsilon - \epsilon^2 > 0$), the statement follows for the functions $\rho_{11}(x)$ and $\rho_{22}(\lambda)$, relying on their monotonicity. For the function $\rho_{12}(\lambda)$, it follows from the Cauchy-Schwarz inequality. The proof is now complete. ∎

To show that, by Lemma 8.4.1 and the familiar theorems on taking the limit in the Stieltjes integrals, the Parseval equation for (4.1) can be derived from the equality (4.5), the following should be proved.

Theorem 8.4.1 *Let $f(x) \in \mathcal{L}^2(0, \infty)$. There are two monotone and bounded functions $\rho_{11}(\lambda)$, $\rho_{22}(\lambda)$, and a function $\rho_{12}(\lambda)$ with variation bounded in each finite interval, none of which depends on the vector-valued function $f(x)$, and such that the Parseval equation*

$$\int_{-\infty}^{\infty} |f(x)|^2\, dx = \int_{-\infty}^{\infty} \sum_{i,j=1}^{2} F_i(\lambda) F_j(\lambda)\, d\rho_{ij}(\lambda) \tag{4.15}$$

holds, where $F_i(\lambda) = \underset{n \to \infty}{\text{l.i.m.}} \int_{-n}^{n} f^T(x) \varphi_i(x, \lambda)\, dx$, $i = 1, 2$.

Proof. First, let the vector-valued function $f_n(x)$ vanish outside an interval $[-n, n]$, $n < \min\{|a|, |b|\}$, have continuous first derivative, and let it satisfy the boundary conditions (4.3). Applying (4.4) to $f_n(x)$, we obtain

$$\int_{-n}^{n} |f_n(x)|^2 \, dx = \sum_{k=-\infty}^{\infty} \frac{1}{\alpha_k^2} \left\{ \int_a^b f_n^T(x) y_k(x) \, dx \right\}^2. \tag{4.16}$$

Since $f_n(x)$ is identically zero in the vicinity of the points a, b, integrating by parts, we have

$$\begin{aligned}
\int_a^b f_n^T(x) y_k(x) \, dx &= \frac{1}{\lambda_k} \int_a^b f_{n1}(x)[y_{k2}'(x) - q_1(x) y_{k1}(x)] \, dx \\
&\quad - \frac{1}{\lambda_k} \int_a^b f_{n2}(x)[y_{k1}'(x) + q_2(x) y_{k2}(x)] \, dx \\
&= \frac{1}{\lambda_k} \int_a^b \{[f_{n2}'(x) - q_1(x) f_{n1}(x)] y_{k1}(x) \\
&\quad + [-f_{n1}'(x) - q_2(x) f_{n2}(x)] y_{k2}(x) \} \, dx.
\end{aligned}$$

By the Parseval equation, we derive

$$\begin{aligned}
\sum_{|\lambda_k| \geq \mu} \frac{1}{\alpha_k^2} \left\{ \int_a^b f_n^T(x) y_k(x) \, dx \right\}^2 &\leq \frac{1}{\mu^2} \sum_{|\lambda_k| \geq \mu} \frac{1}{\alpha_k^2} \left\{ \int_a^b [\{f_{n2}'(x) - q_1(x) f_{n1}(x)\} y_{k1}(x) \right. \\
&\quad \left. + \{-f_{n1}'(x) - q_2(x) f_{n2}(x)\} y_{k2}(x)] \, dx \right\}^2 \\
&\leq \frac{1}{\mu^2} \sum_{k=-\infty}^{\infty} \frac{1}{\alpha_k^2} \left\{ \int_a^b [\{f_{n2}'(x) - q_1(x) f_{n1}(x)\} y_{k1}(x) \right. \\
&\quad \left. + \{-f_{n1}'(x) - q_2(x) f_{n2}(x)\} y_{k2}(x)] \, dx \right\}^2 \\
&= \frac{1}{\mu^2} \int_{-n}^{n} \{[f_{n2}'(x) - q_1(x) f_{n1}(x)]^2 \\
&\quad + [f_{n1}'(x) + q_2(x) f_{n2}(x)]^2 \} \, dx.
\end{aligned}$$

Hence, also relying on (4.16),

$$\left| \int_{-n}^{n} |f_n(x)|^2 \, dx - \sum_{-\mu \leq \lambda_k \leq \mu} \frac{1}{\alpha_k^2} \left\{ \int_a^b f_n^T(x) y_k(x) \, dx \right\}^2 \right|$$

$$\leq \frac{1}{\mu^2} \int_{-n}^{n} \{[f_{n2}'(x) - q_1(x) f_{n1}(x)]^2 + [f_{n1}'(x) + q_2(x) f_{n2}(x)]^2\} \, dx. \tag{4.17}$$

Furthermore, we have

$$\sum_{-\mu \leq \lambda_k \leq \mu} \frac{1}{\alpha_k^2} \left\{ \int_a^b f^T(x) y_k(x) \, dx \right\}^2$$

$$= \sum_{-\mu \le \lambda_k \le \mu} \frac{1}{\alpha_k^2} \left\{ \int_a^b f_n^T(x)[\beta_n \varphi_1(x, \lambda_k) + \gamma_k \varphi_2(x, \lambda_k)] \, dx \right\}^2$$

$$= \int_{-\mu}^{\mu} \sum_{i,j=1}^2 F_{ni}(\lambda) F_{nj}(\lambda) \, d\rho_{ij,[a,b]}(\lambda),$$

where $F_{ni}(\lambda) = \int_a^b f_n^T(x)\varphi_i(x, \lambda) \, dx$, $i = 1, 2$. Therefore, the equality (4.17) can be rewritten in the form

$$\left| \int_{-n}^n |f_n(x)|^2 \, dx - \int_{-\mu}^{\mu} \sum_{i,j=1}^2 F_{ni}(\lambda) F_{nj}(\lambda) \, d\rho_{ij,[a,b]}(\lambda) \right|$$

$$\le \frac{1}{\mu^2} \int_{-n}^n \{[f_{n2}'(x) - q_1(x)f_{n1}(x)]^2 + [f_{n1}'(x) + q_2(x)f_{n2}(x)]^2\} \, dx. \quad (4.18)$$

By Lemma 8.4.1, the set of monotone functions $\rho_{ii,[a,b]}(\lambda)$, $-\mu \le \lambda \le \mu$, $i = 1, 2$ is bounded, whereas the set of $\rho_{12,[a,b]}(\lambda)$ is bounded, and has bounded variation in each finite interval, uniformly with respect to a, b. Therefore, we can select two sequences $\{a_k\}$ and $\{b_k\}$ for which $\rho_{ij,[a,b]}(\lambda)$ converge to $\rho_{ij}(\lambda)$, respectively, as $a_k \to -\infty$ and $b_k \to +\infty$.

Passing to the limit with respect to $\{a_k\}$ and $\{b_k\}$ in (4.18), we get

$$\left| \int_n^n |f_n(x)|^2 \, dx - \int_{-\mu}^{\mu} \sum_{i,j=1}^2 F_{ni}(\lambda) F_{nj}(\lambda) \, d\rho_{ij}(\lambda) \right|$$

$$\le \frac{1}{\mu^2} \int_{-n}^n \{[f_{n2}'(x) - q_1(x)f_{n1}(x)]^2 + [f_{n1}'(x) + q_2(x)f_{n2}(x)]^2\} \, dx.$$

Finally, letting $\mu \to \infty$, we obtain

$$\int_{-n}^n |f_n(x)|^2 \, dx = \int_{-\infty}^{\infty} \sum_{i,j=1}^2 F_{ni}(\lambda) F_{nj}(\lambda) \, d\rho_{ij}(\lambda),$$

or (4.15) for the vector-valued functions $f_n(x)$ satisfying the above conditions.

The extension of the Parseval equation to arbitrary vector-valued functions of class $\mathcal{L}^2(0, \infty)$ can be done by the usual methods. The proof is thus complete. ∎

Let $f(x), g(x) \in \mathcal{L}^2(-\infty, \infty)$, and let the functions $G_i(\lambda)$, $i = 1, 2$, be constructed from $g(x)$ as $F_i(\lambda)$ is from $f(x)$. It is obvious that the vector-valued functions $f(x) \pm g(x)$ have $F_i(\lambda) \pm G_i(\lambda)$, $i = 1, 2$, as their Fourier transforms. Therefore,

$$\int_{-\infty}^{\infty} |f(x) + g(x)|^2 \, dx = \int_{-\infty}^{\infty} \sum_{i,j=1}^2 [F_i(\lambda) + G_i(\lambda)][F_j(\lambda) + G_j(\lambda)] \, d\rho_{ij}(\lambda),$$

$$\int_{-\infty}^{\infty} |f(x) - g(x)|^2 \, dx = \int_{-\infty}^{\infty} \sum_{i,j=1}^2 [F_i(\lambda) - G_i(\lambda)][F_j(\lambda) - G_j(\lambda)] \, d\rho_{ij}(\lambda).$$

Subtracting the lower equality from the upper, we obtain

$$\int_{-\infty}^{\infty} f^T(x)g(x)\,dx = \int_{-\infty}^{\infty} \sum_{i,j=1}^{2} F_i(\lambda)G_j(\lambda)\,d\rho_{ij}(\lambda), \qquad (4.19)$$

which is called the *generalized Parseval equation*.

Theorem 8.4.2 *(Expansion Theorem.)* *Let $f(x) \in \mathcal{L}^2(-\infty, \infty)$ be a continuous vector-valued function $(-\infty \le x \le \infty)$, let $\varphi_i(x, \lambda)$, $F_i(\lambda)$, $i = 1, 2$, be as before, and let the integrals*

$$\int_{-\infty}^{\infty} F_i(\lambda)\varphi_j(x, \lambda)\,d\rho_{ij}(\lambda), \quad i, j = 1, 2, \qquad (4.20)$$

converge absolutely and uniformly in x in each finite interval. Then

$$f(x) = \int_{-\infty}^{\infty} \sum_{i,j=1}^{2} F_i(\lambda)\varphi_j(x, \lambda)\,d\rho_{ij}(\lambda).$$

Proof. Let $g(x)$ be a continuous vector-valued function vanishing outside the interval $[-n, n]$. Applying the generalized Parseval equation (4.19) to the vector-valued functions $f(x)$ and $g(x)$, we obtain

$$\int_{-n}^{n} f^T(x)g(x)\,dx = \int_{-\infty}^{\infty} \sum_{i,j=1}^{2} F_i(\lambda) \left\{ \int_{-n}^{n} g^T(x)\varphi_j(x, \lambda)\,dx \right\} d\rho_{ij}(\lambda).$$

The uniform convergence allows us to change the order of integration on the right-hand side. We then have

$$\int_{-n}^{n} g^T(x)f(x)\,dx = \int_{-n}^{n} g^T(x) \left\{ \int_{-\infty}^{\infty} \sum_{i,j=1}^{2} F_i(\lambda)\varphi_j(x, \lambda)\,d\rho_{ij}(\lambda) \right\} dx.$$

The statement follows, also relying on the arbitrariness of $g(x)$, and the continuity of $f(x)$ of (4.20) (which, for the latter, follows from the assumed uniform convergence). ∎

8.5 Floquet (Bloch) solutions

Consider the system

$$y_2' + p(x)y_1 = \lambda y_1, \qquad -y_1' + r(x)y_2 = \lambda y_2. \qquad (5.1)$$

Let $\varphi(x, \lambda)$ and $\vartheta(x, \lambda)$ be two solutions of (5.1), satisfying the initial conditions

$$\varphi_1(0, \lambda) = \vartheta_2(0, \lambda) = 0, \qquad \varphi_2(0, \lambda) = \vartheta_1(0, \lambda) = 1.$$

It is obvious that they are linearly independent; therefore, any solution $y(x, \lambda)$ of (5.1) can be represented in the form $y(x, \lambda) = c_1 \vartheta(x, \lambda) + c_2 \varphi(x, \lambda)$, where c_1 and c_2 are two arbitrary constants.

Thus, the solutions of (5.1) and the two-dimensional vectors $C = \begin{pmatrix} c_1 \\ c_2 \end{pmatrix}$ are in a one-to-one correspondence $y(x, \lambda) \leftrightarrow \bar{C}$.

Assume that the coefficients $p(x)$ and $r(x)$ are periodic functions with the same period a. Then the vector-valued function $y_a(x, \lambda) = y(x + a, \lambda)$ is also a solution of (5.1). By the definition of $y_a(x, \lambda)$,

$$y_a(0, \lambda) = y(a, \lambda) = c_1 \vartheta(a, \lambda) + c_2 \varphi(a, \lambda),$$

i.e.,

$$y_1(a, \lambda) = c_1 \vartheta_1(a, \lambda) + c_2 \varphi_1(a, \lambda),$$
$$y_2(a, \lambda) = c_1 \vartheta_2(a, \lambda) + c_2 \varphi_2(a, \lambda),$$

or

$$y_a(x, \lambda) = y(x + a, \lambda) \leftrightarrow T\bar{C},$$

where

$$T = T(a, \lambda) = \begin{pmatrix} \vartheta_1(a, \lambda) & \varphi_1(a, \lambda) \\ \vartheta_2(a, \lambda) & \varphi_2(a, \lambda) \end{pmatrix}.$$

T is called the *monodromy matrix* (as in the case of the Sturm-Liouville equation). It is obvious that

$$y(x + na, \lambda) = T^n \bar{C}$$

for any natural n.

We now find the eigenvalues and eigenvectors of T. We have

$$|t - \kappa I| = \begin{vmatrix} \vartheta_1(a, \lambda) - \kappa & \varphi_1(a, \lambda) \\ \vartheta_2(a, \lambda) & \varphi_2(a, \lambda) - \kappa \end{vmatrix} = \kappa^2 - F(\lambda)\kappa + 1 = 0,$$

where $F(\lambda) = \vartheta_1(a, \lambda) + \varphi_2(a, \lambda)$. Therefore, the eigenvalues of T are

$$\kappa^{\mp} = \frac{1}{2}\{F(\lambda) \pm \sqrt{F^2(\lambda) - 4}\}. \tag{5.2}$$

We specify the eigenvectors of T in the form

$$e^{\pm} = \begin{pmatrix} 1 \\ m^{\pm}(\lambda) \end{pmatrix}.$$

We then derive from the equality $Te^+ = \kappa^+ e^+$ that

$$\vartheta_1(a, \lambda) + m^+(\lambda)\varphi_1(a, \lambda) = \kappa^+;$$

hence, using the expression for κ^+ from (5.2), we also find[1] that

$$m^+(\lambda) = \frac{\varphi_2 - \vartheta_1 - \sqrt{F^2(\lambda) - 4}}{2\varphi_1}, \qquad m^-(\lambda) = \frac{\varphi_2 - \vartheta_1 + \sqrt{F^2(\lambda) - 4}}{2\varphi_1}.$$

Furthermore, since $Te^\pm = \kappa^\pm e^\pm$,

$$T^n e^\pm = (\kappa^\pm)^n e^\pm \tag{5.3}$$

for any n.

Consider

$$\psi^+(x, \lambda) = \vartheta(x, \lambda) + m^+(\lambda)\varphi(x, \lambda)$$
$$\psi^-(x, \lambda) = \vartheta(x, \lambda) + m^-(\lambda)\varphi(x, \lambda)$$

called the *Floquet (Bloch) solutions* of (5.1). Relying on (5.3), we have

$$T^n \psi^\pm(x, \lambda) = \psi^\pm(x + na, \lambda) = (\kappa^\pm)^n \psi(x, \lambda). \tag{5.4}$$

Therefore, $\psi^\pm(x, \lambda)$ are bounded on the whole number axis if and only if $|\kappa^\pm| = 1$; this holds due to (5.2) for real λ and $|F(\lambda)| \leq 2$. However, if $|F(\lambda)| \geq 2$, then one of $\kappa^\pm$ is greater than 1 in absolute value, whereas the other is less than 1; therefore, $\psi^\pm(x, \lambda)$ are not bounded.

Put $\chi^\pm(x, \lambda) = (\kappa^\pm)^{-x/a}\psi^\pm(x, \lambda)$. By (5.4),

$$\chi^\pm(x + na, \lambda) = (\kappa^\pm)^{-x/a-n}\psi^\pm(x + na, \lambda) = (\kappa^\pm)^{-x/a}\psi^\pm(x, \lambda).$$

Hence, the functions $\chi^\pm(x, \lambda)$ are periodic with period a, and

$$\psi^\pm(x, \lambda) = (\kappa^\pm)^{x/a}\chi^\pm(x, \lambda).$$

It follows that if λ is such that $|\kappa^+| < 1$ and, therefore, $|\kappa^-| > 1$, then the vector-valued function $\psi^+(x, \lambda)$ is in the space $\mathcal{L}^2(0, \infty)$ while $\psi^-(x, \lambda) \in \mathcal{L}^2(-\infty, 0)$, i.e., $\psi^\pm(x, \lambda)$ are Weyl solutions of (5.1), and $m^\pm(\lambda)$ are Weyl-Titchmarsh functions on $(0, \infty)$, $(-\infty, 0)$, respectively.

8.6   The self-adjointness of the Dirac systems

8.6.1

Consider the operator defined on vector-valued functions $y(x) = \begin{pmatrix} y_1(x) \\ y_2(x) \end{pmatrix}, 0 \leq x < \infty$, and generated by the differential expression of first order

$$l(y) = \begin{cases} \dfrac{dy_2}{dx} + p(x)y_1 + q(x)y_2 \\[2mm] -\dfrac{dy_1}{dx} + q(x)y_1 + r(x)y_2 \end{cases} \tag{6.1}$$

---

[1]Where $\vartheta_1(a, \lambda)$, $\vartheta_2(a, \lambda)$, $\varphi_1(a, \lambda)$ and $\varphi_2(a, \lambda)$ are denoted by ϑ_1, ϑ_2, φ_1 and φ_2, respectively.

and the boundary condition

$$y_1(0)\cos\alpha + y_2(0)\sin\alpha = 0. \tag{6.2}$$

We assume that the coefficients $p(x)$, $q(x)$ and $r(x)$ are continuous in each finite interval. As the domain $\mathcal{D}(l)$ of the operator l, we take the continuous vector-valued functions with bounded support and continuous first derivatives, satisfying (6.2).

It immediately follows from the definitions of an adjoint and a self-adjoint operator that, to prove the self-adjointness of a symmetric operator A, it suffices to show that

$$\mathcal{D}(A^*) = \mathcal{D}(A), \tag{6.3}$$

i.e., if an element $x \in \mathcal{D}(A^*)$, then $x \in \mathcal{D}(A)$, to prove which, we should show the domain $\mathcal{D}(A^*)$.

This can be done for a Dirac operator l of (6.1), giving a sufficient condition for $\bar{l}$ to be self-adjoint on its basis (where $\bar{l}$ is the closure of l) in terms of $p(x)$, $q(x)$ and $r(x)$.

8.6.2

The following is important for the sequel.

Lemma 8.6.1 *Let $f(x)$ be a continuous function satisfying the conditions :*

1. $f(x)$ is absolutely continuous in an interval (a, b), and

2. $f'(x) \in \mathcal{L}^2(a, b)$,

3. $f(a) = f(b) = 0$.

Let $h(x)$ be an arbitrary function in the class $\mathcal{L}^2(a, b)$. If, for an arbitrary function $f(x)$ satisfying Conditions 1, 2 and 3,

$$\int_a^b f'(x)h(x)\,dx = 0, \tag{6.4}$$

then $h(x) = $ const. almost everywhere.

Proof. We first specify the class of functions fulfilling Conditions 1, 2 and 3. Put $f'(t) = \lambda(t)$. Then

$$f(x) = \int_a^x \lambda(t)\,dt. \tag{6.5}$$

It therefore follows from Condition 3 that

$$\int_a^b \lambda(t)\,dt = 0. \tag{6.5'}$$

Conversely, if a function $\lambda(t)$ from $\mathcal{L}^2(a, b)$ fulfils the condition (6.5'), then the function $f(x)$ determined by (6.5) satisfies Conditions 1, 2 and 3.

Now, if $h(x) \in \mathcal{L}^2(a, b)$ does not satisfy (6.5'), we can select a constant number c_0 such that the function

$$h_1(x) = h(x) - c_0 \qquad (6.6)$$

satisfies the condition. In fact, to determine c_0, we obtain

$$\int_a^b h_1(t) \, dt = \int_a^b h(t) \, dt - c_0(b - a) = 0. \qquad (6.7)$$

We now come back to (6.4), which can be written as

$$\int_a^b \lambda(x) h(x) \, dx = 0.$$

Using (6.5), we rewrite the equality as

$$\int_a^b \lambda(x) h_1(x) \, dx = 0, \qquad (6.8)$$

where $h_1(x)$ is of the form (6.6) and c_0 is determined by (6.7).

Now, putting $\lambda(x) = h_1(x)$ in (6.8), which is possible, since $h_1(\lambda)$ satisfies (6.5'), we obtain

$$\int_a^b h_1^2(x) \, dx = 0;$$

hence, $h_1(x) = 0$, i.e., $h(x) = c_0$ almost everywhere, as required.   ∎

8.6.3

Let l be the operator determined in Item 1. The following supplies a description of the domain $\mathcal{D}(l^*)$ of the adjoint operator l^*.

Lemma 8.6.2 *The domain $\mathcal{D}(l^*)$ of l^* consists of the vector-valued functions $z(x) = \begin{pmatrix} z_1(x) \\ z_2(x) \end{pmatrix}$ satisfying the conditions*

1. *$z_1(x), z_2(x) \in \mathcal{L}^2(0, \infty)$,*

2. *$z_1(x), z_2(x)$ are absolutely continuous in each finite interval,*

3. *$l(z) = z^* = \begin{pmatrix} z_1^* \\ z_2^* \end{pmatrix} \in \mathcal{L}^2(0, \infty)$, i.e.,*

$$z_2' + p(x)z_1 + q(x)z_2 = z_1^* \in \mathcal{L}^2(0, \infty)$$

$$-z_1' + q(x)z_1 + r(x)z_2 = z_2^* \in \mathcal{L}^2(0, \infty)$$

and

4. $z_1(0) \cos \alpha + z_2(0) \sin \alpha = 0$.

Proof. By the definition of l^*, if $y(x) \in \mathcal{D}(l^*)$ and $z(x) \in \mathcal{D}(l^*)$, then there exits a vector-valued function $z^*(x) \in \mathcal{L}^2(0, \infty)$ such that the equality

$$\int_0^\infty (ly, z)\, dx = \int_0^\infty (y, z^*)\, dx$$

holds $((y, z) \equiv y^T z = y_1 z_1 + y_2 z_2)$, or, in expanded form,

$$\int_0^\infty \{[y_2' + p(x)y_1 + q(x)y_2]z_1 + [-y_1' + q(x)y_1 + r(x)y_2]z_2\}\, dx = \int_0^\infty (y_1 z_1^* + y_2 z_2^*)\, dx. \tag{6.9}$$

We first assume that $y_2(x) = 0$. The equality (6.9) then is reduced to

$$\int_0^\infty \{p(x)y_1 z_1 + [-y_1' + q(x)y_1]z_2\}\, dx = \int_0^\infty y_1 z_1^*\, dx. \tag{6.10}$$

Integrating by parts, we obtain

$$\int_0^\infty p(x)y_1(x)z_1(x)\, dx = -\int_0^\infty y_1'(x)\left[\int_0^x p(t)z_1(t)\, dt\right] dx$$

$$\int_0^\infty q(x)y_1(x)z_2(x)\, dx = -\int_0^\infty y_1'(x)\left[\int_0^x q(t)z_2(t)\, dt\right] dx$$

$$\int_0^\infty y_1(x)z_1^*(x)\, dx = -\int_0^\infty y_1'(x)\left[\int_0^x z_1^*(t)\, dt\right] dx.$$

Consequently, (6.10) can be rewritten as

$$\int_0^\infty y_1'(x)\left\{z_2(x) + \int_0^x [p(t)z_1(t) + q(t)z_2(t) - z_1^*(t)]\, dt\right\} dx = 0;$$

hence, by Lemma 8.6.1,

$$z_2(x) = \int_0^x [-p(t)z_1(t) - q(t)z_2(t) + z_1^*(t)]\, dt + c_2. \tag{6.11}$$

Now, assuming that $y_1(x) = 0$, we derive the equality

$$z_1(x) = \int_0^x [p(t)z_1(t) + r(t)z_2(t) - z_2^*(t)]\, dt + c_1 \tag{6.12}$$

from (6.9) similarly.

It follows from the representations (6.11) and (6.12) that the vector-valued function $z(x) = \begin{pmatrix} z_1(x) \\ z_2(x) \end{pmatrix}$ satisfies Conditions 2 and 3, and also that the limits $z_1(+0) = c_1$, $z_2(+0) = c_2$ exist.

Consider arbitrary vector-valued functions satisfying the boundary condition (6.2). Integrating the left-hand side of (6.9) by parts, we get

$$\int_0^\infty \{[y_2' + p(x)y_1 + q(x)y_2]z_1 + [-y_1' + q(x)y_1 + r(x)y_2]z_2\}dx$$

$$= \int_0^\infty \{[z_2' + p(x)z_1 + q(x)z_2]y_1 + [-z_1' + q(x)z_1 + r(x)z_2]y_2\}\,dx + y_2(0)z_1(0) - y_1(0)z_2(0)$$

$$= \int_0^\infty (y_1z_1^* + y_2z_2^*)\,dx. \tag{6.13}$$

Since

$$l^*(z) = \begin{cases} z_2' + p(x)z_1 + q(x)z_2 \\ \\ -z_1' + q(x)z_1 + r(x)z_2 \end{cases}$$

(proved), it follows from (6.13) that

$$y_2(0)z_1(0) - y_1(0)z_2(0) = 0. \tag{6.14}$$

By (6.2), $y_1(0)/y_2(0) = -\tan\alpha$, which, substituted in (6.14), yields Condition 4 for the vector-valued function $z(x)$. The proof is thus complete. ∎

Using Lemma 8.6.2, we can obtain the following remarkable proposition.

Theorem 8.6.1 *If the coefficients of the system (6.1) are continuous in each finite interval, then the operator $\bar{l}$, or the closure of the operator l, is self-adjoint.*

Proof. We show that the operator l^* is symmetric on its domain $\mathcal{D}(l^*)$. It will then follow that

$$l^* \subseteq (l^*)^* = l^{**}. \tag{6.15}$$

On the other hand, a symmetric operator l always satisfies the inclusion

$$l^{**} \subseteq l^*. \tag{6.16}$$

It follows from (6.15) and (6.16) that $l^{**} = l^*$, or that l^* is self-adjoint. Furthermore, for any symmetric operator A (with a dense domain), $A^{**} = \bar{A}$, where $\bar{A}$ is the closure of A (see N. I. Akhiezer, I. M. Glazman (1961; pp. 127, 148)). Therefore, $\bar{l} = l^*$ and $\bar{l}$ is self-adjoint. Thus, it remains to prove the symmetry property of l^*.

Let $y = \begin{pmatrix} y_1 \\ y_2 \end{pmatrix}$, and let $z = \begin{pmatrix} z_1 \\ z_2 \end{pmatrix} \in \mathcal{D}(l^*)$. By Lemma 8.6.2,

$$y_1(x)z_2(x), y_2(x)z_1(x) \in \mathcal{L}^2(0, \infty).$$

Hence, there is a sequence $a_n \to \infty$ such that

$$\lim_{n\to\infty} y_1(a_n)z_2(a_n) = 0, \qquad \lim_{n\to\infty} y_2(a_n)z_1(a_n) = 0. \tag{6.17}$$

By Green's identity, we have

$$(y, l^*(z))_{x=a_n} - (l^*(y), z)_{x=a_n} = \int_0^{a_n} \left[\frac{d}{dx}(y_1, z_2) - \frac{d}{dx}(z_1, y_2) \right] dx$$
$$= y_1(a_n)z_2(a_n) - z_1(a_n)y_2(a_n)$$

(also invoking Lemma 8.6.2). Passing to the limit, we obtain

$$(y, l^*(z)) = (l^*(y), z)$$

relying on (6.17) for any $y, z \in \mathcal{D}(l^*)$, as required.   ∎

Theorem 8.6.1 shows that we cannot be in the limit-circle case for (6.1), with coefficients which are continuous in each finite interval (i.e., nullity index $(0,0)$).

Notes

Section 1

The fundamental result, i.e., Theorem 8.1.1, was first obtained by S. Conte and W. Sangren (1953). Our proof is due to I. S. Sargsjan.

Section 2

The results are due to E. Titchmarsh (1950) and R. Macleod (1966). However, it should be noted that if the coefficients $p(x)$ and $r(x)$ are continuous in each finite interval, then, as was shown in Section 6, we can only be in the limit-point case.

Section 3

The integral representation of the resolvent was obtained by B. M. Levitan and I. S. Sargsjan (1975). The other results are due to E. Titchmarsh (1950) and R. Macleod (1966).

Section 4

Theorem 8.4.1 is due to E. Titchmarsh (1950). Lemma 8.4.1 and the proof of Theorem 8.4.1, given here, are due to B. M. Levitan and I. S. Sargsjan (1975).

Section 5

The results are similar to those for the Sturm-Liouville operator, and are published for the first time.

Section 6

Theorem 8.6.1 is due to B. M. Levitan. A similar theorem was proved by V. V. Martynov (1965) independently.

Chapter 9

THE STUDY OF THE SPECTRUM

9.1 The spectrum in the case of summable coefficients

Consider the problem

$$y_1' - \{\lambda + q_1(x)\}y_2 = 0 \tag{1.1}$$

$$y_2' + \{\lambda + q_2(x)\}y_1 = 0 \tag{1.2}$$

$$y_1(0)\cos\alpha + y_2(0)\sin\alpha = 0 \tag{1.3}$$

on the half-axis $[0, \infty)$, assuming that the coefficients $q_1(x)$ and $q_2(x)$ are in the class $\mathcal{L}(0, \infty)$.

Let $\varphi(x, \lambda)$ be the solution of the system (1.1), (1.2), satisfying the initial conditions

$$\varphi_1(0, \lambda) = -\sin\alpha, \qquad \varphi_2(0, \lambda) = \cos\alpha. \tag{1.4}$$

We rewrite (1.1), (1.2) in the form

$$\varphi_1' - \lambda\varphi_2 = q_1(x)\varphi_2 \tag{1.1'}$$

$$\varphi_2' + \lambda\varphi_1 = -q_2(x)\varphi_1 \tag{1.2'}$$

(replacing y by φ). Applying the method of variation of the constants, we find

$$\varphi_1(x, \lambda) = \sin(\lambda x - \alpha) + \int_0^x \{\varphi_2(s, \lambda)q_1(s)\cos\lambda(x - s) - \varphi_1(s, \lambda)q_2(s)\sin\lambda(x - s)\}\, ds \tag{1.5}$$

$$\varphi_2(x, \lambda) = \cos(\lambda x - \alpha) - \int_0^x \{\varphi_2(s, \lambda)q_1(s)\sin\lambda(x - s) + \varphi_1(s, \lambda)q_2(s)\cos\lambda(x - s)\}\, ds. \tag{1.6}$$

Let $\lambda = \sigma + i\tau$ $(\tau > 0)$. Put

$$\varphi_1(x, \lambda) = h_1(x)e^{\tau x}, \qquad \varphi_2(x, \lambda) = h_2(x)e^{\tau x}. \tag{1.7}$$

It then follows from the formulas (1.5) and (1.6) that

$$h_1(x) = e^{-\tau x}\sin(\lambda x - \alpha) + \int_0^x \{h_2(s)q_1(s)\cos\lambda(x - s) - h_1(s)q_2(s)\sin\lambda(x - s)\}e^{\tau(s - x)}\, ds$$

$$h_2(x) = e^{-\tau x}\cos(\lambda x - \alpha) - \int_0^x \{h_2(s)q_1(s)\sin\lambda(x-s) + h_1(s)q_2(s)\cos\lambda(x-s)\}e^{\tau(s-x)}\,ds.$$

Since $|\sin(\lambda x - \alpha)| \le e^{\tau x}$, $|\cos(\lambda x - \alpha)| \le e^{\tau x}$, we obtain, in terms of the absolute values,

$$|h_1(x)|, |h_2(x)| \le 1 + \int_0^x \{|h_1(s)|\,|q_2(s)| + |h_2(s)|\,|q_1(s)|\}\,ds.$$

Therefore, Lemma 3.2.1 is applicable, and we get

$$|h_1(s)|, |h_2(s)| \le \exp\left\{\int_0^x [|q_1(s)| + |q_2(s)|]\,ds\right\}.$$

Since $q_1(x), q_2(x) \in \mathcal{L}(0, \infty)$ (given), it follows from the last estimate that the vector-valued function $h(x)$ is bounded for $x \in [0, \infty)$, $|\lambda| \ge \rho > 0$ and $\tau \ge 0$. Therefore, for large x,

$$\varphi_1(x, \lambda) = O(e^{\tau x}), \qquad \varphi_2(x, \lambda) = O(e^{\tau x}), \tag{1.8}$$

relying on (1.7).

First, we consider real λ. The functions $\varphi_1(x, \lambda)$ and $\varphi_2(x, \lambda)$ are bounded according to (1.8). Hence, we derive from the formulas (1.5) and (1.6), as $x \to \infty$, that

$$\varphi_1(x) = \sin(\lambda x - \alpha) + \int_0^\infty \{\varphi_2(s, \lambda)q_1(s)\cos\lambda(x-s) - \varphi_1(s, \lambda)q_2(s)\sin\lambda(x-s)\}\,ds$$
$$+ O\left\{\int_x^\infty [|q_1(s)| + |q_2(s)|]\,ds\right\} = \mu(\lambda)\cos\lambda x + \nu(\lambda)\sin\lambda x + o(1) \tag{1.9}$$

$$\varphi_2(x, \lambda) = \cos(\lambda x - \alpha) - \int_0^\infty \{\varphi_2(s, \lambda)q_1(s)\sin\lambda(x-s) + \varphi_1(s, \lambda)q_2(s)\cos\lambda(x-s)\}\,ds$$
$$+ O\left\{\int_x^\infty [|q_1(s)| + |q_2(s)|]\,ds\right\} = \nu(\lambda)\cos\lambda x - \mu(\lambda)\sin\lambda x + o(1), \tag{1.10}$$

where

$$\mu(\lambda) = -\sin\alpha + \int_0^\infty \{\varphi_2(s, \lambda)q_1(s)\cos\lambda s + \varphi_1(s, \lambda)q_2(s)\sin\lambda s\}\,ds, \tag{1.11}$$

$$\nu(\lambda) = \cos\alpha + \int_0^\infty \{\varphi_2(s, \lambda)q_1(s)\sin\lambda s - \varphi_1(s, \lambda)q_2(s)\cos\lambda s\}\,ds. \tag{1.12}$$

Since the integrals in λ converge uniformly, the functions $\mu(\lambda)$ and $\nu(\lambda)$ are continuous.

Similarly, if $\psi(x, \lambda)$ is the solution of the system (1.1), (1.2), satisfying the conditions

$$\psi_1(0, \lambda) = -\cos\alpha, \quad \psi_2(0, \lambda) = -\sin\alpha, \tag{1.4'}$$

then

$$\psi_1(x, \lambda) = \xi(\lambda)\cos\lambda x + \eta(\lambda)\sin\lambda x + o(1) \tag{1.13}$$

$$\psi_2(x, \lambda) = \eta(\lambda) \cos \lambda x - \xi(\lambda) \sin \lambda x + o(1), \tag{1.14}$$

where

$$\xi(\lambda) = -\cos \alpha + \int_0^\infty \{\psi_2(s, \lambda)q_1(s) \cos \lambda s + \psi_1(s, \lambda)q_2(s) \sin \lambda s\}\, ds, \tag{1.13'}$$

$$\eta(\lambda) = -\sin \alpha + \int_0^\infty \{\psi_2(s, \lambda)q_1(s) \sin \lambda s - \psi_1(s, \lambda)q_2(s) \cos \lambda s\}\, ds. \tag{1.14'}$$

By (1.9), (1.10), (1.13) and (1.14), we have

$$W\{\varphi, \psi\} = \mu(\lambda)\eta(\lambda) - \nu(\lambda)\xi(\lambda) + o(1). \tag{1.15}$$

Since, by (1.4) and (1.4'), $W\{\varphi, \psi\}_{x=0} = 1$, it follows from (1.15) that

$$\mu(\lambda)\eta(\lambda) - \nu(\lambda)\xi(\lambda) = 1. \tag{1.16}$$

Therefore, $\mu(\lambda)$ and $\nu(\lambda)$ cannot vanish for the same value of λ.

We now consider complex λ. It follows from (1.5) for fixed positive τ, as $x \to \infty$, that

$$\begin{aligned}
\varphi_1(x, \lambda) &= -\frac{e^{-i\lambda x}}{2i} \cos \alpha - \frac{e^{-i\lambda x}}{2} \sin \alpha + O(e^{-\tau x}) \\
&\quad + \frac{1}{2} \int_0^x e^{-i\lambda(x-s)} \{\varphi_2(s, \lambda)q_1(s) - i\varphi_1(s, \lambda)q_2(s)\}\, ds \\
&\quad + O\left\{ \int_0^x e^{-\tau(x-s)} |\varphi_2(s, \lambda)q_1(s) - i\varphi_1(s, \lambda)q_2(s)|\, ds \right\}.
\end{aligned} \tag{1.17}$$

Furthermore, on the one hand, relying on (1.8), as $x \to \infty$,

$$O\left\{ \int_0^x e^{-\tau(x-s)} |\varphi_2(s, \lambda)q_1(s) - i\varphi_1(s, \lambda)q_2(s)|\, ds \right\}$$
$$= O\left\{ \int_0^x e^{-\tau(x-2s)} [|q_1(s)| + |q_2(s)|]\, ds \right\}$$
$$= O\left\{ e^{\tau(x-2\delta)} \int_0^{x-\delta} [|q_1(s)| + |q_2(s)|]\, ds \right\} + O\left\{ e^{\tau x} \int_{x-\delta}^x [|q_1(s)| + |q_2(s)|]\, ds \right\} = o(e^{\tau x}),$$

and, on the other hand,

$$\int_x^\infty e^{-i\lambda(x-s)} \{\varphi_2(s, \lambda)q_1(s) - \varphi_1(s, \lambda)q_2(s)\}\, ds$$
$$= O\left\{ e^{\tau x} \int_x^\infty [|q_1(s)| + |q_2(s)|]\, ds \right\} = o(e^{\tau x}).$$

Therefore, we derive from the equality (1.17) that

$$\varphi_1(x, \lambda) = e^{-i\lambda x} \{M_1(\lambda) + o(1)\} \tag{1.18}$$

as $x \to \infty$, where

$$M_1(\lambda) = -\frac{\sin \alpha}{2} - \frac{\cos \alpha}{2i} + \frac{1}{2} \int_0^\infty e^{i\lambda x} \{\varphi_2(s, \lambda)q_1(s) - i\varphi_1(s, \lambda)q_2(s)\} \, ds. \quad (1.19)$$

Similarly, we obtain the estimates

$$\varphi_2(x, \lambda) = e^{-i\lambda x} \{M_2(\lambda) + o(1)\}$$

$$\psi_1(x, \lambda) = e^{-i\lambda x} \{N_1(\lambda) + o(1)\} \quad (1.20)$$

$$\psi_2(x, \lambda) = e^{-i\lambda x} \{N_2(\lambda) + o(1)\}$$

from (1.6) and also similar formulas for the solution $\psi(x, \lambda)$, where

$$
\begin{aligned}
M_2(\lambda) &= -\frac{\sin \alpha}{2i} - \frac{\cos \alpha}{2} - \frac{1}{2} \int_0^\infty e^{i\lambda s} \{i\varphi_2(s, \lambda)q_1(s) + \varphi_1(s, \lambda)q_2(s)\} \, ds \\
N_1(\lambda) &= -\frac{\cos \alpha}{2} + \frac{\sin \alpha}{2i} + \frac{1}{2} \int_0^\infty e^{i\lambda s} \{\psi_2(s, \lambda)q_1(s) - i\psi_1(s, \lambda)q_2(s)\} \, ds (1.21) \\
N_2(\lambda) &= -\frac{\sin \alpha}{2} - \frac{\cos \alpha}{2i} - \frac{1}{2} \int_0^\infty e^{i\lambda s} \{i\psi_2(s, \lambda)q_1(s) + \psi_1(s, \lambda)q_2(s)\} \, ds.
\end{aligned}
$$

Theorem 9.1.1 *If the coefficients $q_1(x)$ and $q_2(s)$ are in the class $\mathcal{L}(0, \infty)$, then the spectrum of the problem (1.1), (1.2), (1.3) is continuous, and fills up the whole axis $(-\infty, \infty)$.*

Proof. Let $\vartheta(x, \lambda) = \psi(x, \lambda) + m(\lambda)\varphi(x, \lambda)$ be that solution of the system (1.1), (1.2), which is in the class $\mathcal{L}^2(0, \infty)$. Using the formulas (1.18) and (1.20), we then have

$$\vartheta_1(x, \lambda) = e^{-i\lambda x} \{N_1(\lambda) + m(\lambda)M_1(\lambda) + o(1)\}$$

$$\vartheta_2(x, \lambda) = e^{-i\lambda x} \{N_2(\lambda) + m(\lambda)M_2(\lambda) + o(1)\}.$$

Since there is only one solution of (1.1), (1.2) in $\mathcal{L}^2(0, \infty)$ for any nonreal λ, it is obvious that $\varphi(x, \lambda)$ and $\psi(x, \lambda)$ do not belong to this class. Because $\vartheta(x, \lambda) \in \mathcal{L}^2(0, \infty)$ by assumption, we should have

$$m(\lambda) = -\frac{N_1(\lambda)}{M_1(\lambda)} = -\frac{N_2(\lambda)}{M_2(\lambda)}.$$

Furthermore, note that if λ tends to a real limit (i.e., as $\tau \to 0$), then

$$M_1(\lambda) \to \frac{1}{2}\{\mu(\lambda) + i\nu(\lambda)\}, \quad N_1(\lambda) \to \frac{1}{2}\{\xi(\lambda) + i\eta(\lambda)\}$$

$$M_2(\lambda) \to \frac{1}{2}\{\nu(\lambda) - i\mu(\lambda)\}, \quad N_2(\lambda) \to \frac{1}{2}\{\eta(\lambda) - i\xi(\lambda)\}$$

relying on (1.19), (1.21), (1.11), (1.12), (1.13'), (1.14').

Therefore, $\lim_{\tau \to 0} m(\lambda) = -\dfrac{\xi(\lambda) + i\eta(\lambda)}{\mu(\lambda) + i\nu(\lambda)}$ and

$$\mathrm{Im}\, m(\lambda) = -\frac{1}{\mu^2(\lambda) + \nu^2(\lambda)}$$

for real λ. Since $\mu(\lambda)$, $\nu(\lambda)$ do not vanish for the same value of λ, $\mathrm{Im}\, m(\lambda)$ is a continuous function on the whole axis $(-\infty, \infty)$. By the formula (3.16), Chapter 8, we then have

$$\rho(\lambda + \Delta) - \rho(\lambda) = \frac{1}{\pi} \int_\lambda^{\lambda + \Delta} \frac{d\lambda}{\mu^2(\lambda) + \nu^2(\lambda)},$$

and the proof is thus complete.   ∎

9.2   Transformation of the basic system

In some cases, the system

$$y_1' - \{\lambda + q_1(x)\}y_2 = 0 \tag{2.1}$$

$$y_2' + \{\lambda + q_2(x)\}y_2 = 0 \tag{2.2}$$

with indefinitely increasing coefficients $q_1(x)$, $q_2(x)$ can, by a convenient change of variables, be transformed to another form admitting the derivation of asymptotic formulas.

Assume that λ is a real number, and that $q_1(x)$, $q_2(x)$ are two twice continuously differentiable functions. Put

$$\alpha(x) = \int_0^x \{[\lambda + q_1(s)][\lambda + q_2(s)]\}^{1/2}\, ds \tag{2.3}$$

$$u(x) = F(x, \lambda)y_2(x)$$
$$v(x) = G(x, \lambda)y_2(x) - F^{-1}(x, \lambda)y_1(x) \tag{2.4}$$

where

$$F(x, \lambda) = \left\{\frac{\lambda + q_1(x)}{\lambda + q_2(x)}\right\}^{1/4}, \qquad G(x, \lambda) = \frac{F_x'(x, \lambda)}{\alpha'(x)}. \tag{2.5}$$

Then

$$\frac{du}{d\alpha} = \frac{du}{dx} \cdot \frac{dx}{d\alpha} = \{F_x'y_2 + Fy_2'\}\frac{dx}{d\alpha} = \frac{F_x'}{\alpha'}y_2 + \frac{F}{\alpha'}y_2'$$

and $y_2' = \left\{\dfrac{du}{d\alpha} - \dfrac{F_x'}{\alpha'}y_2\right\}\dfrac{\alpha'}{F}$, which, substituted in (2.2), yields

$$\frac{dx}{d\alpha}\frac{\alpha'}{F} - \frac{F_x'}{F}y_2 + \{\lambda + q_2(x)\}y_1 = 0,$$

or, by (2.3), (2.4) and (2.5),

$$\frac{du}{d\alpha} = \frac{F'_x}{\alpha'}y_2 - \frac{F}{\alpha'}\{\lambda + q_2(x)\}y_1 = Gy_2 - F^{-1}y_1 = v(x). \tag{2.6}$$

Furthermore, we have

$$\frac{dv}{d\alpha} = \frac{dv}{dx} \cdot \frac{dx}{d\alpha} = \{G'_x y_2 + G y'_2 + F^{-2} F'_x y_1 - F^{-1} y'_1\}\frac{1}{\alpha'}$$

and

$$y'_1 = -\frac{dv}{d\alpha}F\alpha' + G'_x F y^2 - \{GF[\lambda + q_2(x)] - F'_x F^{-1}\}y_1,$$

relying on (2.2). Substituting y'_1 in (2.1) and taking into account the first equation in (2.4), we get

$$\frac{dv}{d\alpha} = \left\{-\frac{\lambda + q_1(x)}{\alpha' F'_x} + \frac{G_x}{\alpha' F}\right\}u - \left\{\frac{G[\lambda + q_2(x)]}{\alpha'} - \frac{F_x}{\alpha' F^2}\right\}y_1.$$

Since the coefficient of u is -1 (by (2.3) and (2.5)), and since the coefficient of y_1 is zero, (2.7) takes the form

$$\frac{dv}{d\alpha} = -u + \frac{G'_x}{\alpha' F}u. \tag{2.8}$$

Therefore, the system (2.1), (2.2) is reduced to

$$\frac{du}{d\alpha} = v, \qquad \frac{dv}{d\alpha} = -u + R(x,\lambda)u \tag{2.9}$$

according to the equalities (2.6) and (2.8), where $R(x,\lambda) = \dfrac{G'_x(x,\lambda)}{\alpha'(x)F(x,\lambda)}$, or, in more detail,

$$R(x,\lambda) = \frac{q''_1(x)}{4[\lambda + q_1(x)]^2[\lambda + q_2(x)]} - \frac{q''_2(x)}{4[\lambda + q_1(x)][\lambda + q_2(x)]^2}$$
$$- \frac{5}{16}\frac{q'^2_1(x)}{[\lambda + q_1(x)]^3[\lambda + q_2(x)]} - \frac{1}{8}\frac{q'_1(x)q'_2(x)}{[\lambda + q_1(x)]^2[\lambda + q_2(x)]}$$
$$+ \frac{7}{16}\frac{q'^2_2(x)}{[\lambda + q_1(x)][\lambda + q_2(x)]}, \tag{2.10}$$

using (2.3) and (2.4).

Lemma 9.2.1 *If a vector-valued function $\eta(x)$ is a solution of (2.9), i.e.,*

$$\frac{d\eta_1}{d\alpha} = \eta_2, \qquad \frac{d\eta_2}{d\alpha} = -\eta_1 + R(x,\lambda)\eta_1, \tag{2.11}$$

then it satisfies the system of integral equations

$$\eta_1(x, \lambda) = \eta_1(0) \cos \alpha(x) + \eta_2(0) \sin \alpha(x) - \int_0^x \eta_1(s, \lambda) P(s, \lambda) \sin[\alpha(s) - \alpha(x)] \, ds \tag{2.12}$$

$$\eta_2(x, \lambda) = -\eta_1(0) \sin \alpha(x) + \eta_2(0) \cos \alpha(x) + \int_0^x \eta_1(s, \lambda) P(s, \lambda) \cos[\alpha(s) - \alpha(x)] \, ds \tag{2.13}$$

where

$$P(x, \lambda) = R(x, \lambda)\alpha'(x) = G'_x(x, \lambda)/F(x, \lambda). \tag{2.14}$$

Proof. Indeed, it follows from (2.13), (2.11) and (2.14) that

$$\int_0^x \eta_1(s, \lambda) P(s, \lambda) \sin[\alpha(s) - \alpha(x)] \, ds$$

$$= \int_0^x \left\{ \frac{d\eta_2}{d\alpha} \alpha'(s) + \eta_1(s, \lambda)\alpha'(s) \right\} \sin[\alpha(s) - \alpha(x)] \, ds$$

$$= \int_0^x \frac{d\eta_2}{d\alpha} \alpha'(s) \sin[\alpha(s) - \alpha(x)] \, ds + \int_0^x \eta_1(s, \lambda)\alpha'(s) \sin[\alpha(s) - \alpha(x)] \, ds. \tag{2.15}$$

Integrating by parts the first addend on the right, using the first equation in (2.9), and integrating by parts again, we have

$$\int_0^x \frac{d\eta_2}{d\alpha} \sin[\alpha(s) - \alpha(x)] \, d\alpha(s) = \eta_2(0) \sin \alpha(x) - \int_0^x \frac{d\eta_1}{d\alpha} \cos[\alpha(s) - \alpha(x)] \, d\alpha(s)$$

$$= \eta_2(0) \sin \alpha(x) - \eta_1(x, \lambda) + \eta_1(0) \cos \alpha(x) - \int_0^x \eta_1(s, \lambda) \sin[\alpha(s) - \alpha(x)] \, d\alpha(s).$$

The equation (2.12) follows from (2.15), relying on the last relation; the equation (2.13) is proved similarly. ∎

Lemma 9.2.2 *Let $q_1(x)$, $q_2(x)$ be two monotone functions, and let one of the following hold :*

1. *$q_1(x) \geq 0$, $q_2(x) \leq 0$, $q_1'(x) > 0$, $q_2'(x) < 0$ and, as $x \to \infty$, $q_1(x) \to +\infty$, $q_2(x) \to -\infty$,*

2. *$q_1(x) \leq 0$, $q_2(x) \geq 0$, $q_1'(x) < 0$, $q_2'(x) > 0$ and, as $x \to \infty$, $q_1(x) \to -\infty$, $q_2(x) \to +\infty$,*

3. *$q_1(x), q_2(x) \geq 0$, $q_1'(x), q_2'(x) > 0$ and, as $x \to \infty$, $q_1(x), q_2(x) \to +\infty$,*

4. *$q_1(x), q_2(x) \leq 0$, $q_1'(x), q_2'(x) < 0$ and, as $x \to \infty$, $q_1(x), q_2(x) \to -\infty$,*

and the conditions

5.

$$q_1'(x) = O\{|q_1(x)|^c\}, \qquad q_2'(x) = O\{|q_2(x)|^c\}, \tag{2.16}$$

   where $0 < c < 3/2$,

6. $q_1''(x)$ and $q_2''(x)$ preserve sign for large x

hold. Then the integral

$$\int_0^\infty |P(x, \lambda)|\, dx \tag{2.17}$$

converges uniformly in λ in each domain $|\lambda + q_1(x)|, |\lambda + q_2(x)| \geq \delta > 0$, for $0 \leq x < \infty$.

Proof. By (2.10), (2.14), (2.3), we have

$$P(x, \lambda) = \frac{q_1''(x)}{4[\lambda + q_1(x)]^{3/2}[\lambda + q_2(x)]^{1/2}} - \frac{q_2''(x)}{4[\lambda + q_1(x)]^{1/2}[\lambda + q_2(x)]^{3/2}}$$

$$- \frac{5}{16} \frac{q_1'^2(x)}{[\lambda + q_1(x)]^{5/2}[\lambda + q_2(x)]^{1/2}} - \frac{1}{8} \frac{q_1'(x)q_2'(x)}{[\lambda + q_1(x)]^{3/2}[\lambda + q_2(x)]^{3/2}}$$

$$+ \frac{7}{16} \frac{q_2'^2(x)}{[\lambda + q_1(x)]^{1/2}[\lambda + q_2(x)]^{5/2}}. \tag{2.18}$$

It obviously suffices to show the integrability of each addend on the half-axis $[0, \infty)$.

Let Conditions 1, 5 and 6 hold (the proof is similar in the other cases). By (2.16), we have

$$\int_{X_0}^X \frac{\{q_1'(x)\}^2}{\{q_1(x)\}^{5/2}\{q_2'(x)\}^{1/2}}\, dx = O\left\{ \int_{X_0}^X \frac{q_1'(x)}{\{q_1(x)\}^{5/2-c}\{q_2(x)\}^{1/2}}\, dx \right\}.$$

Since $q_2(x)$ is a monotone function (given), applying the mean-value theorem to the last integral, we get

$$\int_{X_0}^X \frac{\{q_1'(x)\}^2}{\{q_1(x)\}^{5/2}\{q_2(x)\}^{1/2}}\, dx = O\left[\{q_2(\xi)\}^{-1/2} \int_{X_0}^X \frac{dq_1(x)}{q_1^{5/2-c}(x)} \right]$$

$$= O[q_2^{-1/2}(\xi)q_1^{c-3/2}(x)]_{X_0}^X = O(1) \tag{2.19}$$

$(X_0 < \xi < X)$. Similarly, we derive

$$\int_{X_0}^X \frac{\{q_2'(x)\}^2}{q_1^{1/2}(x)q_2^{5/2}(x)}\, dx = O(1). \tag{2.20}$$

Furthermore, again by the mean-value theorem and integrating by parts, we have

$$\int_{X_0}^X \frac{q_1''(x)}{q_1^{3/2}(x)q_2^{1/2}(x)}\, dx = q_2^{-1/2}(\xi) \int_{X_0}^X \frac{q_1''(x)}{q_1^{3/2}(x)}\, dx$$

$$= q_2^{-1/2}(\xi) \left[\frac{q_1'(x)}{q_1^{3/2}(x)} \right]_{X_0}^X + \frac{3}{2} \int_{X_0}^X \frac{\{q_1'(x)\}^2}{q_1^{5/2}(x)}\, dx.$$

Therefore, it follows from (2.16) and (2.19) that

$$\int_{X_0}^{X} \frac{q_1''(x)}{q_1^{3/2}(x)q_2^{1/2}(x)}\, dx = O(1). \tag{2.21}$$

Similarly, we see that

$$\int_{X_0}^{X} \frac{q_2''(x)}{q_1^{1/2}(x)q_2^{3/2}(x)}\, dx = O(1). \tag{2.22}$$

Finally, using (2.16) and the mean-value theorem, we obtain

$$\int_{X_0}^{X} \frac{q_1'(x)q_2'(x)}{q_1^{3/2}(x)q_2^{3/2}(x)}\, dx = O\left[\int_{X_0}^{X} \frac{q_1'(x)}{q_1^{3/2}(x)q_2^{3/2-c}(x)}\, dx\right]$$
$$= O[\{q_2(\xi)\}^{c-3/2}q_1^{-1/2}(x)|_{X_0}^{X}] = O(1). \tag{2.23}$$

The convergence of (2.17) follows from the estimates (2.19–23) because of (2.18). The proof is thus complete. ∎

9.3 The case of a pure point spectrum

Our goal here is to establish certain sufficient conditions for the spectrum of

$$y_1' - \{\lambda + q_1(x)\}y_2 = 0 \tag{3.1}$$

$$y_2' + \{\lambda + q_2(x)\}y_1 = 0 \tag{3.2}$$

$$y_1(0)\cos\alpha + y_2(0)\sin\alpha = 0 \tag{3.3}$$

to be discrete on the half-line $[0, \infty)$.

Theorem 9.3.1 *If $q_1(x)$, $q_2(x)$ satisfy Conditions 1, 5, 6 or 2, 5, 6 of Lemma 9.2.2, then the problem (3.1), (3.2), (3.3) has a pure point spectrum.*

Proof. If $\mathrm{Im}\,\lambda > 0$, we assume that $0 < \arg\lambda < \pi$, while, if $q_1(x)$ or $q_2(x) \to -\infty$, we assume that both $\arg\{\lambda + q_1(x)\}$ and $\arg\{\lambda + q_2(x)\}$ increase from $\arg\lambda$ to π. Therefore,

$$\frac{1}{2}\arg\lambda \le \arg\{\lambda + q_1(x)\}^{1/2}, \qquad \arg\{\lambda + q_2(x)\}^{1/2} \le \frac{\pi}{2},$$

Hence, the imaginary part of the function

$$\alpha(x) = \int_0^x \{[\lambda + q_1(s)][\lambda + q_2(s)]\}^{1/2}\, ds$$

is positive. Furthermore, if Condition 1 or 2 is fulfilled, λ bounded and $x \to \infty$, we have

$$\alpha(x) \sim i \int_0^x \{q_1(s)q_2(s)\}^{1/2}\, ds,$$

so that $e^{-i\alpha(x)} \to \infty$. Let $\xi(x, \lambda) = \eta(x, \lambda)e^{i\alpha(x)}$. Then it follows from (2.12) and (2.13) that

$$\xi_1(x, \lambda) = \eta_1(0)e^{i\alpha(x)}\cos\alpha(x) + \eta_2(0)e^{i\alpha(x)}\sin\alpha(x)$$
$$-\int_0^x \xi_1(s, \lambda)e^{i(\alpha(x)-\alpha(s))}P(s, \lambda)\sin\{\alpha(s) - \alpha(x)\}\, ds$$

$$\xi_2(x, \lambda) = -\eta_1(0)e^{i\alpha(x)}\sin\alpha(x) + \eta_2(0)e^{i\alpha(x)}\cos\alpha(x)$$
$$+\int_0^x \xi_1(x, \lambda)e^{i(\alpha(x)-\alpha(s))}P(s, \lambda)\cos\{\alpha(s) - \alpha(x)\}\, ds.$$

Therefore, by the inequalities $|e^{i\alpha(x)}\sin\alpha(x)| \leq 1$, $|e^{i\alpha(x)}\cos\alpha(x)| \leq 1$,

$$|\xi_1|, |\xi_2| \leq |\eta_1(0)| + |\eta_2(0)| + \int_0^x |\xi_1(s, \lambda)|\,|P(s, \lambda)|\, ds,$$

and, by Lemma 3.2.1, which is applicable according to Lemma 9.2.2,

$$|\xi_1|, |\xi_2| \leq \{|\eta_1(0)| + |\eta_2(0)|\} \exp\left\{\int_0^x |P(s, \lambda)|\, ds\right\}.$$

Hence,

$$|\eta_1(x, \lambda)|, |\eta_2(x, \lambda)| \leq \{|\eta_1(0)| + |\eta_2(0)|\} \exp\left\{\int_0^x |P(s, \lambda)|\, ds\right\} |e^{-i\alpha(x)}| = O(e^{-i\alpha(x)})$$
$$= O(e^{-\operatorname{Im}\alpha(x)}). \quad (3.4)$$

Furthermore, for fixed λ, $\operatorname{Im}\lambda > 0$ and as $x \to \infty$, it follows from (2.12) that

$$\eta_1(x, \lambda) = \frac{1}{2}\eta_1(0)e^{-i\alpha(x)} - \frac{1}{2i}\eta_2(0)e^{-i\alpha(x)} + O(e^{-\operatorname{Im}\alpha(x)})$$
$$+\frac{1}{2i}\int_0^x \eta_1(s, \lambda)P(s, \lambda)e^{i\{\alpha(s)-\alpha(x)\}}\, ds$$
$$+O\left\{\int_0^x |\eta_1(x, \lambda)|\,|P(s, \lambda)|e^{-\operatorname{Im}\{\alpha(s)-\alpha(x)\}}\, ds\right\}. \quad (3.5)$$

By (3.4),

$$O\left\{\int_0^x |\eta_1|\,|P|e^{-\operatorname{Im}\{\alpha(x)-\alpha(s)\}}\, ds\right\} = O\left\{\int_0^x |P|e^{-\operatorname{Im}\{\alpha(x)-2\alpha(s)\}}\, ds\right\}$$
$$= O\left\{\int_0^x e^{-\operatorname{Im}\{\alpha(x)-2\alpha(x-\delta)\}}\int_0^{x-\delta} |P|\, ds\right\} + O\left\{e^{\operatorname{Im}\alpha(x)}\int_{x-\delta}^x |P|\, ds\right\}. \quad (3.6)$$

Since

$$\alpha(x - \delta) = \alpha(x) - \int_{x-\delta}^{x} \{[\lambda + q_1(s)][\lambda + q_2(s)]\}^{1/2} ds,$$

we have

$$- \text{Im}[\alpha(x) - 2\alpha(x - \delta)] = \text{Im}\,\alpha(x) - 2\,\text{Im}\int_{x-\delta}^{x} \{[\lambda + q_1(s)][\lambda + q_2(s)]\}^{1/2} ds.$$

Therefore,

$$\exp\{- \text{Im}[\alpha(x) - 2\alpha(x - \delta)]\} = o(e^{\text{Im}\,\alpha(x)}), \quad x \to \infty,$$

and, using Lemma 9.2.2, we obtain from (3.6) that

$$O\left\{\int_0^x |\eta_1(s, \lambda)|\,|P(s, \lambda)|\,e^{-\text{Im}[\alpha(x) - \alpha(s)]}\,ds\right\} = o(e^{\text{Im}\,\alpha(x)}) \tag{3.6'}$$

as $x \to \infty$. On the other hand, by Lemma 9.2.2, as $x \to \infty$, we get

$$\int_0^\infty \eta_1(s, \lambda)P(s, \lambda)e^{-i[\alpha(x) - \alpha(s)]}\,ds = O\left\{e^{\text{Im}\,\alpha(x)}\int_x^\infty |P(s, \lambda)|\,ds\right\} = o(e^{\text{Im}\,\alpha(x)}). \tag{3.7}$$

Owing to the estimates (3.6') and (3.7), the equality (3.5) can be rewritten in the form

$$\eta_1(x, \lambda) = e^{-i\alpha(x)}\{M^+(\lambda) + o(1)\}, \tag{3.8}$$

where

$$M^+(\lambda) = \frac{\eta_1(0)}{2} - \frac{\eta_2(0)}{2i} + \frac{1}{2i}\int_0^\infty \eta_1(s, \lambda)P(s, \lambda)e^{i\alpha(s)}\,ds. \tag{3.9}$$

Similarly, we derive from the equalities (2.13) that

$$\eta_2(x, \lambda) = e^{-i\alpha(x)}\{N^+(\lambda) + o(1)\}, \tag{3.10}$$

where

$$N^+(\lambda) = \frac{\eta_1(0)}{2i} + \frac{\eta_2(0)}{2} + \frac{1}{2}\int_0^\infty \eta_1(s, \lambda)P(s, \lambda)e^{i\alpha(s)}\,ds.$$

Now, if $\text{Im}\,\alpha(x) \to -\infty$ as $x \to \infty$ and, therefore, $e^{i\alpha(x)} \to \infty$, then, proceeding as above, (2.12) and (2.13) are given the form

$$\eta_1(x, \lambda) = e^{i\alpha(x)}\{M^-(\lambda) + o(1)\} \tag{3.12}$$

$$\eta_2(x, \lambda) = e^{i\alpha(x)}\{N^-(\lambda) + o(1)\} \tag{3.13}$$

where

$$M^-(\lambda) = \frac{\eta_1(0)}{2} + \frac{\eta_2(0)}{2i} - \frac{1}{2i}\int_0^\infty \eta_1(s, \lambda)P(s, \lambda)e^{-i\alpha(s)}\,ds \tag{3.14}$$

$$N^-(\lambda) = \frac{\eta_2(0)}{2} - \frac{\eta_1(0)}{2i} + \frac{1}{2}\int_0^\infty \eta_1(s, \lambda)P(s, \lambda)e^{-i\alpha(s)}\,ds. \tag{3.15}$$

.

Let $\varphi(x, \lambda)$ be the solution of the system (3.1), (3.2), satisfying the conditions

$$\varphi_1(0, \lambda) = -\sin \alpha$$

$$\varphi_2(0, \lambda) = \cos \alpha$$

(3.16)

and $\psi(x, \lambda)$ the one satisfying

$$\psi_1(0, \lambda) = -\cos \alpha$$

$$\psi_2(0, \lambda) = -\sin \alpha.$$

(3.17)

Furthermore, since it follows from (2.4) that

$$\eta_1(0) = F(0)y_2(0)$$

$$\eta_2(0) = G(0)y_2(0) - F^{-1}(0)y_1(0)$$

(3.18)

relying on (3.16), we have

$$\eta_1(0) = F(0) \cos \alpha$$

$$\eta_2(0) = F^{-1} \sin \alpha + G(0) \cos \alpha.$$

(3.18′)

On the other hand, thanks to (2.4),

$$\varphi_1(x, \lambda) = G(x, \lambda)\eta_1(x, \lambda) - F(x, \lambda)\eta_2(x, \lambda)$$

$$\varphi_2(x, \lambda) = F^{-1}(x, \lambda)\eta_1(x, \lambda).$$

(3.19)

Therefore, using (3.8) and (3.10), we obtain the asymptotic formulas for $\varphi(x, \lambda)$ as $x \to \infty$ (the plus sign meaning that $\operatorname{Im} \alpha(x) \to +\infty$ as $x \to \infty$)

$$\varphi_1^+(x, \lambda) = e^{-i\alpha(x)}\{GM_\varphi^+(\lambda) - FN_\varphi^+(\lambda) + o(1)\}$$

(3.20)

$$\varphi_2^+(x, \lambda) = e^{-i\alpha(x)}\{F^{-1}M_\varphi^+(\lambda) + o(1)\}$$

(3.21)

where, relying on (3.9), (3.11), (3.18′) and (3.19),

$$M_\varphi^+(\lambda) = \frac{1}{2i}\bigg\{ -F^{-1}(0)\sin \alpha + [iF(0) - G(0)]\cos \alpha$$
$$+ \int_0^\infty \varphi_2^+(s, \lambda)P(s, \lambda)F(s, \lambda)e^{i\alpha(s)}\,ds\bigg\}, \quad (3.20')$$

$$N_\varphi^+(\lambda) = \frac{1}{2}\bigg\{ F^{-1}(0)\sin \alpha - [iF(0) - G(0)]\cos \alpha$$
$$+ \int_0^\infty \varphi_2^+(s, \lambda)P(s, \lambda)F(s, \lambda)e^{i\alpha(s)}\,ds\bigg\}. \quad (3.21')$$

Similarly, we obtain the asymptotic formulas

$$\psi_1(x, \lambda) = e^{-i\alpha(x)}\{GM_\psi^+(\lambda) - FN_\psi^+(\lambda) + o(1)\} \tag{3.22}$$

$$\psi_2(x, \lambda) = e^{-i\alpha(x)}\{F^{-1}M_\psi^+(\lambda) + o(1)\} \tag{3.23}$$

for $\psi(x, \lambda)$, as $x \to \infty$, where $M_\psi^+(\lambda)$ and $N_\psi^+(\lambda)$ are obtained from $M_\varphi^+(\lambda)$ and $N_\varphi^+(\lambda)$, respectively, by replacing $\sin\alpha$, $\cos\alpha$, $\varphi_2^+(x, \lambda)$ with $\cos\alpha$, $-\sin\alpha$, $\psi_2(x, \lambda)$, respectively. Let $\vartheta(x, \lambda) = \psi(x, \lambda) + m^+(\lambda)\varphi(x, \lambda)$ be the solution of the system (3.1), (3.2), of class $\mathcal{L}^2(0, \infty)$. Using (3.20), (3.21), (3.22) and (3.23), we get

$$\vartheta_1(x, \lambda) = \psi_1^+(x, \lambda) + m^+(\lambda)\varphi_1^+(x, \lambda)$$
$$= e^{-i\alpha(x)}\{GM_\psi^+ - FN_\psi^+ + m^+(\lambda)[GM_\varphi^+ - FN_\varphi^+] + o(1)\}$$

$$\vartheta_2(x, \lambda) = \psi_2^+(x, \lambda) + m^+(\lambda)\varphi_2^+(x, \lambda) = e^{-i\alpha(x)}\{F^{-1}M_\psi^+ + m^+(\lambda)F^{-1}M_\varphi^+ + o(1)\}.$$

Since $\vartheta(x, \lambda) \in \mathcal{L}^2(0, \infty)$, it follows from the last two formulas that

$$m^+(\lambda) = -\frac{M_\psi^+}{M_\varphi^+} = -\lim_{x\to\infty} \frac{G(x, \lambda)M_\psi^+(\lambda) - F(x, \lambda)N_\psi^+(\lambda)}{G(x, \lambda)M_\varphi^+(\lambda) - F(x, \lambda)N_\varphi^+(\lambda)}. \tag{3.24}$$

Similar formulas also hold as $\text{Im}\,\alpha(x) \to -\infty$, replacing $m^+(\lambda)$, $M_{\varphi,\psi}^+(\lambda)$, $N_{\varphi,\psi}^+(\lambda)$ in (3.24) by $m^-(\lambda)$, $M_{\varphi,\psi}^-(\lambda)$, $N_{\varphi,\psi}^-(\lambda)$, respectively, where the last two functions are obtained from (3.14), (3.15) in the way $M_{\varphi,\psi}^+(\lambda)$, $N_{\varphi,\psi}^+(\lambda)$ are from (3.9) and (3.11).

Let λ be a fixed real number. By Condition 1 in Lemma 9.2.2, $q_1(x)$ tends to $+\infty$ monotonically, while $q_2(x) \to -\infty$. First, we assume that, for all x, $\lambda + q_1(x) > 0$ and $\lambda + q_2(x) < 0$. Then the function $\alpha(x)$ is purely imaginary, so that $\alpha(x) = |\alpha(x)|e^{i\pi/2}$. We assume the same when Condition 2 holds. As can easily be seen, we then have

$$F(x, \lambda) = |F(x, \lambda)|e^{i\pi/4}, \qquad G(x, \lambda) = |G(x, \lambda)|e^{-i\pi/4}.$$

Therefore, it follows from (3.20′) that the function $e^{-i\pi/4}M_\varphi^+(\lambda)$ is real-valued. It is proved similarly that $M_\psi^+(\lambda)$, $N_{\varphi,\psi}^\pm(\lambda)$, $M_{\varphi,\psi}^-(\lambda)$ are real-valued if each function is multiplied by $e^{i\pi/4}$ or $e^{-i\pi/4}$.

Thus, we have established that $e^{-i\pi/4}/M_\varphi^+(\lambda)$ is actually real-valued, and continuous on the whole real axis. Furthermore, since $M^+(\lambda)$ is regular in the upper half-plane, by the symmetry principle, $M_\varphi^+(\lambda)$ is an entire function. We can infer similarly that $M_\psi^+(\lambda)$ is entire, too. Therefore, by (3.24), $m^+(\lambda)$ is meromorphic. The eigenvalues coincide with its poles; hence, we have a point spectrum.

The other cases are considered similarly.

The preceding argument can also be applied when, as $q_1(x) \to +\infty$, $q_2(x) \to -\infty$ or $q_1(x) \to -\infty$, $q_2(x) \to +\infty$, the inequalities $\lambda + q_1(x) > 0$, $\lambda + q_2(x) < 0$ or $\lambda + q_1(x) < 0$, $\lambda + q_2(x) > 0$ do not hold for all x. It is obvious that we can then choose X_0 such that, for all $x > X_0$ and fixed real λ, $\lambda + q_1(x) > 0$, $\lambda + q_2(x) < 0$ (for Condition 1) and $\lambda + q_1(x) < 0$, $\lambda + q_2(x) > 0$ (for Condition 2); the above conclusions are also valid for the interval (X_0, ∞).

The proof is thus complete. ∎

9.4   Other cases

Here, we study the spectrum of the problem (3.1), (3.2), (3.3) further. Let the vector-valued functions $\varphi(x, \lambda)$ and $\psi(x, \lambda)$ be the same as above, and let $\eta_\varphi(x, \lambda)$, $\eta_\psi(x, \lambda)$ be two solutions of the system (2.11), obtained from the formula (2.4) by replacing $y(x)$ with $\varphi(x, \lambda)$, $\psi(x, \lambda)$, respectively. It then follows from the formulas (3.18) that

$$\eta_{1\varphi}(0) = F(0) \cos \alpha$$

$$\eta_{2\varphi}(0) = G(0) \cos \alpha + F^{-1}(0) \sin \alpha$$

$$(4.1)$$

$$\eta_{1\psi}(0) = -F(0) \sin \alpha$$

$$\eta_{2\psi} = -G(0) \sin \alpha + F^{-1}(0) \cos \alpha$$

$$(4.2)$$

relying on (3.16), (3.17).

Furthermore, we derive

$$\varphi_1(0, \lambda) = G(0)\eta_{1\varphi}(0) - F(0)\eta_{2\varphi}(0)$$

$$\varphi_2(0, \lambda) = F^{-1}(0)\eta_{1\varphi}(0)$$

$$(4.3)$$

$$\psi_1(0, \lambda) = G(0)\eta_{1\psi}(0) - F(0)\eta_{2\psi}(0)$$

$$\psi_2(0, \lambda) = F^{-1}(0)\eta_{1\psi}(0)$$

$$(4.4)$$

from the formulas (3.19) for $\varphi(0, \lambda)$, $\psi(0, \lambda)$. By the conditions (3.16) and (3.17), we have $W\{\varphi, \psi\} = 1$. Therefore, using the expressions (4.3), (4.4), we obtain

$$W\{\varphi, \psi\} = \eta_{1\varphi}(0)\eta_{2\psi}(0) - \eta_{2\varphi}(0)\eta_{1\psi}(0) = W\{\eta_\varphi, \eta_\psi\} = 1. \qquad (4.5)$$

Theorem 9.4.1 *If $q_1(x)$, $q_2(x)$ fulfil Conditions 3, 5 and 6 or 4, 5 and 6 of Lemma 9.2.2, then the spectrum of the problem (3.1), (3.2), (3.3) is continuous and fills up the whole axis $(-\infty, \infty)$.*

Proof. Let λ be a fixed real number. Since, according to Conditions 3 and 4 of Lemma 9.2.2, $q_1(x)$, $q_2(x)$ both tend to $+\infty$ or $-\infty$ as $x \to \infty$, we first assume that, for all x, either both $\lambda + q_1(x)$ and $\lambda + q_2(x) > 0$ or both $\lambda + q_1(x)$ and $\lambda + q_2(x) < 0$. Anyway, the functions $\alpha(x)$ and $F(x, \lambda)$ (see the formulas (2.3), (2.5)) are real-valued. Therefore, $\operatorname{Im} \alpha(x) = 0$, and it follows from the estimate (3.4) that $|\eta_1(x, \lambda)|$, $|\eta_2(x, \lambda)|$ are bounded. On the other hand, the conditions of the theorem provide for the convergence of the integral (2.17). The formulas (2.12), (2.13) can then be rewritten as

$$\eta_1(x, \lambda) = \mu(\lambda) \sin \alpha(x) + \nu(\lambda) \cos \alpha(x) + o(1) \qquad (4.6)$$

$$\eta_2(x,\lambda) = \mu(\lambda)\cos\alpha(x) - \nu(\lambda)\sin\alpha(x) + o(1) \tag{4.7}$$

$(x \to \infty)$, where

$$\mu(\lambda) = \eta_2(0) + \int_0^\infty \eta_1(s,\lambda)P(s,\lambda)\cos\alpha(s)\,ds \tag{4.8}$$

$$\nu(\lambda) = \eta_1(0) - \int_0^\infty \eta_1(s,\lambda)P(s,\lambda)\sin\alpha(s)\,ds. \tag{4.9}$$

Since the integrals in (4.8), (4.9) converge uniformly in λ, $\mu(\lambda)$, $\nu(\lambda)$ are both continuous and bounded functions of λ.

Furthermore, by the asymptotic formulas (4.6), (4.7), it follows from the equality (4.5) that

$$W\{\eta_\varphi, \eta_\psi\} = \mu_\psi(\lambda)\nu_\varphi(\lambda) - \mu_\varphi(\lambda)\nu_\psi(\lambda) + o(1) = 1$$

as $x \to \infty$, where $\mu_{\varphi,\psi}(\lambda)$, $\nu_{\varphi,\psi}(\lambda)$ are obtained from (4.8), (4.9) by replacing $\eta_1(x)$, $\eta_2(x)$ by

$$F(x,\lambda)\varphi_2(x,\lambda), \qquad F(x,\lambda)\psi_2(x,\lambda)$$

and

$$\{G(x,\lambda)\varphi_2(x,\lambda) - F^{-1}(x,\lambda)\varphi_1(x,\lambda)\}, \qquad \{G(x,\lambda)\psi_2(x,\lambda) - F^{-1}(x,\lambda)\psi_1(x,\lambda)\},$$

respectively. Therefore,

$$\mu_\psi(\lambda)\nu_\varphi(\lambda) - \mu_\varphi(\lambda)\nu_\psi(\lambda) = 1$$

as $x \to \infty$, and neither $\mu_\varphi(\lambda)$, $\nu_\varphi(\lambda)$ nor $\mu_\psi(\lambda)$, $\nu_\psi(\lambda)$ can vanish for the same value of λ. Furthermore, we derive from the formula (3.20') that

$$M_\varphi^+(\lambda) = \frac{1}{2i}\left\{i\eta_{1\varphi}(0) - \eta_{2\psi}(0) + \int_0^\infty \eta_{1\varphi}(s,\lambda)P(s,\lambda)e^{i\alpha(s)}\,ds\right\}$$

in accordance with (4.1), (4.2) or, using the expressions (4.8), (4.9),

$$M_\varphi^+(\lambda) = -\frac{1}{2i}\{\mu_\varphi(\lambda) - i\nu_\varphi(\lambda)\}.$$

Similarly,

$$N_\varphi^+(\lambda) = \frac{1}{2}\{\mu_\varphi(\lambda) - i\nu_\varphi(\lambda)\}$$

$$M_\psi^+(\lambda) = -\frac{1}{2i}\{\mu_\psi(\lambda) - i\nu_\psi(\lambda)\}$$

$$N_\varphi^+(\lambda) = \frac{1}{2}\{\mu_\psi(\lambda) - i\vartheta_\psi(\lambda)\}.$$

Then, by (3.24), we get

$$m^+(\lambda) = -\frac{\mu_\psi(\lambda)\mu_\varphi(\lambda) - \nu_\psi(\lambda)\nu_\varphi(\lambda)}{\mu_\varphi^2(\lambda) + \nu_\varphi^2(\lambda)} - \frac{i}{\mu_\varphi^2(\lambda) + \nu_\varphi^2(\lambda)}.$$

Therefore, $\operatorname{Im} m^+(\lambda) = -1/(\mu_\varphi^2(\lambda)+\nu_\varphi^2(\lambda))$, and, similarly, $\operatorname{Im} m^-(\lambda) = 1/(\mu_\varphi^2(\lambda)+\nu_\varphi^2(\lambda))$. In both cases, $\operatorname{Im} m(\lambda)$ is a continuous and bounded function of λ. Therefore, the statement of the theorem concerning the continuity of the spectrum follows from the formula (3.16), Chapter 8, since

$$\rho(\lambda + \Delta) - \rho(\lambda) = \pm\frac{1}{\pi} \int_\lambda^{\lambda+\Delta} \frac{d\lambda}{\mu_\varphi^2(\lambda) + \nu_\varphi^2(\lambda)}.$$

Notes

Section 1

The results are due to S. Conte and W. Sangren (1954). Their work is the basis for this section.

Sections 2, 3 and 4

These sections are written on the basis of the paper of W. Roos and W. Sangren (1961). The treatment here of their results is due to B. M. Levitan and I. S. Sargsjan (1975).

Chapter 10

THE SOLUTION OF THE CAUCHY PROBLEM FOR THE NONSTATIONARY DIRAC SYSTEM

10.1 Derivation of the formula for the solution of the Cauchy problem

Here and onwards, in studying various problems in the spectral theory of a Dirac system, we will use the canonical form (1.5) from Chapter 7, or

$$\begin{pmatrix} 0 & 1 \\ -1 & 0 \end{pmatrix} \frac{dy}{dx} + \begin{pmatrix} p(x) & q(x) \\ q(x) & -p(x) \end{pmatrix} y = \lambda y.$$

Consider the Cauchy problem for the nonstationary Dirac system

$$i\frac{\partial u}{\partial t} + \begin{pmatrix} 0 & 1 \\ -1 & 0 \end{pmatrix} \frac{\partial u}{\partial x} + \begin{pmatrix} p(x) & q(x) \\ q(x) & -p(x) \end{pmatrix} u = 0 \qquad (1.1)$$

$$u(x,t) = \begin{pmatrix} u_1(x,t) \\ u_2(x,t) \end{pmatrix}$$

$$(1.2)$$

$$u(x,0) = f(x)$$

where $p(x)$, $q(x)$ are two real-valued functions defined on the whole number line and summable in each finite interval[1] while the vector function $f(x)$ is real-valued and continuously differentiable.

We introduce the notation

$$B \equiv \begin{pmatrix} 0 & 1 \\ -1 & 0 \end{pmatrix}, \qquad Q(x) \equiv \begin{pmatrix} p(x) & q(x) \\ q(x) & -p(x) \end{pmatrix}.$$

The Cauchy problem (1.1), (1.2) can then be rewritten in the form

$$i\frac{\partial u}{\partial t} + B\frac{\partial u}{\partial x} + Q(x)u = 0 \qquad (1.1')$$

---

[1]Depending on the circumstances, the functions $p(x)$, $q(x)$ are in the sequel subject to various smoothness conditions.

$$u(x,0) = f(x). \tag{1.2'}$$

We first consider (1.1'), (1.2'), assuming that $p(x) = q(x) \equiv 0$ ($Q(x) \equiv 0$), i.e.,

$$i\frac{\partial u^0(x,t)}{\partial t} + B\frac{\partial u^0(x,t)}{\partial x} = 0, \qquad u^0(x,0) = f(x), \tag{1.3}$$

which is equivalent to two independent problems relative to the components $u_1^0(x,t)$, $u_2^0(x,t)$ of the vector-valued function $u^0(x,t)$, viz.

$$\frac{\partial^2 u_1^0}{\partial t^2} = \frac{\partial^2 u_1^0}{\partial x^2}, \qquad u_1^0(x,0) = f_1(x), \qquad \left.\frac{\partial u_1^0}{\partial t}\right|_{t=0} = if_2'(x) \tag{1.4}$$

$$\frac{\partial^2 u_2^0}{\partial t^2} = \frac{\partial^2 u_2^0}{\partial x^2}, \qquad u_2^0(x,0) = f_2(x), \qquad \left.\frac{\partial u_2^0}{\partial t}\right|_{t=0} = if_1'(x). \tag{1.5}$$

Solving (1.4), (1.5) by the method of d'Alembert, we obtain the solutions

$$u_1^0(x,t) = \frac{1}{2}\{f_1(x+t) + if_2(x+t)\} + \frac{1}{2}\{f_1(x-t) - if_2(x-t)\}$$

$$u_2^0(x,t) = \frac{1}{2}\{-if_1(x+t) + f_2(x+t)\} + \frac{1}{2}\{if_1(x-t) + f_2(x-t)\}.$$

Therefore, the solution of the problem (1.3) is given by the formula

$$u^0(x,t) = \frac{1}{2}\{Hf(x+t) + H^T f(x-t)\}, \tag{1.6}$$

where

$$H = \begin{pmatrix} 1 & i \\ -i & 1 \end{pmatrix}, \qquad H^T = \begin{pmatrix} 1 & -i \\ i & 1 \end{pmatrix}.$$

We now consider the nonhomogeneous problem with the homogeneous initial conditions

$$i\frac{\partial \tilde{u}(x,t)}{\partial t} + B\frac{\partial \tilde{u}(x,t)}{\partial x} = g(x,t), \qquad \tilde{u}(x,0) = 0, \tag{1.7}$$

where $g(x,t)$ is a differentiable vector-valued function. This also splits up into two independent problems with respect to the components $\tilde{u}_1(x,t)$, $\tilde{u}_2(x,t)$, viz.

$$\frac{\partial^2 \tilde{u}_1}{\partial t^2} - \frac{\partial^2 \tilde{u}_1}{\partial x^2} = -i\frac{\partial g_1}{\partial t} + \frac{\partial g_2}{\partial x}$$

$$\tilde{u}_1(x,0) = 0, \qquad \left.\frac{\partial \tilde{u}_1}{\partial t}\right|_{t=0} = -ig_1(x,0) \tag{1.8}$$

$$\frac{\partial^2 \tilde{u}_2}{\partial t^2} - \frac{\partial^2 \tilde{u}_2}{\partial x^2} = -i\frac{\partial g_2}{\partial t} - \frac{\partial g_1}{\partial x}$$

$$\tilde{u}_2(x,0) = 0, \qquad \left.\frac{\partial \tilde{u}_2}{\partial t}\right|_{t=0} = -ig_2(x,0). \tag{1.9}$$

Since the solution of the problem

$$\frac{\partial^2 v}{\partial t^2} - \frac{\partial^2 v}{\partial x^2} = h(x,t), \qquad v(x,0) = 0, \qquad \frac{\partial v}{\partial t}\bigg|_{t=0} = k(x)$$

is given by the formula

$$v(x,t) = \frac{1}{2}\int_{x-t}^{x+t} k(\tau)\,d\tau + \frac{1}{2}\int_0^t d\tau \int_{x-(t-\tau)}^{x+(t-\tau)} h(s,\tau)\,ds,$$

we apply the latter to the problems (1.8), (1.9), and obtain

$$\tilde{u}_1(x,t) = \frac{1}{2}\int_0^t \{-ig_1(x+t-\tau,\tau)+g_2(x+t-\tau,\tau)-ig_1(x-t+\tau,\tau)-g_2(x-t+\tau,\tau)\}\,d\tau$$

$$\tilde{u}_2(x,\tau) = \frac{1}{2}\int_0^t \{-g_1(x+t-\tau,\tau)-ig_2(x+t-\tau,\tau)+g_1(x-t+\tau,\tau)-ig_2(x-t+\tau,\tau)\}\,d\tau$$

for their solutions. Therefore, the solution of the problem (1.7) is specified by the formula

$$\tilde{u}(x,t) = -\frac{i}{2}\int_0^t \{Hg(x+t-\tau,\tau) + H^T g(x-t+\tau,\tau)\}\,d\tau.$$

We now retrace our steps to the problem (1.1'), (1.2'), and rewrite it in the form

$$i\frac{\partial u}{\partial t} + B\frac{\partial u}{\partial x} = -Q(x)u(x,t), \qquad u(x,0) = f(x). \tag{1.11}$$

Putting $g(x,t) = -Q(x)u(x,t)$, considering $g(x,t)$ as known and taking into account that the solution of the problem (1.11) is the sum of those of the problems (1.3) and (1.7), we have

$$u(x,t) = \frac{1}{2}\{Hf(x+t) + H^T f(x-t)\}$$
$$+ \frac{i}{2}\int_0^t \{HQ(x+t-\tau)u(x+t-\tau,\tau) + H^T Q(x-t+\tau)u(x-t+\tau,\tau)\}\,d\tau,$$

relying on (1.6), (1.10), or, replacing the integration variable,

$$u(x,t) = \frac{1}{2}\{Hf(x+t) + H^T f(x-t)\}$$
$$+ \frac{i}{2}\int_x^{x+t} HQ(s)u(s,x+t-s)\,dx + \frac{i}{2}\int_{x-t}^x H^T Q(s)u(s,s-x+t)\,ds. \tag{1.12}$$

Thus, to solve the problem (1.1'), (1.2'), its equivalent system of integral equations (1.12) is constructed. They are Volterra integral equations and, therefore, can be solved by the method of successive approximations.

Let $u_0(x, t)$ be determined from the formula (1.6). For $k \geq 1$, $k = 1, 2, 3, \ldots$, we put

$$u_k(x, t) = \frac{i}{2} \int_x^{x+t} HQ(s)u_{k-1}(s, x + t - s)\, ds + \frac{i}{2} \int_{x-t}^x H^T Q(s)u_{k-1}(s, s - x + t)\, ds.$$

$$\tag{1.13}$$

It then follows from the uniform convergence of the successive approximations (whose proof we omit) that the solution of the equation (1.12) is determined by the equality

$$u(x, t) = \sum_{k=0}^{\infty} u_k(x, t). \tag{1.14}$$

We show that each vector-valued function $u_k(x, t)$, $k = 1, 2, 3, \ldots$, can be represented as

$$u_k(x, t) = \frac{1}{2} \int_{x-t}^{x+t} W_k(x, t, s)f(s)\, ds, \tag{1.15}$$

where $W_k(x, t, s)$ is a matrix of order two.

In fact, if $k = 1$, we have

$$u_1(x, t) = \frac{i}{4} \int_x^{x+t} HQ(s)\{Hf(x + t) + H^T f(2s - x - t)\}\, ds$$

$$+ \frac{i}{4} \int_{x-t}^x H^T Q(s)\{Hf(2s - x + t) + H^T f(x - t)\}\, ds,$$

relying on (1.6), (1.13). Changing the variable and taking into account that $HQH = H^T Q H^T = 0$, we get

$$u_1(x, t) = \frac{1}{2} \int_{x-t}^{x+t} \frac{i}{4} \left\{ HQ \left(\frac{x + t + s}{2} \right) H^T + H^T Q \left(\frac{x - t - s}{2} \right) H \right\} f(s)\, ds,$$

i.e., the formula (1.15), if

$$W_1(x, t, s) = \frac{i}{4} \left\{ HQ \left(\frac{x + t + s}{2} \right) H^T + H^T Q \left(\frac{x - t - s}{2} \right) H \right\}.$$

We now assume that, for $k = 1, 2, \ldots, n - 1$, the formula (1.15) is proved, and prove it for $k = n$. Putting $k = n - 1$ in (1.15), and substituting the expression for $u_{n-1}(x, t)$ in (1.13), we obtain

$$u_n(x, t) = \frac{i}{4} \int_x^{x+t} HQ(s) \left\{ \int_{2s-(x+t)}^{x+t} W_{n-1}(s, x + t - s, \tau)f(\tau)\, d\tau \right\} ds$$

$$+ \frac{i}{4} \int_{x-t}^x H^T Q(s) \left\{ \int_{x-t}^{2s-(x-t)} W_{n-1}(s, s - x + t, \tau)f(\tau)\, d\tau \right\} ds$$

while, changing the order of integration, that

$$u_n(x,t) = \frac{1}{2}\int_{x-t}^{x+t}\left\{\frac{i}{2}\int_x^{(x+t+\tau)/2} HQ(s)W_{n-1}(s,x+t-s,s)\,ds\right.$$

$$\left. + \frac{i}{2}\int_{x-t}^{(x-t+\tau)/2} H^T Q(s)W_{n-1}(s,s-x+t,s)\,ds\right\} f(\tau)\,d\tau,$$

i.e., (1.15) again, if the expression in braces is denoted by $W_n(x,t,\tau)$.

Thus, (1.15) is proved for any n. ∎

Now, substituting $u_k(x,t)$, $k = 0,1,2,\ldots$, in the equality (1.14), we get that the solution of the problem (1.1), (1.2) is specified by the formula

$$u(x,t) = \frac{1}{2}\{Hf(x+t) + H^T f(x-t)\} + \frac{1}{2}\int_{x-t}^{x+t} W(x,t,s)f(s)\,ds, \qquad (1.16)$$

where $W(x,t,s) = \sum_{k=1}^{\infty} W_k(x,t,s)$.

10.2   The Goursat problem for the solution kernel of the Cauchy problem

We again consider the Cauchy problem (1.1), (1.2), and clarify the conditions which the matrix kernel $W(x,t,s)$ should satisfy for the vector-valued function $u(x,t)$ determined by (1.16) to yield the solution of (1.1), (1.2). We show that the Goursat problem is obtained for $W(x,t,s)$.

Let the solution of (1.1), (1.2) be given by (1.16). It follows from the latter that the initial condition (1.2) holds automatically.

Let the vector-valued function $u(x,t;f)$ be the solution of (1.1), (1.2), represented as in (1.16), and let

$$T_t \equiv -iI\frac{\partial}{\partial t}, \qquad B_x \equiv B\frac{\partial}{\partial x} + Q(x)$$

(where I is the identity 2×2 matrix). We prove that

$$B_x u(x,t;f) = u(x,t;B_x f). \qquad (2.1)$$

Indeed, it is easy to see that both the left- and right-hand sides satisfy the same Cauchy problem

$$B_x v(x,t) = T_t v(x,t), \qquad v(x,0) = B_x f(x). \qquad (2.2)$$

Therefore, (2.1) follows from the uniqueness of the solution (2.2), and the equation (1.1) can be written in the form

$$T_t u(x,t;f) = u(x,t;B_x f). \qquad (2.3)$$

It follows from the definition of the operator T_t and from the formula (1.16) that

$$T_t u(x, t; f) \equiv -iI\frac{\partial}{\partial t} u(x, t; f) = -\frac{i}{2}\{Hf'(x+t) - H^T f'(x-t)\}$$

$$-\frac{i}{2}\{W(x, t, x+t)f(x+t) - W(x, t, x-t)f(x-t)\} - \frac{i}{2}\int_{x-t}^{x+t} W_t'(x, t, s)f(s)\,ds. \quad (2.4)$$

On the other hand, by the definition of the operator B_x, relying on (1.16), we have

$$u(x, t; B_x f) = \frac{1}{2}\{HBf'(x+t) + H^T Bf(x-t)\}$$

$$+\frac{1}{2}\{HQ(x+t)f(x+t) + H^T Q(x-t)f(x-t)\}$$

$$+\frac{1}{2}\{W(x, t, x+t)Bf(x+t) - W(x, t, x-t)Bf(x-t)\}$$

$$-\frac{1}{2}\int_{x-t}^{x+t}\{W_s'(x, t, s)B - W(x, t, s)Q(s)\}f(s)\,ds. \quad (2.5)$$

Since the vector-valued function $f(x)$ is arbitrary, the coefficients of $f(x+t)$, $f(x-t)$ in the expressions (2.4), (2.5) and the integrands should coincide because of (2.3) (the coefficients of the derivatives $f'(x+t)$, $f'(x-t)$ are eliminated). Thus, equalizing the coefficients of $f(x+t)$, $f(x-t)$, respectively, we obtain

$$-iW(x, t, x+t) = HQ(x+t) + W(x, t, x+t)B$$

$$-iW(x, t, x-t) = H^T Q(x-t) - W(x, t, x-t)B$$

or, taking into account that $B + iI = H^T$, $B - iI = -iH$,

$$W(x, t, x+t)H^T = iHQ(x+t),$$

$$W(x, t, x-t)H = iH^T Q(x-t). \qquad (2.6)$$

Now, equalizing the integrand of one function to the other, we obtain

$$iW_t'(x, t, s) - W_s'(x, t, s)B = W(x, t, s)Q(s), \qquad (2.7)$$

which, along with the conditions (2.6) for the kernel matrix $W(x, t, s)$, determines the Goursat problem.

It follows from the equalities (2.6) that $W(x, 0, x) = iQ(x)$. Therefore, if the matrix $Q(x)$ is normalized, so that $Q(0) = 0$, then

$$W(0, 0, 0) = 0. \qquad (2.8)$$

10.3   The transformation matrix operator

Let E be the space of 2-component, complex-valued continuous vector functions $f(x)$, $0 \leq x < \infty$, with continuous first derivative. No restrictions are placed on their growth at infinity. The topology on E is defined via the uniform convergence of the vector-valued functions and of their first derivatives, in each finite interval.

Furthermore, let A_1, A_2 be two linear (not necessarily continuous) 2×2 matrix operators from E to E, and let E_1, E_2 be two closed subspaces of E.

Definition 10.3.1 *An invertible linear operator (resp. 2×2-matrix) X defined on the whole space E and acting from E_1 to E_2 is called the* transformation *(resp. matrix) operator for two operators (resp. matrices) A_1, A_2 if X satisfies the conditions:*

1. *The operator (resp. matrix) X and its inverse X^{-1} are continuous in E.*

2. *The matrix-operator identity*

$$A_1 X = X A_2 \tag{3.1}$$

holds.

Here, we consider the three cases.

Case 1   Let E be as above, and let $E_1 = E_2$ be the subspace of vector-valued functions $f(x) = \begin{pmatrix} f_1(x) \\ f_2(x) \end{pmatrix}$ fulfilling the boundary condition

$$f_2(0) = h f_1(0), \tag{3.2}$$

where h is an arbitrary finite complex number. Consider the matrix operators

$$A_1 = \begin{pmatrix} 0 & 1 \\ -1 & 0 \end{pmatrix} \frac{d}{dx} + \begin{pmatrix} p(x) & q(x) \\ q(x) & -p(x) \end{pmatrix} \equiv B \frac{d}{dx} + Q(x),$$

$$A_2 = B \frac{d}{dx}, \quad 0 \leq x < \infty,$$

where $p(x)$, $q(x)$ are two continuous complex-valued functions.

Theorem 10.3.1 *The operator (resp. matrix) X mapping E_1 onto E_2 can be realized in the form*

$$X f(x) = f(x) + \int_0^x K(x, t) f(t) \, dt. \tag{3.3}$$

The matrix kernel $K(x, t)$ of (3.3) is the solution of the differential equation

$$B K_x'(x, t) + K_t'(x, t) B = -Q(x) K(x, t), \tag{3.4}$$

and fulfils the conditions

$$K(x,x)B - BK(x,x) = Q(x) \tag{3.5}$$

$$K(x,0)BH = 0, \qquad H = \begin{pmatrix} 1 \\ h \end{pmatrix}. \tag{3.6}$$

Conversely, if $K(x,t)$ is the solution of the problem (3.4), (3.5), (3.6), then the operator (resp. matrix) X determined by the formula (3.3) is the transformation operator (resp. matrix) for the two operators (resp. matrices) A_1, A_2 and acts from E_1 to E_2.

Proof. By the definition of the matrix operators A_1, X, we have

$$A_1\{Xf(x)\} = Bf'(x) + Q(x)f(x) + BK(x,x)f(x)$$
$$+ \int_0^x \{BK'_x(x,t) + Q(x)K(x,t)\}f(t)\,dt. \tag{3.7}$$

On the other hand, it follows from the definition of X and A_2 that

$$X\{A_2f(x)\} = Bf'(x) + \int_0^x K(x,t)Bf'(t)\,dt.$$

Integrating by parts, we obtain

$$X\{A_2f(x)\} = Bf'(x) + K(x,x)Bf(x) - K(x,0)Bf(0) - \int_0^x K'_t(x,t)Bf(t)\,dt. \tag{3.8}$$

Equalizing (3.7) to (3.8) (by (3.1)), by the arbitrariness of the vector-valued function $f(x)$, we obtain

$$BK'_x(x,t) + Q(x)K(x,t) = -K'_t(x,t)B,$$

i.e., (3.4) (making the integrands equal).

Now, from the equality of the coefficients of $f(x)$, we get $Q(x) + BK(x,x) = K(x,x)B$, or

$$BK(x,x) - K(x,x)B = -Q(x). \tag{3.9}$$

Finally, the term involving $f(0)$ in the expression (3.8) should vanish, because there is no similar term in (3.7). Thus,

$$K(x,0)Bf(0) = 0. \tag{3.10}$$

It is obvious that $f(0) = \begin{pmatrix} 1 \\ h \end{pmatrix} \equiv H$ satisfies the boundary condition (3.2).

Therefore, it follows from (3.10) that $K(x,0)BH = 0$, i.e., the condition (3.6).

We have shown that if an operator X can be realized as in (3.3), then its kernel, or the matrix $K(x,t)$, is a solution of the problem (3.4-6). Conversely, if $K(x,t)$ is a solution of (3.4–6), then the operator X constructed with its help by (3.3) is the transformation operator for the pair A_1, A_2, and maps E_1 onto E_2.

Thus, to complete the proof, we should show that (3.4–6) is solvable. However, we first somewhat change the initial condition (3.6).

If we put

$$K(x,0) = \begin{pmatrix} K_{11}(x,0) & K_{12}(x,0) \\ K_{21}(x,0) & K_{22}(x,0) \end{pmatrix},$$

then (3.6) means that

$$K(x,0)BH = \begin{pmatrix} K_{11}(x,0) & K_{12}(x,0) \\ K_{21}(x,0) & K_{22}(x,0) \end{pmatrix} \begin{pmatrix} 0 & 1 \\ -1 & 0 \end{pmatrix} \begin{pmatrix} 1 \\ h \end{pmatrix}$$

$$= \begin{pmatrix} -K_{12}(x,0) + hK_{11}(x,0) \\ K_{22}(x,0) + hK_{21}(x,0) \end{pmatrix} = 0,$$

i.e., $K_{12}(x,0) = hK_{11}(x,0)$, $K_{22}(x,0) = -hK_{21}(x,0)$.

Therefore, if we put $K_{11}(x,0) = \varphi_1(x)$, $K_{21}(x,0) = \varphi_2(x)$, where $\varphi(x) = \begin{pmatrix} \varphi_1(x) \\ \varphi_2(x) \end{pmatrix}$ is (for the present) an unknown vector-valued function, then the initial condition (3.6) takes the form

$$K(x,0) = \begin{pmatrix} \varphi_1(x) \\ \varphi_2(x) \end{pmatrix}(1\ h) \equiv \varphi(x)H^T = \begin{pmatrix} \varphi_1(x) & h\varphi_1(x) \\ \varphi_2(x) & h\varphi_2(x) \end{pmatrix} \equiv H(x), \qquad (3.11)$$

since

$$K(x,0)BH = \begin{pmatrix} \varphi_1(x) & h\varphi_1(x) \\ \varphi_2(x) & h\varphi_2(x) \end{pmatrix} \begin{pmatrix} 0 & 1 \\ -1 & 0 \end{pmatrix} \begin{pmatrix} 1 \\ h \end{pmatrix} = \begin{pmatrix} -h\varphi_1(x) + h\varphi_1(x) \\ -h\varphi_2(x) + h\varphi_2(x) \end{pmatrix} = 0.$$

Retracing our steps to (3.4–6), we first consider the problem (3.4), (3.6). If we take into account (3.11), then we should solve the problem

$$BK'_x(x,t) + K'_t(x,t)B = -Q(x)K(x,t) \qquad (3.12)$$

$$K(x,0) = H(x) \qquad (3.13)$$

for which we proceed as in solving the problem (1.1'), (1.2'), Section 1. First, we consider the problem (3.12), (3.13), assuming that $Q(x) \equiv 0$, i.e.,

$$BK^{0'}_x(x,t) + K^{0'}_t(x,t)B = 0, \qquad K^0(x,0) = H(x). \qquad (3.14)$$

This is equivalent to four independent problems relative to the elements $K^0_{i,j}(x,t)$, $(i,j = 1,2)$ of the matrix $K^0(x,t)$, all of which are of the same form, viz.

$$u''_{xx} - u''_{tt} = 0, \qquad u(x,0) = f(x), \qquad u'_t(x,0) = g(x).$$

Solving each of them by the d'Alembert method, we obtain

$$K^0(x,t) = \frac{1}{2}\{H(x+t) + BH(x+t)B\} + \frac{1}{2}\{H(x-t) - BH(x-t)B\} \quad (3.15)$$

for the problem (3.14), the validity of which can be seen also by direct verification.

We now consider the nonhomogeneous problem with the homogeneous initial condition

$$B\tilde{K}'_x(x,t) + \tilde{K}'_t(x,t)B = G(x,t), \quad \tilde{K}(x,0) = 0, \quad (3.16)$$

splitting up into four independent problems relative to the elements $\tilde{K}_{ij}(x,t)$, $(i,j = 1,2)$ of the matrix $\tilde{K}(x,t)$, all of which are of the same kind, viz.

$$u''_{xx} - u''_{tt} = g(x,t), \quad u(x,0) = 0, \quad u'_t(x,0) = f(x).$$

Solving them by the familiar formula (see the problems (1.8), (1.9), Section 1), we obtain

$$\tilde{K}(x,t) = \frac{1}{2}\int_x^{x+t}\{-BG(\tau, x+t-\tau) + G(\tau, x+t-\tau)B\}\, d\tau$$

$$+\frac{1}{2}\int_{x-t}^x\{G(\tau, t-x+\tau)B - BG(\tau, t-x+\tau)\}\, d\tau \quad (3.17)$$

for the solution of (3.16).

Retracing our steps to the problem (3.12), (3.13), putting $G(x,t) = -Q(x)K(x,t)$, considering $G(x,t)$ as known, and then taking into account that the solution of (3.12), (3.13) is the sum of those of the problems (3.14), (3.16), we obtain

$$K(x,t) = \frac{1}{2}\{H(x+t) + BH(x+t)B\} + \frac{1}{2}\{H(x-t) - BH(x-t)B\}$$

$$+\frac{1}{2}\int_x^{x+t} Q(s)\{BK(s, x+t-s) + K(s, x+t-s)B\}\, ds$$

$$+\frac{1}{2}\int_{x-t}^x Q(s)\{K(s, t-x+s)B - BK(s, t-x+s)\}\, ds, \quad (3.18)$$

relying on (3.15), (3.17).

Thus, to solve (3.12), (3.13), the equivalent Volterra-type integral equation (3.18) is constructed, and, therefore, can be solved by the method of successive approximations.

Let $K_0(x,t)$ be determined by the formula (3.15). For $n = 1,2,3,\ldots$, we put

$$K_n(x,t) = \frac{1}{2}\int_x^{x+t} Q(s)\{BK_{n-1}(s, x+t-s) + K_{n-1}(s, x+t-s)B\}\, ds$$

$$+\frac{1}{2}\int_{x-t}^x Q(s)\{K_{n-1}(s, t-x+s)B - BK_{n-1}(s, t-x+s)\}\, ds. \quad (3.19)$$

We show that each matrix $K_n(x,t)$, $n = 1, 2, 3, \ldots$, can be represented in the form

$$K_n(x,t) = \frac{1}{2} \int_{x-t}^{x+t} \{M_n(x,t,s)BH(s) + N_n(x,t,s)H(s)B\} \, ds, \qquad (3.20)$$

where $M_n(x,t,s)$, $N_n(x,t,s)$ are two matrices only depending on $Q(x)$.

Indeed, for $n = 1$, we have

$$K_1(x,t) = \frac{1}{2} \int_{x}^{x+t} Q(s)\{BH(2s - x - t) + H(2s - x - t)B\} \, ds$$
$$+ \frac{1}{2} \int_{x-t}^{x} Q(s)\{H(2s + x - t)B - BH(2s + x - t)\} \, ds$$

because of (3.15), (3.20). Replacing the integration variable, we reduce the integrals to

$$K_1(x,t) = \frac{1}{2} \int_{x-t}^{x+t} \left\{ Q\left(\frac{x+t+s}{2}\right) - Q\left(\frac{x-t+s}{2}\right) \right\} BH(s) \, ds$$
$$+ \frac{1}{2} \int_{x-t}^{x+t} \left\{ Q\left(\frac{x+t+s}{2}\right) + Q\left(\frac{x-t+s}{2}\right) \right\} H(s)B \, ds,$$

i.e., (3.20) for $n = 1$, if we put

$$M_1(x,t,s) = Q\left(\frac{x+t+s}{2}\right) - Q\left(\frac{x-t+s}{2}\right)$$

$$N_1(x,t,s) = Q\left(\frac{x+t+s}{2}\right) + Q\left(\frac{x-t+s}{2}\right).$$

Assuming that the formula (3.20) is proved for $n = 1, 2, \ldots, k - 1$, we can also prove it for $n = k$; however, we will not do so here, because the calculations are tedious. It follows from the uniform convergence of the successive approximations that the solution of (3.18) is determined by the equality

$$K(x,t) = \sum_{n=0}^{\infty} K_n(x,t).$$

Now, substituting the expressions $K_n(x,t)$, $n = 0, 1, 2, \ldots$, we obtain that the solution of the problem (3.12), (3.13) is given by the formula

$$K(x,t) = \frac{1}{2}\{H(x+t) + BH(x+t)B\} + \frac{1}{2}\{H(x-t) - BH(x-t)B\}$$
$$+ \frac{1}{2} \int_{x-t}^{x+t} \{M(x,t,s)BH(s) + N(x,t,s)H(s)B\} \, ds, \quad (3.21)$$

where

$$M(x,t,s) = \sum_{n=1}^{\infty} M_n(x,t,s), \qquad N(x,t,s) = \sum_{n=1}^{\infty} N_n(x,t,s).$$

The matrix $H(x)$, or, more precisely, the vector-valued function $\varphi(x)$, still remains unknown in the solution (3.21) of (3.12), (3.13). To determine the matrix, we use the condition (3.9), i.e., the equality

$$K(x,x)B - BK(x,x) = Q(x).$$

Substituting the value of $K(x,x)$ from (3.21), we get

$$\frac{1}{2}\{H(2x) + BH(2x)B\}B - \frac{1}{2}B\{H(2x) + BH(2x)B\} + \frac{1}{2}\{H(0) - BH(0)B\}B$$

$$-\frac{1}{2}B\{H(0) - BH(0)B\} + \frac{1}{2}\int_0^{2x}\{M(x,x,s)BH(s) + N(x,x,s)H(s)B\}\,ds = Q(x).$$

Since $B^2 = -I$ (I being the identity matrix), the terms involving $H(0)$ are eliminated, and, therefore, the last equation takes the form

$$H(2x)B - BH(2x) = Q(x) + \int_0^{2x}\{W_1(x,s)BH(s) + \tilde{W}_1(x,s)H(s)B\}\,ds, \quad (3.22)$$

where $W_1(x,s) = \frac{1}{2}M(x,x,s)$, $\tilde{W}_1(x,s) = \frac{1}{2}N(x,x,s)$.

Replacing $2x$ by x in (3.22), we have

$$H(x)B - BH(x) = Q\left(\frac{x}{2}\right) + \int_0^x\{W(x,s)BH(s) + \tilde{W}(x,s)H(s)B\}\,ds,$$

and the system of integral equations for the components $\varphi_1(x)$ and $\varphi_2(x)$ of the unknown vector-valued function $\varphi(x)$ follows by the definition of the matrix $H(x)$, viz.

$$\varphi_1(x) = p\left(\frac{x}{2}\right) + h\varphi_2(x) + \int_0^x\{a(x,s)\varphi_1(s) + b(x,s)\varphi_2(s)\}\,ds$$

$$\varphi_2(x) = -q\left(\frac{x}{2}\right) - h\varphi_1(x) + \int_0^x\{c(x,s)\varphi_1(s) + d(x,s)\varphi_2(s)\}\,ds.$$

∎

Case 2 The space E is as in the first case, and $E_1 = E_2$ is the subspace of E, containing all vector-valued functions $f(x) = \begin{pmatrix} f_1(x) \\ f_2(x) \end{pmatrix}$ satisfying the boundary condition $f_1(0) = 0$. Formally, this condition is associated with (3.2) for $h = \infty$, which obviously is not covered by the above.

Let A_1, A_2 be two matrix operators of the above form. The transformation matrix operator X is again sought for in the form

$$Xf(x) = f(x) + \int_0^x K(x,t)f(t)\,dt. \tag{3.23}$$

Reasoning similarly to the above, we obtain the problem

$$BK_x'(x,t) + K_t'(x,t)B = -Q(x)K(x,t) \tag{3.24}$$

$$BK(x,x) - K(x,x)B = -Q(x) \tag{3.25}$$

$$K_{11}(x,0) = K_{21}(x,0) = 0 \tag{3.26}$$

for the kernel $K(x,t)$.

Case 3 The space E is the same as in Cases 1 and 2, E_1 and E_2 are subspaces of E, containing the vector-valued functions $f(x)$ satisfying the boundary conditions

$$f_2(0) = h_1 f_1(0), \qquad f_2(0) = h_2 f_1(0),$$

respectively, where h_1, h_2 are two arbitrary finite complex numbers.

Let the matrix operators A_1, A_2 be of the form

$$A_1 = \begin{pmatrix} 0 & 1 \\ -1 & 0 \end{pmatrix} \frac{d}{dx} + \begin{pmatrix} p_1(x) & 0 \\ 0 & q_1(x) \end{pmatrix} \equiv B\frac{d}{dx} + Q_1(x),$$

$$A_2 = \begin{pmatrix} 0 & 1 \\ -1 & 0 \end{pmatrix} \frac{d}{dx} + \begin{pmatrix} p_2(x) & 0 \\ 0 & q_2(x) \end{pmatrix} \equiv B\frac{d}{dx} + Q_2(x).$$

We then look for the transformation matrix operator X such that

$$Xf(x) = R(x)f(x) + \int_0^x L(x,t)f(t)\,dt, \tag{3.27}$$

where $R(x) = \begin{pmatrix} \alpha(x) & \beta(x) \\ -\beta(x) & \alpha(x) \end{pmatrix}$ and

$$\alpha(x) = \frac{1}{\kappa}\sin\left\{ -\frac{1}{2}\int_0^x r(\tau)\,d\tau + \sin^{-1}\frac{1}{\kappa} \right\}$$

$$\beta(x) = \frac{1}{\kappa}\cos\left\{ -\frac{1}{2}\int_0^x r(\tau)\,d\tau + \sin^{-1}\frac{1}{\kappa} \right\}$$

$$r(x) = p_1(x) + q_1(x) - p_2(x) - q_2(x)$$

$$\kappa = \frac{(1+h_1)(1+h_2)}{1+h_1 h_2}.$$

The transformation matrix operator kernel $L(x,t)$ is the solution of the problem

$$BL'_x(x,t) + L'_t(x,t)B = L(x,t)Q_2(x) - Q_1(x)L(x,t)$$

$$BL(x,x) - L(x,x)B = R(x)Q_2(x) - Q_1(x)R(x) - BR'(x)$$

$$L(x,0)BH = 0, \qquad H = \begin{pmatrix} 1 \\ h_1 \end{pmatrix}.$$

10.4   Solution of the mixed problem on the half-line

Let a matrix $Q(x) = \begin{pmatrix} p(x) & q(x) \\ q(x) & -p(x) \end{pmatrix}$ and a vector-valued function $f(x) = \begin{pmatrix} f_1(x) \\ f_2(x) \end{pmatrix}$
be defined on the half-line $[0, \infty)$, and let their elements be absolutely continuous.
Extend $Q(x)$, i.e., the functions $p(x)$, $q(x)$, to the negative real axis, preserving the
class, arbitrarily in all other respects, and extend the vector-valued function $f(x)$ in
an indeterminate manner (to be sharpened in the sequel).

Consider the mixed problem on the half-line

$$-i\frac{\partial u}{\partial t} = B\frac{\partial u}{\partial x} + Q(x)u \tag{4.1}$$

$$u(x, 0) = f(x) \tag{4.2}$$

$$u_2(0, t) - hu_1(0, t) = 0. \tag{4.3}$$

By the formula (1.16), the solution $u(x, t)$ of the problem (4.1), (4.2) can be
represented as

$$u(x, t) = \frac{1}{2}\{Hf(x + t) + H^T f(x - t)\} + \frac{1}{2}\int_{x-t}^{x+t} W(x, t, s)f(s)\, ds. \tag{4.4}$$

We now apply the transformation operators to extend the solution of the equation
(4.1) to the negative real axis. We show how $u(x, t)$ can be expressed at the point
$-x$ as a linear operator of $u(s, t)$ for $0 \le s \le x$.

Let two operators L_1, L_2 be of the form

$$L_1 = \begin{pmatrix} 0 & 1 \\ -1 & 0 \end{pmatrix}\frac{\partial}{\partial x} + \begin{pmatrix} p(-x) & q(-x) \\ q(-x) & -p(-x) \end{pmatrix} \equiv B\frac{\partial}{\partial x} + Q(-x)$$

$$L_2 = \begin{pmatrix} 0 & 1 \\ -1 & 0 \end{pmatrix}\frac{\partial}{\partial x} + \begin{pmatrix} p(x) & q(x) \\ q(x) & -p(x) \end{pmatrix} \equiv B\frac{\partial}{\partial x} + Q(x)$$

for $x \ge 0$. Assume that the matrix $Q(x)$ is extended continuously at zero, i.e.,

$$Q(-0) = Q(+0). \tag{4.5}$$

Denote by X the transformation operator (for the pair L_1, L_2) mapping the space
$\mathcal{E}_h$ of absolutely continuous vector-valued functions $f(x)$ satisfying the boundary
condition $f_2(0) - hf_1(0) = 0$ to itself.

Let the vector-valued function $u^+(x, t)$ be the solution of the mixed problem (4.1–
3) on the half-line $[0, \infty)$. Owing to the condition (4.3), $u^+(x, t) \in \mathcal{E}_h$ for each fixed
t; therefore, the operator X can be applied to $u^+(x, t)$. The extension of $u^+(x, t)$ to
the negative real axis is determined by the formula

$$u(-x, t) \equiv u^-(x, t) = X\{u^+(x, t)\}, \tag{4.6}$$

i.e., relying on (3.3),

$$u^-(x,t) = u^+(x,t) + \int_0^x K(x,s)u^+(s,t)\,ds. \tag{4.7}$$

We prove that $u^-(x,t)$ satisfies the equation (4.1) on the negative real axis. By the definition of L_1, we have to show that

$$L_1\{u^-(x,t)\} = -i\frac{\partial u^-}{\partial t}. \tag{4.8}$$

Since $u^+(x,t) \in \mathcal{E}_h$ for each fixed t, by the definition of X and by the condition $L_1X = XL_2$, we have

$$L_1X\{u^+(x,t)\} = XL_2\{u^+(x,t)\}. \tag{4.9}$$

Owing to (4.6), the left-hand side of (4.9) coincides with that of (4.8). To calculate the right-hand side, consider the equality

$$XL_2\{u^+(x,t)\} = X\left\{-i\frac{\partial u^+}{\partial t}\right\} = -iX\left\{\frac{\partial u^+}{\partial t}\right\}$$

holding because of (4.1) which can be rewritten as $L_2\{u^+(x,t)\} = -i\partial u^+/\partial t$. It obviously follows from (4.8) that X and the differentiation operator $\partial/\partial t$ commute; therefore, we derive from the last equality that

$$XL_2\{u^+(x,t)\} = -i\frac{\partial}{\partial t}X\{u^+(x,t)\} = -i\frac{\partial u^-}{\partial t},$$

also relying on (4.6), i.e., the right-hand sides of (4.9), (4.8) coincide; this proves the validity of (4.8).

We now show that the extension of $u^+(x,t)$ to the negative real axis is continuous along with the first derivative with respect to x. In fact, putting $x = 0$ in (4.7), we obtain

$$u^-(-0,t) = u^+(+0,t), \tag{4.10}$$

i.e., the extension is continuous.

Furthermore, using the explicit form of L_1, L_2 and X, the equality (4.9) can be rewritten as

$$B\frac{\partial u^-}{\partial x} + Q(-x)u^- = B\frac{\partial u^+}{\partial x} + Q(x)u^+ + \int_0^x K(x,s)\left\{B\frac{\partial u(s,t)}{\partial s} + Q(s)u(s,t)\right\}ds$$

on the basis of (4.6). If we put $x = 0$, then

$$B\left.\frac{\partial u(-x,t)}{\partial x}\right|_{x=0} + Q(-0)u(-0,t) = B\left.\frac{\partial u(x,t)}{\partial x}\right|_{x=0} + Q(+0)u(+0,t);$$

hence,

$$\frac{\partial u(-x,t)}{\partial x}\bigg|_{x=0} = \frac{\partial u(x,t)}{\partial x}\bigg|_{x=0} \tag{4.11}$$

because of (4.5), (4.10), i.e., the first derivative with respect to x is also continuous at zero.

The formula (4.7) can be used to extend the original vector-valued function $f(x)$. Indeed, setting $x = 0$ and taking into account the initial condition (4.2), we have

$$f(-x) = f(x) + \int_0^x K(x,s)f(s)\,ds. \tag{4.12}$$

The continuity of the extension of $f(x)$ to the negative real axis follows if we put $x = 0$, while that of its derivative follows from (4.1) if we put $t = 0$.

We now retrace our steps to the mixed problem (4.1–3).

If $x > t \geq 0$, the solution coincides with that of (4.1), (4.2), and, therefore, is specified by (4.4).

If $0 < x < t$, we have to extend the vector-valued function $f(x)$ to the negative real axis by (4.12). Break the integration interval $(x-t, x+t)$ in (4.4) into $(x-t, 0)$, $(0, x+t)$, and then replace the variable s by $-s$ in the integral along the interval $(x-t, 0)$, while $f(-s)$ is replaced by (4.12). Similarly, we replace $f(x-t)$ in (4.4). After simple transformations, we then obtain

$$u(x,t) = \frac{1}{2}\{Hf(x+t) + H^T f(x-t)\}$$
$$+ \frac{1}{2}\int_0^{x+t} W(x,t,s)f(s)\,ds + \frac{1}{2}\int_0^{t-x} \tilde{W}(x,t,s)f(s)\,ds$$

for the solution of the problem (4.1–3), where $0 < x < t$,

$$\tilde{W}(x,t,s) = H^T K(t-x,s) + W(x,t-s) + \int_s^{t-x} W(x,t,-\tau)K(\tau,s)\,d\tau.$$

10.5   Solution of the problem (1.1), (1.2) for $t < 0$

Consider the problem (1.1), (1.2) again; however, assuming now that $t < 0$. Put $t = -\tau, \tau > 0$. The equation (1.1) then takes the form

$$i\frac{\partial u(x,-\tau)}{\partial \tau} = B\frac{\partial u(x,-\tau)}{\partial x} + Q(x)u(x,-\tau). \tag{5.1}$$

Let $E = \begin{pmatrix} 0 & 1 \\ 1 & 0 \end{pmatrix}$. Substituting

$$u(x,-\tau) = Ev(x,\tau) \tag{5.2}$$

in (5.1), we obtain

$$iE\frac{\partial v(x,\tau)}{\partial \tau} = BE\frac{\partial v(x,\tau)}{\partial x} + Q(x)Ev(x,\tau).$$

If we pre-multiply by $-E$ and take into account that $E^2 = I$, $-EBE = B$, then

$$-i\frac{\partial v}{\partial \tau} = B\frac{\partial v}{\partial x} + Q_1(x)v, \qquad Q_1 = -EQE. \tag{5.3}$$

The initial condition (1.2) is transformed into

$$v(x,0) = Ef(x), \tag{5.4}$$

relying on (5.2), which, along with (5.3), determines the Cauchy problem for $v(x,\tau)$. Since the problem (5.3), (5.4) is of the same form as (1.1), (1.2), applying (4.4), we have

$$v(x,\tau) = \frac{1}{2}\{HEf(x+\tau) + H^T Ef(x-\tau)\} + \frac{1}{2}\int_{x-\tau}^{x+\tau} W_0(x,\tau,s)Ef(s)\,ds. \tag{5.5}$$

Now, pre-multiplying throughout by E, taking into account that $EHE = H^T$, $EH^T E = H$, putting $\tau = -t$ and retracing our steps to the vector-valued function $u(x,t)$ according to (5.2), we obtain

$$u(x,t) = \frac{1}{2}\{Hf(x+t) + H^T f(x-t)\} + \frac{1}{2}\int_{x+t}^{x-t} \tilde{W}(x,t,s)f(s)\,ds, \tag{5.6}$$

where

$$\tilde{W}(x,t,s) = EW_0(x,-t,s)E. \tag{5.7}$$

We now derive the equation for the matrix kernel $\tilde{W}(x,t,s)$. By (2.7), the matrix $W_0(x,\tau,s)$ (see (5.5)) satisfies the equation

$$i\frac{\partial W_0(x,\tau,s)}{\partial \tau} - \frac{\partial W_0(x,\tau,s)}{\partial s}B = W_0(x,\tau,s)Q_1(s).$$

Replacing τ by $-t$ and taking into account that $Q_1(s) = -EQ(s)E$, we have

$$-i\frac{\partial W_0(x,-t,s)}{\partial t} - \frac{\partial W_0(x,-t,s)}{\partial s}B = -W_0(x,-t,s)EQ(s)E.$$

Hence, pre- and post-multiplying throughout by E once, and taking into account that $BE = -EB$, we obtain

$$i\tilde{W}'_t(x,t,s) - \tilde{W}'_s(x,t,s)B = \tilde{W}(x,t,s)Q(s) \tag{5.8}$$

because of the notation (5.7). Finally, by (2.6), for $W_0(x,\tau,s)$, we have the conditions

$$W_0(x,\tau,x+\tau)H^T = iHQ_1(x+\tau)$$

$$W_0(x, \tau, x - \tau)H = iH^T Q_1(x - \tau)$$

on the characteristics. Replacing τ by $-t$, pre- and post-multiplying throughout by E, taking into account that $Q_1 = -EQE$, and also (5.7), we obtain the conditions

$$\tilde{W}(x, t, x + t)H^T = -iHQ(x + t)$$

$$\tilde{W}(x, t, x - t)H = -iHQ(x - t)$$

(5.9)

on the characteristics. Now, contrasting the Goursat problem (2.6), (2.7) with (5.8), (5.9) and using the uniqueness of the solution, we infer that $\tilde{W}(x, t, s) = -W(x, t, s)$. The relation in (5.6) then takes the form

$$u(x, t) = \frac{1}{2}\{Hf(x + t) + H^T f(x - t)\} + \frac{1}{2}\int_{x-t}^{x+t} W(x, t, s)f(s)\, ds,$$

and we again obtain (1.16).

Summarizing all the results of Sections 1 to 5, we arrive at the following important proposition for the study of the asymptotic behaviour of the so-called *spectral kernel* (in particular, of the spectral function $\rho(\lambda)$) and for the series expansion in terms of vector-valued eigenfunctions of the Dirac operator (sharpening the expansion theorem).

Theorem 10.5.1 *Let $p(x)$, $q(x)$ be two real-valued functions defined on the half-line $[0, \infty)$, with summable first derivatives in each finite interval, and let $f(x)$ be a real-valued vector function of the same class, fulfilling the condition $f_2(0) - hf_1(0) = 0$ at zero, where h is an arbitrary real number. Then the solution $u(x, t) = \begin{pmatrix} u_1(x, t) \\ u_2(x, t) \end{pmatrix}$ of the problem*

$$-i\frac{\partial u}{\partial t} = \begin{pmatrix} 0 & 1 \\ -1 & 0 \end{pmatrix}\frac{\partial u}{\partial x} + \begin{pmatrix} p(x) & q(x) \\ q(x) & -p(x) \end{pmatrix}u \equiv B\frac{\partial u}{\partial x} + Q(x)u \qquad (5.10)$$

$$u(x, 0) = f(x), \qquad u_2(0, t) - hu_1(0, t) = 0 \qquad (5.11)$$

is given by the formula :

1. For $|t| < x$,

$$u(x, t) = \frac{1}{2}\{Hf(x + t) + H^T f(x - t)\} + \frac{1}{2}\int_{x-t}^{x+t} W(x, t, s)f(s)\, ds, \qquad (5.12)$$

where

$$H = \begin{pmatrix} 1 & i \\ -i & 1 \end{pmatrix}, \qquad H^T = \begin{pmatrix} 1 & -i \\ i & 1 \end{pmatrix}, \qquad (5.13)$$

and the matrix $W(x, t, s)$ of order two is the solution of the differential equation

$$iW_t'(x, t, s) - W_s'(x, t, s)B = W(x, t, s)Q(s)$$

fulfilling the conditions

$$W(x, t, x + t)H^T = iHQ(x + t)$$

$$W(x, t, x - t)H = iH^TQ(x - t)$$

on the characteristics.

2. *For $0 < x < t$, the solution is constructed by extending the matrix $Q(x)$ to the negative real axis, preserving the class, and arbitrarily in all other respects, while extending $f(x)$ by the formula*

$$f(-x) = f(x) + \int_0^x K(x, s)f(s)\, ds,$$

where the matrix $K(x, s)$ (of order two) is the solution of the problem

$$BK_x'(x, s) + K_s'(x, s)B = K(x, s)Q(-s) - Q(s)K(x, s)$$

$$K(x, x)B - BK(x, x) = Q(x) - Q(-x), \qquad K(x, 0)B \begin{pmatrix} 1 \\ h \end{pmatrix} = 0.$$

The solution $u(x, t)$ is then determined by the formula

$$u(x, t) = \frac{1}{2}\{Hf(x + t) + H^Tf(x - t)\}$$

$$+ \frac{1}{2}\int_0^{x+t} W(x, t, s)f(s)\, ds + \frac{1}{2}\int_0^{t-x} \tilde{W}(x, t, s)f(s)\, ds,$$

where $\tilde{W}(x, t, s)$ is expressed explicitly in terms of $W(x, t, s)$.

10.6   Asymptotic behaviour of the spectral function

10.6.1

Consider the problem

$$\begin{pmatrix} 0 & 1 \\ -1 & 0 \end{pmatrix} \frac{dy}{dx} + \begin{pmatrix} p(x) & q(x) \\ q(x) & -p(x) \end{pmatrix} y = \lambda y \tag{6.1}$$

$$y(x, \lambda) = \begin{pmatrix} y_1(x, \lambda) \\ y_2(x, \lambda) \end{pmatrix}, \qquad y_2(0, \lambda) - hy_1(0, \lambda) = 0, \tag{6.2}$$

where $p(x)$, $q(x)$ are two real-valued functions defined on the half-line $[0, \infty)$, summable in each finite interval, and h is an arbitrary real number.

Along with (6.1), (6.2), we consider the problem

$$-i\frac{\partial u}{\partial t} = \begin{pmatrix} 0 & 1 \\ -1 & 0 \end{pmatrix}\frac{\partial u}{\partial x} + \begin{pmatrix} p(x) & q(x) \\ q(x) & -p(x) \end{pmatrix} u \equiv B\frac{\partial u}{\partial x} + Q(x)u \qquad (6.3)$$

$$u(x,0) = f(x) \qquad (6.4)$$

where $f(x) = \begin{pmatrix} f_1(x) \\ f_2(x) \end{pmatrix}$ is a real-valued vector function, defined and continuously differentiable on $[0,\infty)$, and fulfilling the condition

$$f_2(0) - hf_1(0) = 0. \qquad (6.5)$$

We extend $p(x)$, $q(x)$ to the whole number axis, preserving the class, and arbitrarily in all other respects, while $f(x)$ is extended by (5.14), i.e.,

$$f(-x) = f(x) + \int_0^x K(x,s)f(s)\,ds.$$

Let $\varphi(x,\lambda) = \begin{pmatrix} \varphi_1(x,\lambda) \\ \varphi_2(x,\lambda) \end{pmatrix}$ be the solution of the equation (6.1), satisfying the initial conditions $\varphi_1(0,\lambda) = 1$, $\varphi_2(0,\lambda) = h$. It obviously fulfils the condition (6.2).

We now put $f(x) = \varphi(x,\lambda)$, and consider the problem (6.3–5). On the one hand, its solution is given by (5.12) for $|t| < x$, i.e.,

$$u(x,t) = \frac{1}{2}\{H\varphi(x+t,\lambda) + H^T\varphi(x-t,\lambda)\} + \frac{1}{2}\int_{x-t}^{x+t} W(x,t,s)\varphi(s,\lambda)\,ds. \qquad (6.6)$$

On the other hand, by the Fourier method, it can be given by the formula

$$u(x,t) = \varphi(x,\lambda)e^{i\lambda t}. \qquad (6.7)$$

Because of the uniqueness of the solution of (6.3–5), equalizing the right-hand side of (6.6) to (6.7), we have the identity

$$\varphi(x,\lambda)e^{i\lambda t} = \frac{1}{2}\{H\varphi(x+t,\lambda) + H^T\varphi(x-t,\lambda)\} + \frac{1}{2}\int_{x-t}^{x+t} W(x,t,s)\varphi(s,\lambda)\,ds. \qquad (6.8)$$

Let ϵ be an arbitrary positive number, and let $g_\epsilon(t)$ be the function satisfying the conditions :

1. It is even and vanishes outside the interval $(-\epsilon, \epsilon)$.

2. It has piecewise continuous derivative.

Let $\psi_\epsilon(\lambda)$ be the Fourier transform of $g_\epsilon(t)$, viz.

$$\psi_\epsilon(\lambda) = \int_{-\epsilon}^{\epsilon} g_\epsilon(t) e^{i\lambda t}\, dt. \tag{6.9}$$

Assume that $x \geq \epsilon$ (the case $x < \epsilon$ is studied similarly). Multiplying (6.8) throughout by $g(t)$, integrating with respect to t from $-\epsilon$ to ϵ, replacing the variable in the first two integrals on the right-hand side and changing the order of integration in the third, we obtain

$$\varphi(x, \lambda)\psi_\epsilon(\lambda) = \frac{1}{2}\int_{x-\epsilon}^{x+\epsilon} \{Hg_\epsilon(s - x) + H^T g_\epsilon(x - s)\}\varphi(s, \lambda)\, ds$$
$$+ \int_{x-\epsilon}^{x+\epsilon} \left\{ \int_{|x-\epsilon|}^{\epsilon} W(x, t, s)g_\epsilon(t)\, dt \right\} \varphi(s, \lambda)\, ds \tag{6.10}$$

because of the notation (6.9).

Given that $g_\epsilon(t)$ is an even function, using the explicit form of the matrices H, H^T from (5.13), we find

$$\frac{1}{2}\{Hg_\epsilon(s - x) + H^T g_\epsilon(x - s)\} = iBg_\epsilon(x - s).$$

Now, also taking into account the notation

$$\chi(x, s) = \int_{|x-s|}^{\epsilon} W(x, t, s)g_\epsilon(t)\, dt,$$
$$\tag{6.11}$$
$$H(x, s; \epsilon) = iBg_\epsilon(x - s) + \chi_\epsilon(x, s),$$

(6.10) can be rewritten in the form

$$\varphi(x, \lambda)\psi_\epsilon(\lambda) = \int_{x-\epsilon}^{x+\epsilon} H(x, s; \epsilon)\varphi(s, \lambda)\, ds. \tag{6.12}$$

10.6.2

Let $f(x) \in \mathcal{L}^2(0, \infty)$. For given h, a nondecreasing and left-continuous function $\rho(\lambda)$, $\lambda \in (-\infty, \infty)$ is known to exist, bounded in each finite interval and generating an isometric mapping of the space of $\mathcal{L}^2(0, \infty)$ vector-valued functions onto the space $\mathcal{L}^2_{\{\rho(\lambda)\}}(-\infty, \infty)$ in accordance with the formulas

$$F(\lambda) = \int_0^{\infty} f^T(x)\varphi(x, \lambda)\, dx$$
$$\tag{6.13}$$
$$f(x) = \int_{-\infty}^{\infty} F(\lambda)\varphi(x, \lambda)\, d\rho(\lambda)$$

where the integrals converge in the metrics of $\mathcal{L}^2{}_{\{\rho(\lambda)\}}(-\infty, \infty)$, $\mathcal{L}^2(0, \infty)$, respectively, and the Parseval equation

$$\int_0^\infty |f(x)|^2 \, dx = \int_{-\infty}^\infty F^2(\lambda) \, d\rho(\lambda) \qquad (6.14)$$

holds (see Theorems 8.1.1 and 8.1.2).

Moreover, if $g(x) \in \mathcal{L}^2(0, \infty)$ and $G(\lambda)$ is determined from $g(x)$ as $F(\lambda)$ from $f(x)$, then, by the formula (1.13), Chapter 8, the generalized Parseval equation

$$\int_0^\infty f^T(x)g(x) \, dx = \int_{-\infty}^\infty F(\lambda)G(\lambda) \, d\rho(\lambda) \qquad (6.14')$$

holds.

We introduce the following terminology :

1. As before, we call $\rho(\lambda)$ the *spectral function* of the problem (6.1), (6.2).

2. We also call the matrix of order two

$$\vartheta(x, s; \lambda) = \int_0^\lambda \varphi(x, \mu)\varphi^T(s, \mu) \, d\rho(\mu) \qquad (6.15)$$

the *spectral kernel* of (6.1), (6.2).

Here, we obtain the formulas for the asymptotic behaviour as $|\lambda| \to \infty$ of the spectral kernel $\vartheta(x, s; \lambda)$; the asymptotic formula for $\rho(\lambda)$ will follow, because

$$\vartheta(0, 0; \lambda) = \begin{pmatrix} 1 & h \\ h & h^2 \end{pmatrix} \rho(\lambda). \qquad (6.16)$$

If we now consider (6.12) again, and

$$H(x, s; \epsilon) = \begin{pmatrix} H_{11} & H_{12} \\ H_{21} & H_{22} \end{pmatrix} (x, s; \epsilon),$$

then we put $H_i(x, s; \epsilon) = \begin{pmatrix} H_{i1}(x, s; \epsilon) \\ H_{i2}(x, s; \epsilon) \end{pmatrix}$. The identity (6.12) means that, for any $x \geq 0$, the functions $\varphi_i(x, \lambda)\psi_\epsilon(\lambda)$, $i = 1, 2$ (i.e., the components of the vector-valued functions $\varphi(x, \lambda)\psi_\epsilon(\lambda)$) are the Fourier transforms of the vector-valued functions equal to $H_i(x, s; \epsilon)$ for $|x-s| \leq \epsilon$ and vanishing for $|x-s| \geq \epsilon$, respectively. Therefore, if $f(x) \in \mathcal{L}^2(0, \infty)$, then, by the generalized Parseval equation (6.14'), we have

$$\int_{-\infty}^\infty \varphi_i(x, \lambda)\psi_\epsilon(\lambda)F(\lambda) \, d\rho(\lambda) = \int_0^\infty H_i^T(x, s; \epsilon)f(s) \, ds,$$

$i = 1, 2$, or, in matrix form,

$$\int_{-\infty}^\infty \varphi(x, \lambda)F(\lambda)\psi_\epsilon(\lambda) \, d\rho(\lambda) = \int_0^\infty H(x, s; \epsilon)f(s) \, ds. \qquad (6.17)$$

Lemma 10.6.1 *If a matrix $Q(x)$ is summable in each finite interval, and (x_0, x) is an arbitrary interval on the real axis, then there exists a constant $C = C(x_0, x_1)$ such that, for all $x, s \in (x_0, x_1)$ and all a, the estimate*

$$\bigvee_a^{a+1} \{\vartheta_{ik}(x, s; \lambda)\} < C, \quad i, k = 1, 2,$$

holds, where $\vartheta_{ik}(x, s; \lambda)$ are the entries of the matrix $\vartheta(x, s; \lambda)$.

Proof. Let

$$g_\epsilon(t) = \begin{cases} (\epsilon - |t|)/\epsilon, & |t| \le \epsilon \\ 0, & |t| > \epsilon \end{cases}.$$

Then

$$\psi_\epsilon(\lambda) = \int_{-\epsilon}^{\epsilon} g_\epsilon(t) e^{i\lambda t} \, dt = \epsilon \left(\frac{\sin(\epsilon\lambda/2)}{\epsilon\lambda/2} \right)^2.$$

If we put $g_\epsilon(t, a) = g_\epsilon(t) e^{-iat}$ for arbitrary a, then

$$\psi_\epsilon(\lambda, s) = \int_{-\epsilon}^{\epsilon} g_\epsilon(t, a) e^{i\lambda t} \, dt = \epsilon \left(\frac{\sin\{\epsilon(\lambda - a)/2\}}{\epsilon(\lambda - a)/2} \right)^2. \tag{6.18}$$

Therefore, it follows from (6.12), relying on (6.18) and the Parseval equation (6.14), that

$$\epsilon^2 \int_{-\infty}^{\infty} \left\{ \frac{\sin[\epsilon(\lambda - a)/2]}{\epsilon(\lambda - a)/2} \right\}^4 |\varphi(x, \lambda)|^2 \, d\rho(\lambda) = \int_0^{\infty} \|H(x, s; \epsilon, a)\|^2 \, ds, \tag{6.19}$$

where $\|H\|^2$ is the Hilbert-Schmidt norm of the matrix H, or the sum of its elements squared, $H(x, s; \epsilon, a)$ is determined by the equalities in (6.11), and $g_\epsilon(t)$ is replaced by $g(t, a)$. Since $g_\epsilon(t)$ has bounded support, $H(x, s; \epsilon, a)$ also has bounded support, and $\|H(x, s; \epsilon, a)\|^2$ is summable, relying on the summability of $Q(x)$. Therefore, the integral on the right-hand side of (6.19) exists. Putting $\epsilon = 2$ in (6.19), we have

$$\int_{-\infty}^{\infty} \left(\frac{\sin(\lambda - a)}{\lambda - a} \right)^4 |\varphi(x, \lambda)|^2 \, d\rho(\lambda) < C,$$

and, much more so,

$$\int_a^{a+1} \frac{\sin^4(\lambda - a)}{(\lambda - a)^4} |\varphi(x, \lambda)|^2 \, d\rho(\lambda) < C. \tag{6.20}$$

Furthermore, since $(\sin x)/x > 2/\pi$ for $0 \le x \le 1$, it follows that

$$\int_a^{a+1} \varphi_i^2(x, \lambda) \, d\rho(\lambda) < \frac{\pi^4}{16} C = C_i, \quad i = 1, 2, \tag{6.21}$$

which proves the lemma for the entries $\vartheta_{ii}(x, s; \lambda)$, $i = 1, 2$, of $\vartheta(x, s; \lambda)$ for $s = x$. The statement of the lemma for these elements when $s \neq x$ follows from the estimate (6.21) by the Cauchy-Schwarz inequality. The proof is completed similarly for $\vartheta_{ij}(x, s; \lambda)$ when $i \neq j$. ∎

Corollary 10.6.1 *With the same conditions as in Lemma 10.6.1, and for all a, the estimate*

$$\rho(a + 1) - \rho(a) < C$$

is valid.

Proof. This follows from (6.21) if we put $i = 1$, $x = 0$ and take into account that $\varphi_1(0, \lambda) = 1$. ∎

10.6.3

We have already noted that we derive here the asymptotic formulas as $|\lambda| \to \infty$ for the spectral kernel $\vartheta(x, s; \lambda)$ and for the spectral function $\rho(\lambda)$ of the problem (6.1), (6.2). We show that the principal terms in the formulas are the spectral kernel $\vartheta^*(x, s; \lambda)$ and the spectral function $\rho^*(\lambda)$, respectively, of the simpler problem (6.1), (6.2), i.e.,

$$\begin{pmatrix} 0 & 1 \\ -1 & 0 \end{pmatrix} \frac{dy}{dx} = \lambda y, \quad 0 \leq x < \infty \tag{6.22}$$

$$y_2(0) - h y_1(0) = 0 \tag{6.23}$$

for $Q(x) \equiv 0$.

We now calculate $\rho^*(\lambda)$, and thereby also $\vartheta^*(x, s; \lambda)$, using the definition of the spectral function in Section 1 of Chapter 8.

Add

$$y_2(b) - H y_1(b) = 0 \tag{6.24}$$

to the boundary condition (6.23), where b is an arbitrary positive number, H an arbitrary real number, and calculate the eigenvalues of (6.22–24).

Let $\psi(x, \lambda)$ be the solution of the equation (6.22), satisfying the initial conditions

$$\psi_1(0, \lambda) = 1, \qquad \psi_2(0, \lambda) = h. \tag{6.25}$$

As can easily be seen, the problem (6.22), (6.25) splits into two independent ones relative to the components $\psi_1(x, \lambda)$, $\psi_2(x, \lambda)$ of the vector-valued function $\psi(x, \lambda)$, with solutions given by the formulas

$$\psi_1(x, \lambda) = \cos \lambda x - h \sin \lambda x$$

$$\tag{6.26}$$

$$\psi_2(x, \lambda) = \sin \lambda x + h \cos \lambda x$$

Since $\psi(x, \lambda)$ satisfies (6.23), to determine the eigenvalues of the boundary-value problem (6.22-24), we have to substitute the expressions for $\psi_1(x, \lambda)$, $\psi_2(x, \lambda)$ in the boundary condition (6.24). We have

$$\sin \lambda b + h \cos \lambda b - H \cos \lambda b + hH \sin \lambda b = 0,$$

i.e.,

$$(1 + hH) \sin \lambda b + (h - H) \cos \lambda b = 0. \tag{6.27}$$

If we put

$$\cos \omega = \frac{1 + hH}{\sqrt{(1 + h^2)(1 + H^2)}},$$

then the equation (6.27) can be rewritten as $\sin(\lambda b + \omega) = 0$; hence,

$$\lambda_{n,b} b + \omega = n\pi, \quad n = 0, \pm 1, \pm 2, \ldots,$$

or

$$\lambda_{n,b} = n\frac{\pi}{b} + \frac{\omega}{b}, \quad n = 0, \pm 1, \pm 2, \ldots. \tag{6.28}$$

As can easily be seen,

$$\alpha_{n,b}^2 = \int_0^b \{\psi_1^2(x, \lambda_{n,b}) + \psi_2^2(x, \lambda_{n,b})\} \, dx = (1 + h^2)b.$$

Therefore, by the definition of $\rho_b^*(\lambda)$ (see Section 1 of Chapter 8), we have

$$\rho_b^*(\lambda) = \sum_{0 \le \lambda_{n,b} \le \lambda} \frac{1}{(1 + h^2)b}, \quad \lambda \ge 0$$

$$\rho_b^*(\lambda) = -\sum_{\lambda_{n,b} > \lambda} \frac{1}{(1 + h^2)b}, \quad \lambda < 0.$$

We transform $\rho_b^*(\lambda)$ for $\lambda \ge 0$ (for $\lambda < 0$, we proceed similarly). Using the formulas (6.28), we have

$$\rho_b^*(\lambda) = \frac{1}{1 + h^2} \sum_{0 \le \lambda_{n,b} \le \lambda} \frac{1}{b} = \frac{1}{\pi(1 + h^2)} \sum_{0 \le \lambda_{n,b} \le \lambda} \frac{\pi}{b} = \frac{1}{\pi(1 + h^2)} \sum_{0 \le \lambda_{n,b} \le \lambda} (\lambda_{n+1,b} - \lambda_{n,b})$$

$$= \frac{1}{\pi(1 + h^2)} \sum_{0 \le \lambda_{n,b} \le \lambda} \Delta\lambda_{n,b} \to \frac{\lambda}{\pi(1 + h^2)}$$

as $b \to \infty$. Thus, for the boundary-value problem (6.22), (6.23),

$$\rho^*(\lambda) = \frac{\lambda}{\pi(1 + h^2)}, \tag{6.29}$$

and

$$\vartheta^*(x, s; \lambda) = \frac{1}{\pi(1 + h^2)} \int_0^\lambda \psi(x, \mu)\psi^T(s, \mu) \, d\mu. \tag{6.29'}$$

The following is an immediate consequence of the above.

Lemma 10.6.2 *As $a \to \infty$, the estimates*

$$\bigvee_{a}^{a+1} \{\vartheta_{ik}^*(x, s; \lambda)\} < C, \quad i, k = 1, 2,$$

hold uniformly for all positive values of x and s, where $\vartheta_{ik}^*(x, s; \lambda)$ are the entries of the matrix $\vartheta^*(x, s; \lambda)$.

10.6.4

To transform the equality (6.17), we substitute in it the expression for $F(\lambda)$ from (6.13), change the order of integration on the left-hand side, which is possible by the Corollary to Lemma 10.6.1 and the Fubini theorem, and obtain

$$\int_0^\infty \left\{ \int_{-\infty}^\infty \varphi(x, \lambda)\varphi^T(s, \lambda)\psi_\epsilon(\lambda) \, d\rho(\lambda) \right\} f(s) \, ds = \int_0^\infty H(x, s; \epsilon)f(s) \, ds.$$

And, since $f(s)$ is arbitrary,

$$\int_{-\infty}^\infty \varphi(x, \lambda)\varphi^T(s, \lambda)\psi_\epsilon(\lambda) \, d\rho(\lambda) = H(x, s; \epsilon),$$

which can be finally rewritten as

$$\int_{-\infty}^\infty \psi_\epsilon(\lambda) \, d_\lambda \vartheta(x, s; \lambda) = \begin{cases} iBg_\epsilon(x - s) + \chi_\epsilon(x, s), & |x - s| \le \epsilon \\ 0, & |x - s| > \epsilon \end{cases} \tag{6.30}$$

relying on the notation (6.11) and on the definition of $\vartheta(x, s; \lambda)$, i.e., (6.15).

We now write out the similar identity to (6.30) for $\vartheta^*(x, s; \lambda)$. As easily follows from the problem (2.6), (2.7), the solution of the problem, $W(x, t, s)$, is identically zero for $Q(x) \equiv 0$. Therefore, we derive from (6.11) that $\chi_\epsilon(x, s) \equiv 0$, and, in this case, (6.30) takes the form

$$\int_{-\infty}^\infty \psi_\epsilon(\lambda) \, d_\lambda \vartheta^*(x, s; \lambda) = \begin{cases} iBg_\epsilon(x - s), & |x - s| \le \epsilon, \\ 0, & |x - s| > \epsilon. \end{cases} \tag{6.30'}$$

Subtracting the identity (6.30') from (6.30), we have

$$\int_{-\infty}^\infty \psi_\epsilon \, d_\lambda \{\vartheta(x, s; \lambda) - \vartheta^*(x, s; \lambda)\} = \begin{cases} \chi_\epsilon(x, s), & |x - s| \le \epsilon, \\ 0, & |x - s| > \epsilon. \end{cases} \tag{6.31}$$

We now transform the right-hand side of the identity (6.31), for which we put

$$\alpha(x, s; \lambda) = \int_{|x-s|}^1 W(x, t, s)e^{i\lambda t} \, dt, \tag{6.32}$$

and obtain

$$\frac{1}{2\pi} \int_{-\infty}^{\infty} \psi_\epsilon(\lambda)\bar{\alpha}(x,s,\lambda)\,d\lambda = \chi_\epsilon(x,s)$$

from (6.32) and (6.9), relying on the Parseval equation for the classical Fourier cosine transform, also taking into account (6.11).

The equality enables us to rewrite the identity (6.31) in the form

$$\int_{-\infty}^{\infty} \psi_\epsilon(\lambda)\,d_\lambda \Phi(x,s;\lambda) = 0, \tag{6.33}$$

where

$$\Phi(x,s,\lambda) = \vartheta(x,s,\lambda) - \vartheta^*(x,s,\lambda) - \frac{1}{2\pi}\int_0^\lambda \bar{\alpha}(x,s,\lambda)\,d\lambda. \tag{6.34}$$

The identity (6.33) and the Tauberian theorems for Fourier integrals enable us to study the behaviour of the spectral kernel $\vartheta(x,s;\lambda)$ as $\lambda \to \infty$.

We first prove two propositions.

Lemma 10.6.3 *If a matrix $Q(x)$ is summable in each finite interval, then, for any fixed x, s and all a, the estimates*

$$\bigvee_a^{a+1} \{\Phi_{ik}(x,s;\lambda)\} < C, \quad i,k = 1,2, \tag{6.35}$$

are valid, where $\Phi_{ik}(x,s;\lambda)$ are the elements of the matrix $\Phi(x,s;\lambda)$, uniformly in each bounded domain of x and s.

Proof. It suffices to obtain (6.35) for each element of the matrix on the right-hand side of (6.34). The estimate for elements of the matrices $\vartheta(x,s;\lambda)$ and $\vartheta^*(x,s;\lambda)$ follows from Lemmas 10.6.1 and 10.6.2.

We derive from the definition of the matrix $\alpha(x,s;\lambda)$, i.e., the equality (6.32), since the matrix $W(x,t,s)$ is summable (which follows from the summability of $Q(x)$), by the familiar Lebesgue-Riemann lemma, that, in each bounded domain of x and s, the equalities

$$\lim_{k\to\infty} \alpha_{ik}(x,s;\lambda) = 0, \quad i,k = 1,2,$$

hold. Therefore, as $a \to \infty$,

$$\bigvee_a^{a+1} \left\{ \int_0^\lambda \alpha_{ik}(x,s;\nu)\,d\nu \right\} = \int_a^{a+1} |\alpha_{ik}(x,s,\nu)\,d\nu| = o(1),$$

which completes the proof.   ∎

Lemma 10.6.4 *If the matrix $Q(x)$ uniformly satisfies the Dini condition in any finite interval, then, for any fixed x, s, the equalities*

$$\lim_{\lambda \to \infty} \int_{-\lambda}^{\lambda} \alpha_{ik}(x, s; \nu) \, d\nu = 0 \qquad (6.36)$$

hold unformly in any bounded domain of x and s.

Proof. It follows from the definition of the matrix $\alpha(x, s; \lambda)$ that

$$\int_{-\lambda}^{\lambda} \alpha_{ik}(x, s; \nu) \, d\nu = \int_{-\lambda}^{\lambda} \left\{ \int_{|x-s|}^{1} W_{ik}(x, t, s) e^{i\nu t} \, dt \right\} d\nu = 2 \int_{|x-s|}^{1} W_{ik}(x, t, s) \frac{\sin \lambda t}{t} \, dt.$$
$$(6.37)$$

Since $Q(x)$ satisfies the Dini condition, the matrix $W(x, t, s)$ also possesses this property. Therefore, the statement of the lemma, (6.36), follows from (6.37). The proof is thus complete. ∎

We now prove the fundamental proposition of this section.

Theorem 10.6.1 *(The Asymptotic Behavior of the Spectral Kernel.) If the matrix $Q(x)$ satisfies the Dini condition in any finite interval, then the asymptotic formula*

$$\vartheta(x, s; \lambda) - \vartheta(x, s; -\lambda) = \vartheta^*(x, s; \lambda) - \vartheta^*(x, s; -\lambda) + o(1) \qquad (6.38)$$

holds for any fixed x, s as $\lambda \to \infty$.

Proof. Let $(f, g) = f^T g$ be the scalar product of two vectors f and g.

Suppose that f, g are two arbitrary constant two-component vectors. It then follows from the equality (6.33) that

$$\int_{-\infty}^{\infty} \psi_\epsilon(\lambda) d_\lambda (\Phi(x, s; \lambda) f, g) = 0. \qquad (6.39)$$

It is obvious that $(\Phi(x, s; \lambda) f, g)$ is a monotone function of λ for $s = x$, $g = f$. Therefore, the formula

$$\lim_{\lambda \to \infty} \{ (\Phi(x, x; \lambda) f, f) - (\Phi(x, x; -\lambda) f, f) \} = 0 \qquad (6.40)$$

follows from (6.39) by the Tauberian theorem of V. A. Marchenko (see B. M. Levitan and I. S. Sargsjan (1975, Chapter XIV, Theorem 4.2)). Consider the monotone function

$$(\Phi(x, x; \lambda) f, f) + 2(\Phi(x, s; \lambda) f, f) + (\Phi(s, s; \lambda) f, f)$$

of λ for $s \neq x$, and again apply the above Tauberian theorem. We then obtain

$$\lim_{\lambda \to \infty} \{ (\Phi(x, s; \lambda) f, f) - (\Phi(x, s; -\lambda) f, f) \} = 0,$$

relying on (6.40). Now, reducing the quadratic forms to bilinear ones, we derive from the above equality that

$$\lim_{\lambda \to \infty} \{(\Phi(x, s; \lambda)f, g) - (\Phi(x, s; -\lambda)f, g)\} = 0.$$

Since f, g are arbitrary, it follows that

$$\lim_{\lambda \to \infty} \{\Phi(x, s; \lambda) - \Phi(x, s; -\lambda)\} = 0,$$

and, relying on formula (6.34),

$$\vartheta(x, s; \lambda) - \vartheta(x, s; -\lambda) = \vartheta^*(x, s; \lambda) - \vartheta^*(x, s; -\lambda) + \frac{1}{2\pi} \int_{-\lambda}^{\lambda} \bar{a}(x, s; \nu)\, d\nu + o(1)$$

as $\lambda \to \infty$.

The statement of the theorem, or (6.38), follows from the above asymptotic formula by Lemma 10.6.4. The proof is now complete. ∎

Theorem 10.6.2 *With the conditions of Theorem 10.6.1 and as $\lambda \to \infty$, the asymptotic formula*

$$\rho(\lambda) - \rho(-\lambda) = \frac{2\lambda}{\pi(1 + h^2)} + o(1) \tag{6.41}$$

holds.

Proof. The above follows from (6.38), relying on (6.16) and

$$\vartheta^*(0, 0; \lambda) - \vartheta^*(0, 0; -\lambda) = \begin{pmatrix} 1 & h \\ h & h^2 \end{pmatrix} \frac{2\lambda}{\pi(1 + h^2)}.$$

∎

The formula (6.41) specifies the asymptotic behaviour of the odd component of the spectral function $\rho(\lambda)$. To clarify the asymptotic behaviour of $\rho(\lambda)$ itself, it obviously suffices to find that of the even component. The following is true.

Theorem 10.6.3 *If the conditions of Theorem 10.6.1 are fulfilled as $\lambda \to \infty$, the asymptotic formula*

$$\rho(\lambda) + \rho(-\lambda) = o(1) \tag{6.42}$$

holds.

Proof. We derive from the identity (6.31), relying on (6.11), that

$$\int_{-\infty}^{\infty} \psi_\epsilon(\lambda)\, d_\lambda\{\vartheta_{11}(0, 0; \lambda) - \vartheta_{11}^*(0, 0; \lambda)\} = \int_0^\epsilon W(0, t, 0)g_\epsilon(t)\, dt$$

for $x = s = 0$. Since $\vartheta_{11}(0,0;\lambda) = \rho(\lambda)$, $\vartheta_{11}^*(0,0;\lambda) = \lambda/(\pi(1+h^2))$, we have

$$\rho(\lambda) + \rho(-\lambda) = \frac{\lambda}{|\lambda|} \cdot \frac{1}{2}\{W(0,+0,0) - W(0,-0,0)\} + o(1),$$

relying on the Tauberian theorem of V. A. Marchenko, and (6.42) follows, because $W(0,0,0) = 0$ in accordance with (2.8).   ∎

Theorems 10.6.2 and 10.6.3 have the following as a consequence.

Theorem 10.6.4 *Let the conditions of Theorem 10.6.1 be fulfilled. Then the asymptotic formula*

$$\rho(\lambda) = \frac{\lambda}{\pi(1+h^2)} + o(1)$$

holds as $|\lambda| \to \infty$.

10.7 Sharpening the expansion theorem

Here, we sharpen the expansion theorem for the vector-valued eigenfunctions of the problem (6.1), (6.2), i.e., we obtain the proof of the theorem on the equiconvergence of the vector-valued eigenfunction expansions for the problem (6.1), (6.2) and for the simplest problem (6.22), (6.23) for an arbitrary square-integrable vector-valued function on the half-line. Note that if $h = 0$ in (6.22), (6.23), then we derive from the formulas (6.26) that $\psi(x,\lambda) = \begin{pmatrix} \cos\lambda x \\ \sin\lambda x \end{pmatrix}$; therefore, the Dirac operator vector-valued eigenfunction expansion for a vector-valued function of class $\mathcal{L}^2(0,\infty)$ is compared with the expansion into the usual Fourier integral.

10.7.1

The identity (6.12) was obtained in the preceding section, i.e.,

$$\varphi(x,\lambda)\psi_\epsilon(\lambda) = \int_{x-\epsilon}^{x+\epsilon} H(x,s;\epsilon)\varphi(s,\lambda)\,ds, \tag{7.1}$$

where

$$H(x,s;\epsilon) = iBg_\epsilon(x-s) + \int_{|x-s|}^{\epsilon} W(x,t,s)g_\epsilon(t)\,dt. \tag{7.2}$$

Let $f(x)$ be an arbitrary vector-valued function of class $\mathcal{L}^2(0,\infty)$, and let $F(\lambda)$ be the generalized Fourier transform of $f(x)$, i.e.,

$$F(\lambda) = \int_0^\infty f^T(x)\varphi(x,\lambda)\,dx. \tag{7.3}$$

It follows from (7.1) that the functions $\varphi_1(x, \lambda)\psi_\epsilon(\lambda)$ and $\varphi_2(x, \lambda)\psi_\epsilon(\lambda)$, i.e., the components of the vector-valued function $\varphi(x, \lambda)\psi_\epsilon(\lambda)$ are generalized Fourier transforms of the vector-valued functions equal to

$$\begin{pmatrix} H_{11}(x, s; \epsilon) \\ H_{12}(x, s; \epsilon) \end{pmatrix}, \qquad \begin{pmatrix} H_{21}(x, s; \epsilon) \\ H_{22}(x, s; \epsilon) \end{pmatrix}$$

for each fixed x, $x - \epsilon \le s \le x + \epsilon$, and vanishing outside the interval. Therefore, by the generalized Parseval equality (6.14'), the relation

$$\int_{-\infty}^{\infty} \varphi(x, \lambda)F(\lambda)\psi_\epsilon(\lambda)\, d\rho(\lambda) = \int_{x-\epsilon}^{x+\epsilon} H(x, s; \epsilon)f(s)\, ds \qquad (7.4)$$

follows from the identity (7.1) and from (7.3).

Consider the vector-valued function

$$S(x, \lambda) = \int_0^\lambda F(\lambda)\varphi(x, \lambda)\, d\rho(\lambda). \qquad (7.5)$$

The equality (7.4) then takes the form

$$\int_{-\infty}^{\infty} \psi_\epsilon(\lambda)d_\lambda S(x, \lambda) = \int_{x-\epsilon}^{x+\epsilon} H(x, s; \epsilon)f(s)\, ds. \qquad (7.6)$$

We now find a new representation of the vector-valued function $S(x, \lambda)$ for the sequel. Substituting $F(\lambda)$ from (7.3) in (7.5), changing the order of integration, we obtain

$$S(x, \lambda) = \int_0^\infty \left\{ \int_0^\lambda \varphi(x, \lambda)\varphi^T(s, \lambda)\, d\rho(\lambda) \right\} f(s)\, ds,$$

since $(f^T(s)\varphi(s, \lambda))\varphi(x, \lambda) = (\varphi(x, \lambda)\varphi^T(s, \lambda))f(s)$ by straightforward calculation. Using the definition (6.15) of the spectral kernel $\vartheta(x, s; \lambda)$, we now obtain the final representation

$$S(x, \lambda) = \int_0^\infty \vartheta(x, s; \lambda)f(s)\, ds, \qquad (7.7)$$

and write out the formula similar to (7.6) for the simplest problem (6.22), (6.23). The vector-valued function $S^*(x, \lambda)$ similar to $S(x, \lambda)$ for this problem can be determined by

$$S^*(x, \lambda) = \int_0^\infty \vartheta^*(x, s; \lambda)f(s)\, ds, \qquad (7.7')$$

relying on (7.7), where the matrix $\vartheta^*(x, s; \lambda)$ is determined by the equality (6.29').

Furthermore, proceeding from the identity (6.30'), as in deriving (7.6), we obtain

$$\int_{-\infty}^{\infty} \psi_\epsilon(\lambda)\, d_\lambda S^*(x, \lambda) = iB \int_{x-\epsilon}^{x+\epsilon} g_\epsilon(x - s)f(s)\, ds, \qquad (7.8)$$

which, subtracted from (7.6), taking into account the expression (7.2) for the matrix $H(x, s; \epsilon)$, yields the equality

$$\int_{-\infty}^{\infty} \psi_\epsilon(\lambda)d_\lambda\{S(x, \lambda) - S^*(x, \lambda)\} = \int_{x-\epsilon}^{x+\epsilon} \left\{ \int_{|x-s|}^{\epsilon} W(x, t, s)g_\epsilon(t)\, dt \right\} f(s)\, ds. \qquad (7.9)$$

Lemma 10.7.1 *If the matrix $Q(x)$ is summable in any finite interval, then the estimates*

$$\bigvee_a^{a+1} \{S_k(x,\lambda)\} = \int_a^{a+1} |F(\lambda)| \cdot |\varphi_k(x,\lambda)| \, d\rho(\lambda) = o(1), \tag{7.10}$$

$k = 1, 2$, hold for any fixed s as $a \to \infty$, uniformly in any finite interval.

Proof. It follows from the definition of $S(x,\lambda)$, i.e., (7.5), that

$$\bigvee_a^{a+1} \{S_k(x,\lambda)\} \leq \left(\int_a^{a+1} |F(\lambda)|^2 \, d\rho(\lambda) \right)^{1/2} \left(\int_a^{a+1} |\varphi_k(x,\lambda)|^2 \, d\rho(\lambda) \right)^{1/2}$$

by the Cauchy-Schwarz inequality. The estimates (7.10) are consequences of these inequalities, since, on the one hand, by Lemma 10.6.1,

$$\int_a^{a+1} |\varphi_k(x,\lambda)|^2 \, d\rho(\lambda) < C,$$

and, on the other hand, by the Parseval equation,

$$\int_{-\infty}^{\infty} |F(\lambda)|^2 \, d\rho(\lambda) = \int_0^{\infty} |f(x)|^2 \, dx < \infty,$$

it is obvious that

$$\int_a^{a+1} |F(\lambda)|^2 \, d\rho(\lambda) = o(1)$$

as $a \to \infty$, which completes the proof. ∎

Lemma 10.7.2 *The estimates*

$$\bigvee_a^{a+1} \{S_k^*(x,\lambda)\} = o(1), \quad k = 1, 2, \tag{7.11}$$

hold uniformly on the whole half-line $(0, \infty)$ as $a \to \infty$.

Proof. This follows at once from the definition of $S^*(x,\lambda)$, because of the explicit form of $\vartheta^*(x,s;\lambda)$.

10.7.2

We now transform the right-hand side of (7.9). Changing the order of integration, we rewrite the equation in the form

$$\int_{-\infty}^{\infty} \psi_\epsilon(\lambda) \, d_\lambda \{S(x,\lambda) - S^*(x,\lambda)\} = \frac{1}{2} \int_{-\epsilon}^{\epsilon} \left\{ \int_{x-t}^{x+t} W(x,t,s) f(s) \, ds \right\} g_\epsilon(t) \, dt. \tag{7.12}$$

Let

$$\alpha(x,\lambda) = \int_{-\epsilon}^{\epsilon} \left\{ \int_{x-t}^{x+t} W(x,t,s) f(s) \, ds \right\} e^{i\lambda t} \, dt. \tag{7.13}$$

Then, applying the Parseval equation to the classical Fourier integral, we obtain

$$\int_{-\epsilon}^{\epsilon} \left\{ \int_{x-t}^{x+t} W(x,t,s)f(s)\,ds \right\} g_\epsilon(t)\,dt = \frac{1}{2\pi} \int_{-\infty}^{\infty} \psi_\epsilon(\lambda)\bar{\alpha}(x,\lambda)\,d\lambda,$$

deriving it from the equalities (7.13) and (6.9).

Therefore, (7.12) takes the form

$$\int_{-\infty}^{\infty} \psi_\epsilon(\lambda)\,d_\lambda R(x,\lambda) = 0, \tag{7.14}$$

where

$$R(x,\lambda) = S(x,\lambda) - S^*(x,\lambda) - \frac{1}{4\pi} \int_0^\lambda \bar{\alpha}(x,\nu)\,d\nu. \tag{7.15}$$

Lemma 10.7.3 *If the matrix $Q(x)$ is summable in each finite interval, then the estimates*

$$\bigvee_a^{a+1} \{R_k(x,\lambda)\} = o(1), \quad k = 1, 2, \tag{7.16}$$

hold for any fixed x as $a \to \infty$ uniformly in any finite interval.

Proof. Since the variation of a sum does not exceed the sum of the variations of the addends, to prove (7.16), it suffices to show that the estimates hold for each addend of the sum (7.15); (7.16) coincide with the estimates (7.10), (7.11) for the addends $S(x,\lambda)$, $S^*(x,\lambda)$.

Furthermore, it follows from the definition of $\alpha(x,\lambda)$, i.e., (7.13), by the Lebesgue-Riemann lemma, that $\lim_{\lambda\to\infty} \alpha_k(x,\lambda) = 0$, $k = 1, 2$. Therefore, as $a \to \infty$,

$$\bigvee_a^{a+1} \left\{ \int_0^\lambda \alpha_k(x,\nu)\,d\nu \right\} = \int_a^{a+1} |\alpha_k(x,\nu)|\,d\nu = o(1).$$

Lemma 10.7.4 *Let the matrix $Q(x)$ be summable in any finite interval. Then the equalities*

$$\lim_{\lambda\to\infty} \int_{-\lambda}^{\lambda} \alpha_k(x,\nu)\,d\nu = 0, \quad k = 1, 2, \tag{7.17}$$

hold for any fixed x uniformly in each finite interval.

Proof. Put

$$b(x,t) = \int_{x-\epsilon}^{x+\epsilon} W(x,t,s)f(s)\,ds.$$

It then follows from (7.13) that

$$\alpha(x,\lambda) = \int_{-\epsilon}^{\epsilon} b(x,t)e^{i\lambda t}\,dt.$$

Therefore,

$$\int_{-\infty}^{\infty} \alpha(x,\nu) \, d\nu = 2 \int_{-\epsilon}^{\epsilon} b(x,t) \frac{\sin \lambda t}{t} \, dt,$$

and (7.17) follow because of the differentiability of $b(x,t)$ with respect to t, thus completing the proof.   ∎

Theorem 10.7.1 *(Equiconvergence Theorem.) Let $f(x) \in \mathcal{L}^2(0,\infty)$. If the matrix $Q(x)$ is summable in each finite interval, then the equality*

$$\lim_{\lambda \to \infty} \{[S(x,\lambda) - S(x,-\lambda)] - [S^*(x,\lambda) - S^*(x,-\lambda)]\} = 0$$

holds uniformly in each finite interval, i.e., the difference between the representation of the vector-valued function $f(x)$ as the generalized Fourier integral in terms of the vector-valued eigenfunctions of the problem (6.1), (6.2) and (6.22), (6.23) tends to zero uniformly in each finite interval.

Proof. By Lemma 10.7.3, the Tauberian theorem of B. M. Levitan (see B. M. Levitan and I. S. Sargsjan (1975; Chapter XIV, Theorem 4.1)) can be applied to (7.14), so that

$$\lim_{\lambda \to \infty} \{R(x,\lambda) - R(x,-\lambda)\} = 0,$$

i.e., by the equality (7.15),

$$\lim_{\lambda \to 0} \{[S(x,\lambda) - S(x,-\lambda)] - [S^*(x,\lambda) - S^*(x,-\lambda)]\} = \frac{1}{4\pi} \lim_{\lambda \to \infty} \int_{-\lambda}^{\lambda} \bar{\alpha}(x,\nu) \, d\nu,$$

and the statement follows, relying on Lemma 10.7.4, thus completing the proof.   ∎

Theorem 10.7.1 yields the final solution of the question of convergence of the Dirac operator vector-valued eigenfunction expansion for square-integrable vector-valued functions on the half-line $(0,\infty)$. In fact, if we put $h = 0$ in (6.22), (6.23), then the representation as the generalized Fourier integral in terms of the vector-valued eigenfunctions of this problem is the expansion in the usual Fourier integral, for, as is noted at the beginning of the section, when $h = 0$, the vector-valued eigenfunctions are $\psi(x,\lambda) = \begin{pmatrix} \cos \lambda x \\ \sin \lambda x \end{pmatrix}$. Therefore, when $h = 0$, Theorem 10.7.1 can be stated differently.

Theorem 10.7.2 *Let $f(x) \in \mathcal{L}^2(0,\infty)$, and let the conditions of Theorem 10.7.1 hold. Then the difference between the one-dimensional Dirac system vector-valued eigenfunction expansion of the vector-valued function $f(x)$ and the expansion in the usual Fourier integral tends to zero uniformly in each interval.*

In particular, this implies the following.

Theorem 10.7.3 *Let the conditions of Theorem 10.7.1 be fulfilled. Then the equality*

$$\lim_{\lambda \to \infty} \{S(x_0, \lambda) - S(x_0, -\lambda)\} = f(x_0),$$

i.e., the expansion of the vector-valued function $f(x)$ in the generalized Fourier integral in terms of the one-dimensional Dirac system vector-valued eigenfunctions at any point x_0, where the conditions for $f(x)$ to be expanded in the usual Fourier integral hold locally, tends to the value $f(x_0)$.

Notes

Sections 1 and 2

The results are due to I. S. Sargsjan (1966b).

Section 3

The transformation matrix operator for different canonical forms of a Dirac system was constructed in the papers of F. Prats and J. Toll (1959), B. M. Levitan and M. G. Gasymov (1964) and I. S. Sargsjan (1966c).

Section 4

The extension of the solution to the one-dimensional Dirac system was done in the paper of B. M. Levitan and I. S. Sargsjan (1965).

Section 5

The results of this section are due to I. S. Sargsjan (1966c).

Sections 6 and 7

The results of these sections are due to I. S. Sargsjan (1966a,b). The proofs given in Sections 1, 2, 6, 7 in the canonical form of M. G. Gasymov are due to G. V. Dikhamindzhiya (1978).

Chapter 11

THE DISTRIBUTION OF THE EIGENVALUES

11.1 The integral equation for Green's matrix function

11.1.1

Consider the Dirac system

$$Ly \equiv \begin{pmatrix} 0 & 1 \\ -1 & 0 \end{pmatrix} \frac{dy}{dx} + \begin{pmatrix} p(x) & q(x) \\ q(x) & -p(x) \end{pmatrix} y = \lambda y$$

on the whole number line $(-\infty, \infty)$. Let $G(x, \xi; \lambda)$ be Green's matrix function of the operator L, and let $g_0(x, \xi; \lambda)$ be that with 'frozen' coefficients, i.e., of the system

$$\begin{pmatrix} 0 & 1 \\ -1 & 0 \end{pmatrix} \frac{dy}{d\xi} + \begin{pmatrix} p(x) & q(x) \\ q(x) & -p(x) \end{pmatrix} y = \lambda y, \quad -\infty < \xi < \infty. \tag{1.0}$$

The matrix $g_0(x, \xi; \lambda)$ can be calculated explicitly, and is of the form

$$g_0(x, \xi; \lambda) = \begin{pmatrix} p(x) + \lambda & -q(x) - \kappa \operatorname{sgn}(x - \xi) \\ q(x) - \kappa \operatorname{sgn}(x - \xi) & -p(x) + \lambda \end{pmatrix} \frac{e^{-\kappa|x-\xi|}}{2\kappa},$$

where $\kappa = \kappa(x, \lambda) = \{p^2(x) + q^2(x) - \lambda\}^{1/2}$.

We will seek for $G(x, \xi; \lambda)$ by the parametrix method, showing that it is the solution of the integral equation

$$G(x, \xi; \lambda) = g_0(x, \xi; \lambda) + \int_{-\infty}^{\infty} G(\eta, \xi; \lambda)\{Q(x) - Q(\eta)\}g_0(x, \eta; \lambda)\, d\eta, \tag{1.1}$$

where $g_0(x, \xi; \lambda)$ was defined before and $Q(x) = \begin{pmatrix} p & q \\ q & -p \end{pmatrix}$. It is shown below that (1.1) can be solved by the iteration method for large $|\lambda|$.

Let $||A||$ be the Hilbert-Schmidt norm of the matrix $A = (a_{ij})_{i,j=1}^2$, i.e., let us put $||A|| = \left(\sum_{i,j=1}^2 |a_{ij}|^2\right)^{1/2}$. It then follows from the definition of $g_0(x, \xi; \lambda)$ that

$$||g_0(x, \xi; \lambda)|| = e^{-\kappa|x-\xi|}. \tag{1.2}$$

We now need the trace of the matrix $\frac{\partial}{\partial\lambda}g_0(x,\xi;\lambda)|_{\xi=x}$. After elementary calculations, we obtain

$$\text{trace } \frac{\partial}{\partial\lambda}g_0(x,\xi;\lambda)|_{\xi=x} = \frac{p^2(x)+q^2(x)}{\{p^2(x)+q^2(x)-\lambda^2\}^{3/2}}. \tag{1.3}$$

11.1.2

Our nearest goal is to obtain the asymptotic behaviour of $||G(x,\xi;i\mu)||$ as $\mu \to \infty$, uniform with respect to x and ξ on the whole number line $(-\infty,\infty)$, where $G(x,\xi;\lambda)$ is the solution of (1.1).

Assume that the matrix $Q(x)$ satisfies the conditions :

1. For $|x-\xi| \leq 1$,
$$||\{Q(x)-Q(\xi)\}Q^{-a}(x)|| \leq A|x-\xi|, \tag{1.4}$$
   where $A > 0$, $0 < a < 2$ are two constant numbers.

2. For $|x-\xi| \leq 1$,
$$||Q(x)||^2 \leq B||Q(\xi)||^2, \tag{1.5}$$
   where $B > 1$ is a constant number.

3. For $|x-\xi| > 1$,
$$||Q(\xi)|| < K \exp\{c_0|x-\xi|[p^2(x)+q^2(x)]^{1/2}\}, \tag{1.6}$$
   where $K > 0$, $0 < c_0 < 1$ are two constant numbers.

4. At least one of the two integrals
$$\int_{-\infty}^{\infty} |p(x)|^{-1}\, dx, \qquad \int_{-\infty}^{\infty} |q(x)|^{-1}\, dx \tag{1.7}$$
   converges[1].

We have already noted before that the equation (1.1) is solved by the iteration method. For $k \geq 1$, we put

$$g_k(x,\xi;\lambda) = \int_{-\infty}^{\infty} g_{k-1}(\eta,\xi;\lambda)\{Q(x)-Q(\xi)\}g_0(x,\eta;\lambda)\, d\eta. \tag{1.8}$$

It then follows from (1.1) that

$$G(x,\xi;\lambda) = g_0(x,\xi;\lambda) + \sum_{k=1}^{\infty} g_k(x,\xi,;\lambda) \equiv g_0(x,\xi;\lambda) + g(x,\xi;\lambda). \tag{1.9}$$

---

[1]We require that $|p(x)|^{-1}$ or $|q(x)|^{-1}$ should be summable for the sake of simplicity. We could require the summability of $|p(x)|^{-r}$ and $|q(x)|^{-r}$ for certain $r > 0$, in which case we would have to 'iterate'. But the final asymptotic formulas (4.9) would remain valid.

We now estimate $||g_k(x, \xi; \lambda)||$ for $\lambda = i\mu$ as $\mu \to \infty$. Since we shall have to estimate the kernel

$$K_0(x, \eta; \lambda) = \{Q(x) - Q(\xi)\}g_0(x, \eta; \lambda),$$

we first consider

$$||K_0(x, \eta; i\mu)|| = ||\{Q(x) - Q(\eta)\}Q^{-a}(x)Q^a(x)g_0(x, \eta; i\mu)||$$
$$\leq ||\{Q(x) - Q(\eta)\}Q^{-a}(x)|| \, ||Q^a(x)|| \, ||g_0(x, \eta; i\mu)||, \quad (1.10)$$

where the number a is the same as in (1.4). Now, taking into account that $||Q^a(x)|| < 2^a\{p^2(x) + q^2(x) + \mu^2\}^{a/2} = 2^a\kappa^a(x)$, we derive

$$||K_0(x, \eta; i\mu)|| \leq A_1\kappa^a(x)|x - \eta|e^{-\kappa(x)|x-\eta|} \quad (1.11)$$

from (1.10), relying on (1.2), (1.4) for $|x - \eta| \leq 1$. Applying the inequality $ye^{-\delta y} < c$ valid for all $y \geq 0$ (where c is a constant depending on δ, and $\delta > 0$ is an arbitrary small number), we obtain

$$||K_0(x, \eta; i\mu)|| < A_0\kappa^{a-1}(x)e^{(\delta-1)\kappa(x)|x-\eta|}, \quad |x - \eta| \leq 1. \quad (1.11')$$

To estimate $||K_0(x, \eta; i\mu)||$ for $|x - \eta| > 0$, we consider

$$||K_0(x, \eta; i\mu)|| \leq \{||Q(x)|| + ||Q(\eta)||\}||g_0(x, \eta; i\mu)||.$$

Using (1.6) and the inequality $||Q(x)|| < \sqrt{2}\kappa(x)$, we derive

$$||K_0(x, \eta; i\mu)|| \leq K_1e^{(c_0-1)\kappa(x)|x-\eta|}, \quad |x - \eta| > 0. \quad (1.12)$$

Applying the inequality $y^re^{-\delta_1 y} < c$, valid for all $y \geq 0$ and arbitrary r (c being a constant depending on both δ_1 and r), we have

$$||K_0(x, \eta; i\mu)|| \leq K_1\frac{\kappa^r(x)|x - \eta|^r}{\kappa^r(x)|x - \eta|^r}e^{(c_0-1)\kappa(x)|x-\eta|}$$
$$< \frac{K_0}{\kappa^r(x)}e^{(c_0+\delta_1-1)\kappa(x)|x-\eta|}, \quad |x - \eta| > 1, \quad (1.12')$$

using $|x - \eta| > 1$. We derive from the estimates (1.11') and (1.12') for $||K_0(x, \eta; i\mu)||$ that

$$||K_0(x, \eta; i\mu)|| < \begin{cases} \dfrac{A_0}{\kappa^{1-a}(x)}e^{-B_0\kappa(x)|x-\eta|}, & |x - \eta| \leq 1 \\[3mm] \dfrac{K_0}{\kappa^r(x)}e^{-B_0\kappa(x)|x-\eta|}, & |x - \eta| > 1 \end{cases} \quad (1.13)$$

where

$$B_0 = \min\{1 - \delta, 1 - c_0 - \delta_1\} < 1. \quad (1.14)$$

Note that it follows from the inequalities (1.13) that, relying on the arbitrariness of r for all x and η, the kernel $K_0(x, \eta; i\mu)$ satisfies the inequality

$$\|K_0(x, \eta; i\mu)\| < c_0 \kappa^{a-1}(x) e^{-B_0 \kappa(x)|x-\eta|}. \tag{1.15}$$

We now estimate the iterations $g_k(x, \xi; i\mu) \equiv g_k$. By definition, we have

$$g_1 = \int_{|x-\eta|\leq 1} g_0(\eta, \xi; i\mu)\{Q(x) - Q(\eta)\} g_0(x, \eta; i\mu)\, d\eta$$

$$+ \int_{|x-\eta|>1} g_0(\eta, \xi; i\mu)\{Q(x) - Q(\eta)\} g_0(x, \eta; i\mu)\, d\eta \equiv g_1' + g_1'' \tag{1.15'}$$

relying on (1.8). Taking the norms, (1.2) and the first inequality in (1.13) yield

$$\|g_1'\| \leq \frac{A_0}{\kappa^{1-a}(x)} \int_{|x-\eta|\leq 1} e^{-\kappa(\eta)|\eta-\xi|} e^{-B_0\kappa(x)|x-\eta|}\, d\eta. \tag{1.16}$$

By the condition (1.5) for $|x - \eta| \leq 1$,

$$\{p^2(\eta) + q^2(\eta)\}^{1/2} > \sqrt{B'}\{p^2(x) + q^2(x)\}^{1/2}, \qquad B' = B^{-1} < 1.$$

Therefore, $\kappa(\eta) = \{p^2(\eta) + q^2(\eta) + \mu^2\}^{1/2} > \sqrt{B'}\kappa(x)$; it then follows from (1.16) that

$$\|g_1'\| < \frac{A_0}{\kappa^{1-a}(x)} \int_{|x-\eta|\leq 1} e^{-B_1\kappa(x)\{|x-\eta|+|\eta-\xi|\}}\, d\eta. \tag{1.16'}$$

Because $|x - \eta| + |\eta - \xi| \geq |x - \xi|$, we then obtain

$$\|g_1'\| < \frac{A_0}{\kappa^{1-a}(x)} e^{-(B_1-\delta)\kappa(x)|x-\xi|} \int_{|x-\eta|\leq 1} e^{-\delta\kappa(x)|x-\eta|}\, d\eta$$

$$< \frac{A_0}{\kappa^{1-a}(x)} e^{-(B_1-\delta)\kappa(x)|x-\xi|} \int_{-\infty}^{\infty} e^{-\delta\kappa(x)|x-\eta|}\, d\eta$$

$$= \frac{A_0}{\delta\kappa^{2-a}(x)} e^{-(B_1-\delta)\kappa(x)|x-\xi|},$$

for $\int_{-\infty}^{\infty} \exp\{-\delta\kappa(x)|x - \eta|\}\, d\eta = \dfrac{2}{\delta\kappa(x)}$. Thus,

$$\|g_1'(x, \xi; i\mu)\| < \frac{c_1'}{\kappa^{2-a}(x)} e^{-(B_1-\delta)\kappa(x)|x-\xi|}, \tag{1.17}$$

where $B_1 < 1$ and δ is a small positive number.

We now estimate $\|g_1''\|$. According to (1.15), taking the norms, using (1.2) and the second inequality in (1.13), we have

$$\|g_1''\| \leq \frac{K_0}{\kappa^r(x)} \int_{|x-\eta|>1} e^{-\kappa(\eta)|\eta-\xi|} e^{-B_0\kappa(x)|x-\eta|}\, d\eta,$$

whence

$$\|g_1''\| \leq \frac{K_0}{\kappa^r(x)} \int_{|x-\eta|>1} e^{-\sqrt{B'}|\eta-\xi|} e^{-B_0|x-\eta|} \, d\eta,$$

because (without loss of generality) $\kappa(x) > 1 > \sqrt{1/B} = \sqrt{B'}$ (where $B > 0$ by the condition (1.5)). Proceeding as in the estimation of the integral (1.16′), we then obtain the estimate

$$\|g_1''(x,\xi;i\mu)\| \leq \frac{c_1''}{\kappa^r(x)} e^{-(B_1-\delta)|x-\xi|} \tag{1.18}$$

for $\|g_1''\|$. Therefore, relying on (1.17), (1.18), we obtain

$$\|g_1(x,\xi;i\mu)\| \leq \frac{c_1'}{\kappa^{2-a}(x)} e^{-(B_1-\delta)\kappa(x)|x-\xi|} + \frac{c_1''}{\kappa^r(x)} e^{-(B_1-\delta)|x-\xi|} \tag{1.19}$$

from (1.15), where r is an arbitrary positive number, $0 \leq B_1 < 1$ and δ is an arbitrarily small positive number.

To estimate $g_2(x,\xi;i\mu)$, we have, by its definition,

$$g_2(x,\xi;i\mu) = \int_{|x-\eta|\leq1} g_1(\eta,\xi;i\mu)\{Q(x)-Q(\eta)\}g_0(x,\eta;i\mu) \, d\eta$$

$$+ \int_{|x-\eta|>1} g_1(\eta,\xi;i\mu)\{Q(x)-Q(\eta)\}g_0(x,\eta;i\mu) \, d\eta \equiv g_2' + g_2''. \tag{1.20}$$

Taking the norms and using the inequalities (1.13), (1.19), we get

$$\|g_2'\| \leq \frac{A_1}{\kappa^{1-a}(x)} \int_{|x-\eta|\leq1} e^{-B_0\kappa(x)|x-\eta|} \frac{e^{-(B_1-\delta)\kappa(\eta)|\eta-\xi|}}{\kappa^{2-a}(\eta)} \, d\eta$$

$$+ \frac{A_2}{\kappa^{1-a}(x)} \int_{|x-\eta|\leq1} e^{-B_0\kappa(x)|x-\eta|} \frac{e^{-(B_1-\delta)\kappa(\eta)|\eta-\xi|}}{\kappa^r(\eta)} \, d\eta \equiv I_1' + I_2'. \tag{1.21}$$

We estimate I_1' and I_2' separately, noticing for the former that, because $\kappa(\eta) > \sqrt{B'}\kappa(x)$ for $|x-\eta| \leq 1$,

$$I_1' \leq \frac{A_1'}{\kappa^{3-2a}(x)} \int_{|x-\eta|\leq1} e^{-B_0\kappa(x)|x-\eta|} e^{-\sqrt{B'}(B_1-\delta)\kappa(x)|\eta-\xi|} \, d\eta.$$

Also taking into account that $B = \min(B_0, \sqrt{B'})$, we obtain

$$I_1' \leq \frac{A_1''}{\kappa^{3-2a}(x)} \int_{|x-\eta|\leq1} e^{-B_1(B_1-\delta)\kappa(x)(|x-\eta|+|\eta-\xi|)} \, d\eta,$$

since $B_1 > B_1(B_1-\delta)$ on account of $B_1 < 1$. Precisely as in deriving the estimate (1.17), it follows from the above inequality that

$$I_1' \leq \frac{c_2'}{\kappa^{2(2-a)}(x)} e^{-[B_1(B_1-\delta)-\delta]\kappa(x)|x-\xi|}. \tag{1.22}$$

Having $\kappa(\eta) > \sqrt{B'}\kappa(x)$ for $|x - \eta| \leq 1$, we see that

$$I_2' \leq \frac{A_2'}{\kappa^r(x)} \int_{|x-\eta|\leq 1} \frac{\kappa^{a-1}(x)|x-\eta|^{a-1}}{|x-\eta|^{a-1}} e^{-B_0\kappa(x)|x-\eta|} e^{-(B_1-\delta)|\eta-\xi|} \, d\eta;$$

hence, again using the inequality $y^k e^{-\delta y} < c$, we get

$$I_2' \leq \frac{A_2''}{\kappa^r(x)} e^{-[B_1(B_1-\delta)-\delta]|x-\xi|} \int_{-\infty}^{\infty} \frac{e^{-\delta|x-\eta|}}{|x-\eta|^a} \, d\eta = \frac{c_3}{\kappa^r(x)} e^{-[B_1(B_1-\delta)-\delta]|x-\xi|}, \qquad (1.23)$$

and

$$\|g_2'\| \leq \frac{c_2'}{\kappa^{2(2-a)}(x)} e^{-[B_1(B_1-\delta)-\delta]\kappa(x)|x-\xi|} + \frac{c_3}{\kappa^r(x)} e^{-[B_1(B_1-\delta)-\delta]|x-\xi|}, \qquad (1.24)$$

deriving the estimate for $\|g_2'\|$ from (1.22), (1.23) (see (1.21)).

Now, to estimate g_2'' from the equality (1.20), we see that

$$\|g_2''\| \leq \frac{A_1}{\kappa^r(x)} \int_{|x-\eta|>1} e^{-B_0\kappa(x)|x-\eta|} \frac{e^{-(B_1-\delta)\kappa(\eta)|\eta-\xi|}}{\kappa^{2-a}(\eta)} \, d\eta$$

$$+ \frac{A_2}{\kappa^r(x)} \int_{|x-\eta|>1} e^{-B_0\kappa(x)|x-\eta|} \frac{e^{-(B_1-\delta)|\eta-\xi|}}{\kappa^r(\eta)} \, d\eta \equiv I_1'' + I_2'',$$

by the definition and by the estimates (1.13), (1.19), where I_1'', I_2'' are estimated as I_2'. We thus arrive at an estimate of the same form as (1.23). Therefore,

$$\|g_2''\| \leq \frac{c_4}{\kappa^r(x)} e^{-[B_1(B_1-\delta)-\delta]|x-\xi|},$$

which, combined with (1.18), yields the estimate

$$\|g_2(x,\xi;i\mu)\| \leq \frac{c_2'}{\kappa^{2(2-a)}(x)} e^{-[B_1(B_1-\delta)-\delta]\kappa(x)|x-\xi|} + \frac{c_2''}{\kappa^r(x)} e^{-[B_1(B_1-\delta)-\delta]|x-\xi|} \qquad (1.25)$$

for the second iteration.

Applying the argument for (1.19) and (1.25) to the first and second iterations, respectively, we write down the inequalities on the analogy

$$\|g_k(x,\xi;i\mu)\| \leq \frac{c_k' e^{-B^*\kappa(x)|x-\xi|}}{\kappa^{k(2-a)}(x)} + \frac{c_k'' e^{-B^*|x-\xi|}}{\kappa^r(x)} \qquad (1.26)$$

for the kth iteration, where $B^* = B_1^k - (B_1^{k-1} + B_1^{k-2} + \ldots + 1)\delta$ and δ is an arbitrarily small positive number, which can obviously be made so small that $B^* > 0$.

Until now, r was arbitrary. Now, we choose it so that the integral

$$\int_{-\infty}^{\infty} \frac{dx}{\kappa^r(x)} < +\infty \qquad (1.27)$$

converges, and we choose k_0 so large that $k_0(2-a) > r$.

We obtain

$$\|g_{k_0}(x,\xi;i\mu)\| \leq \frac{c_{k_0}}{\kappa^r(x)} e^{-B^*|x-\xi|} \qquad (1.28)$$

for the iteration $g_{k_0}(x,\xi;i\mu)$.

11.1.3

Starting with the subscript k_0, we estimate the iterations differently. However, we first somewhat change the estimate (1.13). For $|x - \eta| \leq 1$, by (1.11), we have

$$\|K_0(x, \eta; i\mu)\| \leq A_1 \frac{\kappa^{a+\sigma}(x)|x - \eta|^{a+\sigma}}{\kappa^{\sigma}(x)|x - \eta|^{a-1+\sigma}} e^{-\kappa(x)|x-\eta|} \leq \frac{A_2}{\kappa^{\sigma}(x)|x - \eta|^{a-1+\sigma}} e^{-(1-\delta)\kappa(x)|x-\eta|}.$$

$$(1.29)$$

Since $\kappa(x) > 1$ by assumption and $\kappa(x) > \mu$ by definition, choosing $\sigma > 0$ to make $\alpha = a - 1 + \sigma < 1$, we derive from (1.29) that

$$\|K_0(x, \eta; i\mu)\| \leq \frac{A^*}{\mu^{\sigma}} \frac{e^{-(1-\delta)|x-\eta|}}{|x - \eta|^{\alpha}}, \quad |x - \eta| \leq 1. \tag{1.30}$$

Furthermore, it follows from the estimate (1.12) that, for $|x - \eta| > 1$,

$$\|K_0(x, \eta; i\mu)\| \leq K_1 \frac{\kappa^{r+\sigma}(x)|x - \eta|^{r+\sigma}}{\kappa^{r+\sigma}(x)|x - \eta|^{r+\sigma}} e^{-(1-c_0)\kappa(x)|x-\eta|}$$

$$< \frac{K_1}{\mu^{\sigma}\kappa^r(x)} \frac{e^{-(1-c_0-\delta_1)\kappa(x)|x-\eta|}}{|x - \eta|^{r+\sigma}} < \frac{K^*}{\mu^{\sigma}} \frac{e^{-(1-c_0-\delta_1)|x-\eta|}}{\kappa^r(x)}.$$

Hence, also taking into account (1.30), we obtain the estimate

$$\|K_0(x, \eta; i\mu)\| \leq \begin{cases} \dfrac{A^*}{\mu^{\sigma}} \dfrac{e^{-B_0|x-\eta|}}{|x - \eta|^{\alpha}}, & |x - \eta| \leq 1 \\[3mm] \dfrac{K^*}{\mu^{\sigma}} \dfrac{e^{-B_0|x-\eta|}}{\kappa^r(x)}, & |x - \eta| > 1 \end{cases} \tag{1.31}$$

where

$$B_0 = \min(1 - \delta, 1 - c_0 - \delta_1) < 1, \tag{1.32}$$

and r is selected in accordance with (1.27), $\sigma > 0$, $\alpha < 1$.

By definition,

$$g_{k_0+1}(x, \xi; i\mu) = \int_{|x-\eta| \leq 1} g_{k_0}(\eta, \xi; i\mu) K_0(x, \eta; i\mu)\, d\eta$$

$$+ \int_{|x-\eta|>1} g_{k_0}(\eta, \xi; i\mu) K_0(x, \eta; i\mu)\, d\eta = I_1 + I_2. \tag{1.33}$$

Taking the norms and invoking the inequalities (1.28), (1.31), we get

$$\|I_1\| \leq \frac{c_{k_0} A^*}{\mu^{\sigma}} \int_{|x-\eta| \leq 1} \frac{e^{-B_0|x-\eta|}}{|x - \eta|^{\alpha}} \frac{e^{-B^*|\eta-\xi|}}{\kappa^r(\eta)}\, d\eta;$$

hence, relying on $\kappa(\eta) > \sqrt{B'}\kappa(x)$,

$$\|I_1\| \leq \frac{A^*(\sqrt{B'})^r c_{k_0}}{\mu^\sigma \kappa^r(x)} \int_{|x-\eta|\leq 1} \frac{e^{-B_0|x-\eta|}}{|x-\eta|^\alpha} e^{-B^*|\eta-\xi|} \, d\eta$$

$$\leq \frac{A^*(\sqrt{B'})^r c_{k_0}}{\mu^\sigma \kappa^r(x)} e^{-B^*|x-\xi|} \int_{|x-\eta|\leq 1} \frac{e^{-\epsilon|x-\eta|}}{|x-\eta|^\alpha} \, d\eta$$

$$\leq \frac{A^*(\sqrt{B'})^r c_{k_0}}{\mu^\sigma \kappa^r(x)} e^{-B^*|x-\xi|} \int_{-\infty}^{\infty} \frac{e^{-\epsilon|x-\eta|}}{|x-\eta|^\alpha} \, d\eta, \quad (1.34)$$

where $\epsilon = B_0 - B^*$.

To derive the final estimate for $\|I_1\|$, we clarify the relations between certain constants involved in (1.34).

First, it follows from the way we obtained the estimate (1.31) that the constant B_0 can be determined just by the formula (1.32), coinciding with the formula (1.14), with the help of the constants A^*, K^*. On the other hand, since $B^* > 0$, $B^* < B_1$ and $B_1 < B_0$, we have $B^* < B_0$. Therefore, $\epsilon = B_0 - B^* > 0$.

Now, putting

$$M = \frac{A^*}{(\sqrt{B'})^r} \int_{-\infty}^{\infty} \frac{e^{-\epsilon|x-\eta|}}{|x-\eta|^\alpha} \, d\eta,$$

the inequality (1.34) can be rewritten in the form

$$\|I_1\| \leq \frac{c_{k_0}}{\kappa^r(x)} \cdot \frac{M}{\mu^\sigma} e^{-B^*|x-\xi|}. \quad (1.35)$$

We estimate I_2 on the basis of the equality (1.33). By the estimates (1.28), (1.31), we have

$$\|I_2\| \leq \frac{K^* c_{k_0}}{\mu^\sigma \kappa^r(x)} \int_{|x-\eta|>1} e^{-B_0|x-\eta|} \frac{e^{-B^*|\eta-\xi|}}{\kappa^r(\eta)} \, d\eta$$

$$\leq \frac{K^* c_{k_0}}{\mu^\sigma \kappa^r(x)} \int_{|x-\eta|>1} e^{-B_0|x-\eta|} e^{-B^*|\eta-\xi|} \, d\eta$$

$$\leq \frac{K^* c_{k_0}}{\mu^\sigma \kappa^r(x)} e^{-B^*|x-\xi|} \int_{-\infty}^{\infty} e^{-(B_0-B^*)|x-\eta|} \, d\eta.$$

Hence, putting $N = K^* \int_{-\infty}^{\infty} e^{-(B_0-B^*)|x-\eta|} \, d\eta$,

$$\|I_2\| \leq \frac{c_{k_0}}{\kappa^r(x)} \frac{N}{\mu^\sigma} e^{-B^*|x-\xi|}, \quad (1.35')$$

which, together with (1.35), implies that

$$\|g_{k_0+1}(x,\xi;i\mu)\| \leq \frac{c_{k_0}}{\kappa^r(x)} \frac{M+N}{\mu^\sigma} e^{-B^*|x-\xi|}, \quad \sigma > 0.$$

Repeating the above estimations, we can easily establish that

$$\|g_{k_0+p}(x,\xi;i\mu)\| \le \frac{c_{k_0}}{\kappa^r(x)}\frac{(M+N)^p}{\mu^{p\sigma}}e^{-B^*|x-\xi|} \tag{1.36}$$

for the $(k_0+p)^{\text{th}}$ iteration.

Consider the series $\sum_{p=0}^{\infty}\|g_{k_0+p}(x,\xi;i\mu)\|$; substituting $\|g_{k_0+p}(x,\xi;i\mu)\|$ from (1.36), we obtain

$$\sum_{p=0}^{\infty}\|g_{k_0+p}(x,\xi;i\mu)\| \le \frac{c_{k_0}}{\kappa^r(x)}e^{-B^*|x-\xi|}\sum_{p=0}^{\infty}\frac{(M+N)^p}{\mu^{p\sigma}},$$

which obviously converges for large μ. Therefore, the estimate

$$\sum_{p=0}^{\infty}\|g_{k_0+p}(x,\xi;i\mu)\| \le \frac{c^*}{\kappa^r(x)}e^{-B^*|x-\xi|} \tag{1.37}$$

holds uniformly with respect to x, ξ on the whole number axis $(-\infty,\infty)$.

To obtain the required estimate of $G(x,\xi;i\mu)$, it remains to consider the sum

$$\sum_{l=1}^{k_0-1}\|g_l(x,\xi;i\mu)\|. \tag{1.38}$$

Since $l < k_0$, due to the choice of k_0, or the inequality $k_0(2-a) < r$, the first term in the estimates (1.26) is principal. In other words,

$$\|g_l(x,\xi;i\mu)\| < \frac{c_l}{\kappa^{l(2-a)}(x)}e^{-B^*\kappa(x)|x-\xi|}, \quad l = 1,2,\ldots,k_0-1,$$

for all x, ξ. Therefore, the estimate

$$\sum_{l=1}^{k_0-1}\|g_l(x,\xi;i\mu)\| \le \frac{c}{\kappa^{2-a}(x)}e^{-B^*\kappa(x)|x-\xi|} \tag{1.39}$$

is uniform with respect to x, ξ on the whole number axis $(-\infty,\infty)$, and,

$$\|G(x,\xi;i\mu) - g_0(x,\xi;i\mu)\| \le \frac{c}{\kappa^{2-a}(x)}e^{-B^*\kappa(x)|x-\xi|} + \frac{c^*}{\kappa^r(x)}e^{-B^*|x-\xi|}$$

as $\mu \to \infty$, uniformly with respect to x, ξ on $(-\infty,\infty)$ by (1.9), (1.37), (1.39). The important estimate is a consequence, viz.

$$\|G(x,\xi;i\mu)\| \le e^{-\kappa(x)|x-\xi|} + \frac{c}{\kappa^{2-a}(x)}e^{-B^*\kappa(x)|x-\xi|} + \frac{c^*}{\kappa^r(x)}e^{-B^*|x-\xi|} \tag{1.40}$$

(using $g_0(x,\xi;i\mu)$ from (1.2)), uniformly with respect to x, ξ on the whole number axis $(-\infty,\infty)$.

Thus, we have proved the following.

Theorem 11.1.1 *If the matrix $Q(x)$ fulfils Conditions 1 to 4, then (1.40) is valid for the solution $G(x,\xi;i\mu)$ of the integral equation (1.1) as $\mu \to \infty$, uniformly with respect to x, ξ on the whole number axis $(-\infty,\infty)$.*

11.2 Asymptotic behaviour of the matrix $G'_\mu(x, \xi; i\mu)$ as $\mu \to \infty$

Our nearest goal is to obtain the asymptotic behaviour for $G'_\mu(x, \xi; i\mu)$ as $\mu \to \infty$, uniformly with respect to x, ξ on the whole number axis $(-\infty, \infty)$, which will easily determine the asymptotic behaviour of the trace of $G'_\mu(x, \xi; i\mu)$; this is exceptionally important in the derivation of the asymptotic formula for the eigenvalue distribution.

For the estimates in the sequel, the value $\|g'_{0,\mu}(x, \xi; i\mu)\|$ is often required. After elementary calculations, we obtain

$$\left\| \frac{\partial}{\partial \mu} g_0(x, \xi; i\mu) \right\| = \frac{\mu}{\kappa(x)} |x - \xi| e^{-\kappa(x)|x-\xi|}.$$

Hence, by the definition of $\kappa(x)$,

$$\left\| \frac{\partial}{\partial \mu} g_0(x, \xi; i\mu) \right\| \leq |x - \xi| e^{-\kappa(x)|x-\xi|}. \tag{2.1}$$

Differentiating the equation (1.1) with respect to μ, we obtain

$$\frac{\partial}{\partial \mu} G(x, \xi; i\mu) = \frac{\partial}{\partial \mu} g_0(x, \xi; i\mu) + \int_{-\infty}^{\infty} G(\eta, \xi; i\mu)\{Q(x) - Q(\eta)\} \frac{\partial}{\partial \mu} g_0(x, \eta; i\mu) \, d\eta$$

$$+ \int_{-\infty}^{\infty} \frac{\partial}{\partial \mu} G(\eta, \xi; i\mu)\{Q(x) - Q(\eta)\} g_0(x, \eta; i\mu) \, d\eta. \tag{2.2}$$

Introducing the notation

$$\mathcal{L}(x, \xi; \mu) = \frac{\partial}{\partial \mu} G(x, \xi; i\mu), \tag{2.3}$$

$$l_0(x, \xi; \mu) = \int_{-\infty}^{\infty} G(\eta, \xi; i\mu)\{Q(x) - Q(\eta)\} \frac{\partial}{\partial \mu} g_0(x, \eta, i\mu) \, d\eta, \tag{2.4}$$

we can write the equation (2.2) in the form

$$\mathcal{L}(x, \xi; \mu) = \frac{\partial}{\partial \mu} g_0(x, \xi; i\mu) + l_0(x, \xi; \mu)$$

$$+ \int_{-\infty}^{\infty} \mathcal{L}(\eta, \xi; \mu)\{Q(x) - Q(\eta)\} g_0(x, \eta; i\mu) \, d\eta, \tag{2.5}$$

which is easily solvable by the iteration method. We put

$$l_1(x, \xi; \mu) = \int_{-\infty}^{\infty} \left\{ \frac{\partial}{\partial \mu} g_0(\eta, \xi; i\mu) + l_0(\eta, \xi; \mu) \right\} \{Q(x) - Q(\eta)\} g_0(x, \eta; i\mu) \, d\eta, \tag{2.6}$$

and, for $k \geq 2$,

$$l_k(x, \xi; \mu) = \int_{-\infty}^{\infty} l_{k-1}(\eta, \xi; \mu)\{Q(x) - Q(\eta)\} g_0(x, \eta; i\mu) \, d\eta. \tag{2.7}$$

Then

$$\mathcal{L}(x,\xi;\mu) = \frac{\partial}{\partial\mu}g_0(x,\xi;i\mu) + \sum_{k=0}^{\infty} l_k(x,\xi;\mu). \tag{2.8}$$

We assume that the matrix $Q(x)$ fulfils Conditions 1 to 4 of the preceding section. By the inequalities (1.4), (2.1) for $|x-\eta| \leq 1$, we have

$$\left\| \{Q(x) - Q(\eta)\}\frac{\partial}{\partial\mu}g_0(x,\eta;i\mu) \right\|$$

$$\leq \|\{Q(x) - Q(\eta)\}Q^{-a}(x)\| \, \|Q^a(x)\| \left\| \frac{\partial}{\partial\mu}g_0(x,\eta;i\mu) \right\|$$

$$\leq A|x-\eta|^2\kappa^a(x)e^{-\kappa(x)|x-\eta|} \leq \frac{A_0}{\kappa^{2-a}(x)}e^{-(1-\delta)\kappa(x)|x-\eta|}. \tag{2.9}$$

Proceeding in the way the estimate (1.12′) was derived, relying on (2.1) and the condition (1.6), for $|x-\eta| > 0$, we obtain

$$\left\| \{Q(x) - Q(\eta)\}\frac{\partial g_0}{\partial\mu} \right\| \leq \frac{K_0}{\kappa^r(x)}e^{-(1-c_0-\delta_1)\kappa(x)|x-\eta|}, \tag{2.10}$$

where r is an arbitrary positive number, and δ_1 and arbitrarily small positive number.

Combining (2.9) with (2.10), we have

$$\left\| \{Q(x) - Q(\eta)\}\frac{\partial}{\partial\mu}g_0(x,\eta;i\mu) \right\| \leq \begin{cases} \dfrac{A_0}{\kappa^{2-a}(x)}e^{-B_0\kappa(x)|x-\eta|}, & |x-\eta| \leq 1 \\[3mm] \dfrac{K_0}{\kappa^r(x)}e^{-B_0\kappa(x)|x-\eta|}, & |x-\eta| > 1 \end{cases} \tag{2.11}$$

where

$$B_0 = \min\{1-\delta, 1-c_0-\delta_1\} < 1. \tag{2.12}$$

By definition,

$$l_0(x,\xi;\mu) = \int_{|x-\eta|\leq 1} G(\eta,\xi;i\mu)\{Q(x) - Q(\eta)\}\frac{\partial}{\partial\mu}g_0(x,\eta;i\mu)\,d\eta$$

$$+ \int_{|x-\eta|>1} G(\eta,\xi;i\mu)\{Q(x) - Q(\eta)\}\frac{\partial}{\partial\mu}g_0(x,\eta;i\mu)\,d\eta \equiv l_0' + l_0''. \tag{2.13}$$

Taking the norms and using (1.40), (2.11) we obtain

$$\|l_0'\| \leq \int_{|x-\eta|\leq 1} e^{-\kappa(\eta)|\eta-\xi|}\frac{A_0}{\kappa^{2-a}(x)}e^{-B_0\kappa(x)|x-\eta|}\,d\eta$$

$$+ \int_{|x-\eta|\leq 1} \frac{C}{\kappa^{2-a}(\eta)}e^{-B^*\kappa(\eta)|\eta-\xi|}\frac{A_0}{\kappa^{2-a}(x)}e^{-B_0\kappa(x)|x-\eta|}\,d\eta$$

$$+ \int_{|x-\eta|\leq 1} \frac{C^*}{\kappa^r(x)}e^{-B^*|\eta-\xi|}\frac{A_0}{\kappa^{2-a}(x)}e^{-B_0\kappa(x)|x-\eta|}\,d\eta \equiv I_1 + I_2 + I_3. \tag{2.14}$$

We now estimate the integrals I_1, I_2 and I_3 separately. I_1 coincides with the integral (1.16); therefore, by (1.17),

$$I_1 \leq \frac{C_0'}{\kappa^{3-a}(x)} e^{-(B_0-\delta)\kappa(x)|x-\xi|}. \tag{2.15}$$

Furthermore, I_2 is different from the first integral on the right-hand side of (1.21) only in replacing the number B^* by $B_1 - \delta$. Hence, as in obtaining the inequality (1.22), we get

$$I_2 \leq \frac{C_0''}{\kappa^{2(2-a)+1}(x)} e^{-(B_1-\delta)\kappa(x)|x-\xi|}, \tag{2.16}$$

where $B_1 = \min\{B_0, B^*\sqrt{B'}\}$.

Finally, we estimate I_3. Since $\kappa(\eta) > 1$, $a < 2$, by definition,

$$I_3 \leq C^* A_0 \int_{|x-\eta|\leq 1} e^{-B_0\kappa(x)|x-\eta|} \frac{e^{-B^*|\eta-\xi|}}{\kappa^r(\eta)} d\eta$$

while the last integral is different from the second on the right of (1.21) only in replacing B^* by $B_1 - \delta$. Therefore, relying on the estimate (1.23),

$$I_3 \leq \frac{C_0'''}{\kappa^r(x)} e^{-(B_1-\delta)|x-\xi|}, \tag{2.17}$$

where B_1 is the same as in (2.16).

Hence, it follows from the inequalities (2.15), (2.16), (2.17) and (2.14) that

$$\|l_0'\| \leq \frac{C_0}{\kappa^{3-a}(x)} e^{-(B-\delta)\kappa(x)|x-\xi|} + \frac{C_0'''}{\kappa^r(x)} e^{-(B-\delta)|x-\xi|}, \tag{2.18}$$

where $B = \min\{B_0, B_1\}$.

To estimate $\|l_0(x, \xi; \mu)\|$, we also have to estimate l_0''. Since $\kappa(x) > 1$, it easily follows from (1.40), (2.11) and from the definition of l_0'' that

$$\|l_0''\| \leq \frac{K_0'}{\kappa^r(x)} \int_{|x-\eta|>1} e^{-B_0|x-\eta|} e^{-B^*|\eta-\xi|} d\eta \leq \frac{K_0''}{\kappa^r(x)} e^{-(B-\delta)|x-\xi|},$$

which implies, together with (2.18), that

$$\|l_0(x, \xi; \mu)\| \leq \frac{C_0}{\kappa^{3-a}(x)} e^{-(B-\delta)\kappa(x)|x-\xi|} + \frac{C_0'}{\kappa^r(x)} e^{-(B-\delta)|x-\xi|}. \tag{2.19}$$

We now estimate the iterations $l_k(x, \xi; \mu)$ for $k \geq l$. By definition,

$$l_1(x, \xi; \mu) = \int_{|x-\eta|\leq 1} \left\{ \frac{\partial g_0}{\partial \mu} + l_0 \right\} (\eta, \xi; \mu)\{Q(x) - Q(\eta)\} g_0(x, \eta; i\mu) d\eta$$

$$+ \int_{|x-\eta|>1} \left\{ \frac{\partial g_0}{\partial \mu} + l_0 \right\} (\eta, \xi; \mu)\{Q(x) - Q(\eta)\} g_0(x, \eta; i\mu) d\eta \equiv l_1' + l_1'' \tag{2.20}$$

(see (2.6)). Taking the norms and using (2.1), (1.13), (2.19), we obtain

$$\|l_1'\| \leq \frac{A_0}{\kappa^{1-a}(x)} \int_{|x-\eta|\leq 1} e^{-B_0\kappa(x)|x-\eta|} |\eta - \xi| e^{-\kappa(\eta)|\eta-\xi|} \, d\eta$$

$$+ \frac{A_0 C_0}{\kappa^{1-a}(x)} \int_{|x-\eta|\leq 1} e^{-B_0\kappa(x)|x-\eta|} \frac{e^{-(B-\delta)\kappa(\eta)|\eta-\xi|}}{\kappa^{3-a}(\eta)} \, d\eta$$

$$+ \frac{A_0 C_0'}{\kappa^{1-a}(x)} \int_{|x-\eta|\leq 1} e^{-B_0\kappa(x)|x-\eta|} \frac{e^{-(B-\delta)|\eta-\xi|}}{\kappa^r(\eta)} \, d\eta \equiv I_1 + I_2 + I_3. \quad (2.21)$$

To estimate I_1, we proceed as follows. Invoking the inequality $ye^{-\delta_1 y} < C$ and $\kappa(\eta) > \sqrt{B'}\kappa(x)$, we have

$$I_1 \leq \frac{A_0'}{\kappa^{2-a}(x)} \int_{|x-\eta|\leq 1} e^{-B_0\kappa(x)|x-\eta|} e^{-(1-\delta_1)\sqrt{B'}\kappa(x)|\eta-\xi|} \, d\eta.$$

It is obvious that δ_1 can be selected so that $(1-\delta_1)\sqrt{B'} \equiv B_0' < B_0$. Then

$$I_1 \leq \frac{A_0'}{\kappa^{2-a}(x)} \int_{|x-\eta|\leq 1} e^{-B_0'\kappa(x)(|x-\eta|+|\eta-\xi|)} e^{-(B_0-B_0')\kappa(x)|x-\eta|} \, d\eta$$

$$\leq \frac{A_0'}{\kappa^{2-a}(x)} e^{-B_0'\kappa(x)|x-\xi|} \int_{-\infty}^{\infty} e^{-(B_0-B_0')\kappa(x)|x-\eta|} \, d\eta \leq \frac{C_1'}{\kappa^{3-a}(x)} e^{-B_0'\kappa(x)|x-\xi|},$$

relying on $|x - \eta| + |\eta - \xi| \leq |x - \xi|$ and

$$\int_{-\infty}^{\infty} e^{-(B_0-B_0')\kappa(x)|x-\eta|} \, d\eta = \frac{2}{B_0 - B_0'} \frac{1}{\kappa(x)}, \quad B_0 - B_0' > 0.$$

Thus,

$$I_1 \leq \frac{C_1'}{\kappa^{3-a}(x)} e^{-B_0'\kappa(x)|x-\xi|}. \quad (2.22)$$

We now estimate I_2. By the inequality $\kappa(\eta) > \sqrt{B'}\kappa(x)$,

$$I_2 \leq \frac{A_2}{\kappa^{2(2-a)}(x)} \int_{|x-\eta|\leq 1} e^{-B_0\kappa(x)|x-\eta|} e^{-(B-\delta)\sqrt{B'}\kappa(x)|\eta-\xi|} \, d\eta.$$

Estimating I_2 as I_1, we obtain

$$I_2 \leq \frac{C_1''}{\kappa^{2(2-a)+1}(x)} e^{-B_0'\kappa(x)|x-\xi|}. \quad (2.23)$$

Finally, as in estimating the integral l_0'' in (2.13), we have

$$I_3 \leq \frac{C_1'''}{\kappa^r(x)} e^{-B_0'|x-\xi|}.$$

Combining (2.22), (2.23) with the last estimate, we obtain

$$\|l_1'\| \leq \frac{C_1}{\kappa^{3-a}(x)}e^{-B_0'\kappa(x)|x-\xi|} + \frac{C_1'''}{\kappa^r(x)}e^{-B_0'|x-\xi|}, \tag{2.24}$$

also relying on (2.21).

To estimate the first iteration $l_1(x,\xi;\mu)$, it remains to estimate $\|l_1''\|$ in accordance with the notation (2.20). Since $\kappa(x) > 1$, it easily follows from (2.1), (2.19), (1.13) and the definition that

$$\|l_1''\| \leq \frac{C_1^{IV}}{\kappa^r(x)}e^{-B_0'|x-\xi|},$$

which, combined with the inequality (2.24), gives

$$\|l_1(x,\xi;\mu)\| \leq \frac{C_1}{\kappa^{3-a}(x)}e^{-B_0'\kappa(x)|x-\xi|} + \frac{C_1'}{\kappa^r(x)}e^{-B_0'|x-\xi|} \tag{2.25}$$

for the first iteration.

To estimate the subsequent iterations $l_k(x,\xi;\mu)$, $k \geq 2$, we notice that the recurrence relations (2.7) for them coincide with (1.8) for $g_k(x,\xi;i\mu)$, $k \geq 1$. Therefore, we can use the estimates (1.26).

It now follows from (1.8), (2.7) that if, in estimating $\|g_1(x,\xi;i\mu)\|$, we proceeded from (1.2) for $\|g_0(x,\xi;i\mu)\|$, then, in estimating $\|l_2(x,\xi;\mu)\|$, we have to proceed from (2.25) for $\|l_1(x,\xi;\mu)\|$. Therefore, applying the same method as in obtaining the inequality (1.25), i.e., estimating $\|g_2(x,\xi;i\mu)\|$, we have

$$\|l_2(x,\xi;\mu)\| \leq \frac{C_2}{\kappa^{2(2-a)+1}}e^{-(B-\delta)\kappa(x)|x-\xi|} + \frac{C_2'}{\kappa^r(x)}e^{-(B-\delta)|x-\xi|},$$

where the number B is obtained from B_0' as B_1 in (1.25) from B_0.

On the analogy with the kth iteration $l_k(x,\xi;\mu)$, we obtain

$$\|l_k(x,\xi;\mu)\| \leq \frac{C_k}{\kappa^{k(2-a)+1}}\exp\{-[B^{k+1} - (B^{k+2} + \ldots + 1)\delta]\kappa(x)|x-\xi|\}$$

$$+ \frac{C_k'}{\kappa^r(x)}\exp\{-[B^{k-1} - (B^{k-2} + \ldots + 1)\delta]|x-\xi|\}.$$

We now select k_0 to be so large that $k_0(2-a)+1 > r$. Furthermore, reasoning precisely as in obtaining (1.40) and bearing in mind the equality (1.9), or

$$G(x,\xi;i\mu) = g_0(x,\xi;i\mu) + g(x,\xi;i\mu),$$

we obtain the important estimate

$$\left\|\frac{\partial}{\partial\mu}G(x,\xi;i\mu) - \frac{\partial}{\partial\mu}g_0(x,\xi;i\mu)\right\| \leq \left\|\frac{\partial}{\partial\mu}g(x,\xi;i\mu)\right\|$$

$$\leq \frac{C_*}{\kappa^{3-a}(x)}e^{-B_*\kappa(x)|x-\xi|} + \frac{C_*'}{\kappa^r(x)}e^{-B_*|x-\xi|}. \tag{2.26}$$

We have thus proved the following.

Theorem 11.2.1 *Let the matrix $Q(x)$ fulfil Conditions 1 to 3 of the preceding section. Then, for the derivative with respect to μ, $G'_\mu(x, \xi; i\mu)$, of the solution $G(x, \xi; \lambda)$ of the integral equation (1.1), as $\mu \to \infty$, the estimate (2.26) holds uniformly with respect to x, ξ on the whole number axis $(-\infty, \infty)$.*

The relation (2.26) enables us to prove the following important proposition.

Lemma 11.2.1 *(Fundamental Lemma.) Let the matrix $Q(x)$ fulfil Conditions 1 to 3 of the preceding section, and suppose that*

5. *for large $|x|$,*
$$cx^\alpha < p^2(x) + q^2(x) < Cx^\alpha,$$
where $\alpha > 2$, $c > 0$ and $C > 0$ are three constant numbers.

Then, for $z = i\mu$ as $\mu \to \infty$, the asymptotic formula
$$\int_{-\infty}^{\infty} \text{trace} \frac{\partial}{\partial \mu} G(x, x; z)\, dx \sim \int_{-\infty}^{\infty} \frac{p^2(x) + q^2(x)}{\{p^2(x) + q^2(x) - z^2\}^{3/2}}\, dx \qquad (2.27)$$
is valid.

Proof. Because of the equality (1.9), we have
$$G(x, \xi; i\mu) = g_0(x, \xi; i\mu) + g(x, \xi; i\mu).$$

Therefore,
$$\text{trace} \frac{\partial}{\partial \mu} G(x, \xi; i\mu) = \text{trace} \frac{\partial}{\partial \mu} g_0(x, \xi; i\mu) + \text{trace} \frac{\partial}{\partial \mu} g(x, \xi; i\mu).$$

Putting $\xi = x$ and integrating with respect to x, we obtain
$$\int_{-\infty}^{\infty} \text{trace} \frac{\partial}{\partial \mu} G(x, x; i\mu)\, dx - \int_{-\infty}^{\infty} \text{trace} \frac{\partial}{\partial \mu} g_0(x, x; i\mu)\, dx$$
$$= \int_{-\infty}^{\infty} \text{trace} \frac{\partial}{\partial \mu} g(x, x; i\mu)\, dx. \quad (2.28)$$

Let $g(x, x; i\mu) = \begin{pmatrix} g_{11}(x, \mu) & g_{12}(x, \mu) \\ g_{21}(x, \mu) & g_{22}(x, \mu) \end{pmatrix}$. Then

$$\left| \int_{-\infty}^{\infty} \text{trace} \frac{\partial}{\partial \mu} g(x, x; i\mu)\, dx \right| = \left| \int_{-\infty}^{\infty} \frac{\partial}{\partial \mu} \{g_{11} + g_{22}\}(x, \mu)\, dx \right|$$
$$\leq \int_{-\infty}^{\infty} \left\{ \left| \frac{\partial}{\partial \mu} g_{11} \right| + \left| \frac{\partial}{\partial \mu} g_{22} \right| \right\} (x, \mu)\, dx \leq \int_{-\infty}^{\infty} \left\| \frac{\partial}{\partial \mu} g(x, x; i\mu) \right\|\, dx,$$

which implies, together with (2.28), that

$$\left| \int_{-\infty}^{\infty} \text{trace}\, \frac{\partial G}{\partial \mu}(x, x; i\mu)\, dx - \int_{-\infty}^{\infty} \text{trace}\, \frac{\partial g_0}{\partial \mu}(x, x; i\mu)\, dx \right|$$

$$\leq \int_{-\infty}^{\infty} \| g'_\mu(x, x; i\mu) \|\, dx. \quad (2.29)$$

According to the equality (1.3),

$$\int_{-\infty}^{\infty} \text{trace}\, \frac{\partial g_0}{\partial \mu}(x, x; i\mu)\, dx = \int_{-\infty}^{\infty} \frac{p^2(x) + q^2(x)}{\{p^2(x) + q^2(x) + \mu^2\}^{3/2}}\, dx. \quad (2.30)$$

On the other hand, because of the estimate (2.26),

$$\left\| \frac{\partial}{\partial \mu} g(x, x; i\mu) \right\| \leq \frac{C_1}{\kappa^{3-a}(x)} + \frac{C_2}{\kappa^r(x)},$$

or, taking into account the expression for $\kappa(x) = \{p^2(x) + q^2(x) + \mu^2\}^{1/2}$,

$$\left\| \frac{\partial g}{\partial \mu}(x, x; i\mu) \right\| \leq \frac{C_1}{\{p^2(x) + q^2(x) + \mu^2\}^{(3-a)/2}} + \frac{C_2}{\{p^2(x) + q^2(x) + \mu^2\}^{r/2}}.$$

Since $0 < a < 2$ and $r > 0$ is arbitrary (given), it follows from the last inequality that

$$\left\| \frac{\partial}{\partial \mu} g(x, x; i\mu) \right\| \leq \frac{C}{\{p^2(x) + q^2(x) + \mu^2\}^{(3-a)/2}} \quad (2.31)$$

with the consequence, because of (2.29), (2.30), that

$$\left| \int_{-\infty}^{\infty} \text{trace}\, \frac{\partial}{\partial \mu} G(x, x; i\mu)\, dx - \int_{-\infty}^{\infty} \frac{p^2(x) + q^2(x)}{\{p^2(x) + q^2(x) + \mu^2\}^{3/2}}\, dx \right|$$

$$\leq C \int_{-\infty}^{\infty} \frac{dx}{\{p^2(x) + q^2(x) + \mu^2\}^{(3-a)/2}}. \quad (2.32)$$

We put

$$M(\mu) = \int_{-\infty}^{\infty} \frac{p^2(x) + q^2(x)}{\{p^2(x) + q^2(x) + \mu^2\}^{3/2}}\, dx,$$

$$N(\mu) = \int_{-\infty}^{\infty} \frac{dx}{\{p^2(x) + q^2(x) + \mu^2\}^{(3-a)/2}},$$

the integrals converging because of Condition 5.

It follows from (2.32) that, to prove the lemma, it suffices to show the validity of the relation

$$N(\mu)/M(\mu) = O(1/\mu^\delta), \quad \delta > 0, \quad (2.33)$$

as $\mu \to \infty$. By Condition 5, we have

$$N(\mu) \le 2 \int_0^\infty \frac{dx}{\{cx^\alpha + \mu^2\}^{(3-a)/2}},$$

and we obtain that

$$N(\mu) \le \frac{4}{\alpha \sqrt[\alpha]{c}\mu^{3-a-2/\alpha}} \int_0^\infty \frac{t^{(2-a)/\alpha}\, dt}{(1+t^2)^{(3-a)/2}} = \frac{C_1}{\mu^{3-a-2/\alpha}} \qquad (2.34)$$

by substituting $\mu^2 t^2$ for cx^α. Proceeding similarly, we get

$$M(\mu) \ge 2 \int_0^\infty \frac{cx^\alpha}{(Cx^\alpha + \mu^2)^{3/2}}\, dx = \frac{C_2}{\mu^{1-2/\alpha}}. \qquad (2.35)$$

The last two inequalities imply that

$$\frac{N(\mu)}{M(\mu)} = O\left(\frac{1}{\mu^{2-a}}\right),$$

which exactly proves (2.33), since $0 < a < 2$.   ∎

11.3   Other properties of the matrix $G(x, \xi; \lambda)$

Our goal is to prove that the solution $G(x, \xi; \lambda)$ of the integral equation (2.1), i.e.,

$$G(x, \xi; \lambda) = g_0(x, \xi; \lambda) + \int_{-\infty}^\infty G(\eta, \xi; \lambda)\{Q(x) - Q(\eta)\}g_0(x, \eta; \lambda)\, d\eta, \qquad (3.1)$$

is Green's matrix function of the operator

$$Ly \equiv B\frac{dy}{dx} + Q(x)y, \quad -\infty < x < \infty, \qquad (3.2)$$

where

$$B = \begin{pmatrix} 0 & 1 \\ -1 & 0 \end{pmatrix}, \qquad Q(x) = \begin{pmatrix} p(x) & q(x) \\ q(x) & -p(x) \end{pmatrix}. \qquad (3.2')$$

The proof is contained in the following three lemmas, assuming that the matrix $Q(x)$ satisfies Conditions 1 to 4 of Section 1.

We need the following in the sequel.

Theorem 11.3.1 *Let X be the Banach space of matrix functions $A(x, \xi)$, $-\infty < x, \xi < \infty$, with norm*

$$\|A(x, \xi)\|_X = \sup_{-\infty < x < \infty} \int_{-\infty}^\infty \|A(x, \xi)\|\, d\xi,$$

where $||A(x,\xi)||$ is the Hilbert-Schmidt norm of the matrix $A(x,\xi)$. Consider the operator N in X, determined by

$$NA(x,\xi) = \int_{-\infty}^{\infty} A(\eta,\xi)\{Q(x) - Q(\eta)\}g_0(x,\eta;i\mu)\,d\eta, \qquad (3.3)$$

where the matrix $g_0(x,\eta;\lambda)$ has the same value as above. If $Q(x)$ fulfils Conditions 1 to 3 of Section 1, then, for sufficiently large μ, N is contracting in X, i.e., $||N|| < 1$.

Proof. Precisely as Lemma 4.1.1.   ∎

Remark Consider the integral equation (3.1) again, rewriting it now as

$$G(x,\xi;\lambda) = g_0(x,\xi;\lambda) + NG(x,\xi;\lambda) \qquad (3.4)$$

in accordance with (3.3).

Since, by Theorem 11.3.1, for sufficiently large μ ($\lambda = i\mu$), N is contracting in X, (3.4) has a unique solution in X (obtainable by the iteration method) only if the matrix function $g_0(x,\xi;\lambda)$ belongs to X.

We now prove the main properties of $G(x,\xi;\lambda)$.

Lemma 11.3.1 *For all $\xi \neq x$, the matrix $G(x,\xi;\lambda)$ satisfies the equation*

$$B\frac{\partial}{\partial\xi}G(x,\xi;\lambda) = \{\lambda I - Q(\xi)\}G(x,\xi;\lambda), \qquad (3.5)$$

where I is the identity matrix of order two.

Proof. We rewrite the integral equation (3.1) in the form

$$G(x,\xi;\lambda) - g_0(x,\xi;\lambda) = \int_{-\infty}^{\infty} g_0(\eta,\xi;\lambda)\{Q(x) - Q(\eta)\}g_0(x,\eta;\lambda)\,d\eta$$

$$+ \int_{-\infty}^{\infty} \{G(\eta,\xi;\lambda) - g_0(\eta,\xi;\lambda)\}\{Q(x) - Q(\eta)\}g_0(x,\eta;\lambda)\,d\eta, \quad (3.6)$$

and introduce the notation

$$\mathcal{L}(x,\xi;\lambda) = G(x,\xi;\lambda) - g_0(x,\xi;\lambda) \qquad (3.7)$$

$$l(x,\xi;\lambda) = \int_{-\infty}^{\infty} g_0(\eta,\xi;\lambda)\{Q(x) - Q(\eta)\}g_0(x,\eta;\lambda)d\eta. \qquad (3.8)$$

The equation (3.6) is then written as

$$\mathcal{L}(x,\xi;\lambda) = l(x,\xi;\lambda) + \int_{-\infty}^{\infty} \mathcal{L}(\eta,\xi;\lambda)\{Q(x) - Q(\eta)\}g_0(x,\eta;\lambda)\,d\eta. \qquad (3.9)$$

To calculate $\frac{\partial}{\partial \xi} l(x, \xi; \lambda)$, we notice that the matrix $g_0(x, \xi; \lambda)$ has a discontinuity when $\xi = x$; furthermore,

$$g_0(x, \xi; \lambda)|_{\xi=x+0} - g_0(x, \xi; \lambda)|_{\xi=x-0} = -B^{-1} = B, \tag{3.10}$$

the matrix B being the same as in (3.5). Therefore, differentiating the equality (3.8) with respect to ξ and taking into account (3.10), we obtain

$$\frac{\partial}{\partial \xi} l(x, \xi; \lambda) = B^{-1}\{Q(x) - Q(\xi)\}g_0(x, \xi; \lambda)$$
$$+ \int_{-\infty}^{\infty} \frac{\partial g_0(\eta, \xi; \lambda)}{\partial \xi}\{Q(x) - Q(\eta)\}g_0(x, \eta; \lambda)\, d\eta. \tag{3.11}$$

We now prove that $\frac{\partial}{\partial \xi} l(x, \xi; \lambda) \in X$, or

$$\sup_{-\infty < x < \infty} \int_{-\infty}^{\infty} \left\| \frac{\partial l}{\partial \xi}(x, \xi; \lambda) \right\| \, d\xi < \infty. \tag{3.12}$$

Since $g_0(x, \xi; \lambda)$ is Green's matrix function of the equation (1.0) with 'frozen' coefficients (by definition),

$$B\frac{\partial}{\partial \xi}g_0(x, \xi; \lambda) + Q(x)g_0(x, \xi; \lambda) = \lambda g_0(x, \xi; \lambda) \tag{3.13}$$

for $\xi \neq x$; hence,

$$\frac{\partial}{\partial \xi}g_0(x, \xi; \lambda) = B^{-1}\{\lambda I - Q(x)\}g_0(x, \xi; \lambda),$$

which allows (3.11) to be rewritten in the form

$$\frac{\partial}{\partial \xi} l(x, \xi; \lambda) = B^{-1}\{Q(x) - Q(\xi)\}g_0(x, \xi; \lambda)$$
$$+ B^{-1} \int_{-\infty}^{\infty} \{\lambda I - Q(\eta)\}g_0(\eta, \xi; \lambda)\{Q(x) - Q(\eta)\}g_0(x, \eta; \lambda)\, d\eta.$$

Taking the norms, we derive

$$\left\| \frac{\partial}{\partial \xi} l(x, \xi; \lambda) \right\| \leq \sqrt{2}\|\{Q(x) - Q(\xi)\}g_0(x, \xi, \lambda)\|$$
$$+ \sqrt{2} \int_{-\infty}^{\infty} \|\{\lambda I - Q(\eta)\}g_0\| \cdot \|\{Q(x) - Q(\eta)\}g_0\| \, d\eta.$$

Now, using the estimate (1.13) and proceeding as in Section 1, we obtain the estimate

$$\left\| \frac{\partial}{\partial \xi} l(x, \xi; \lambda) \right\| \leq \frac{C_1}{\kappa^{1-a}(x)}e^{-B_0\kappa(x)|x-\xi|} + \frac{C_2}{\kappa^r(x)}e^{-B_0|x-\xi|}.$$

Therefore,

$$\int_{-\infty}^{\infty} \left\| \frac{\partial}{\partial \xi} l(x, \xi; \lambda) \right\| d\xi \leq \frac{C_1'}{\kappa^{1-a}(x)} + \frac{C_2'}{\kappa^{r}(x)},$$

which implies (3.12). Thus, $l_{\xi}'(x, \xi; \lambda) \in X$.

Consider the integral equation

$$M(x, \xi; \lambda) = l_{\xi}'(x, \xi; \lambda) + \int_{-\infty}^{\infty} M(\eta, \xi; \lambda)\{Q(x) - Q(\eta)\}g_0(x, \eta; \lambda)\, d\eta. \qquad (3.14)$$

Since the operator

$$NM(x, \xi; \lambda) = \int_{-\infty}^{\infty} M(\eta, \xi; \lambda)\{Q(x) - Q(\eta)\}g_0(x, \eta; \lambda)\, d\eta$$

by Theorem 11.3.1 is contracting for sufficiently large μ ($\lambda = i\mu$), and $l_{\xi}'(x, \xi; \lambda) \in X$ (proved), the Remark after Theorem 11.3.1 implies that the equation (3.14) has a unique solution belonging to the space X for sufficiently large μ.

Thus, the solution of (3.14), $M(x, \xi; \lambda)$, exists and belongs to X.

Furthermore, it follows from the explicit form of the matrix $g_0(x, \xi; \lambda)$ that it tends to zero as $\xi \to -\infty$. It then follows from the definition of the matrix $l(x, \xi; \lambda)$ that it also tends to zero as $\xi \to -\infty$. Therefore, integrating (3.14) with respect to ξ from $-\infty$ to ξ, we have

$$\int_{-\infty}^{\xi} M(x, t; \lambda)\, dt = l(x, \xi; \lambda) + \int_{-\infty}^{\infty} \left\{ \int_{-\infty}^{\xi} M(\eta, t; \lambda)\, dt \right\} \{Q(x) - Q(\eta)\}g_0(x, \eta; \lambda)\, d\eta,$$

which, compared with (3.9) and using the uniqueness of their solutions, leads to

$$\mathcal{L}(x, \xi; \lambda) \equiv \int_{-\infty}^{\xi} M(x, t; \lambda)\, dt.$$

Therefore, the matrix $\mathcal{L}(x, \xi; \lambda)$ is differentiable with respect to all ξ, $\mathcal{L}_{\xi}'(x, \xi; \lambda) = M(x, \xi; \lambda)$, and the matrix $M(x, \xi; \lambda)$ in (3.14) can be replaced by $\mathcal{L}_{\xi}'(x, \xi; \lambda)$. Substituting $l_{\xi}'(x, \xi; \lambda)$ from (3.11) in (3.14) and taking into account (3.7), (3.8), we then arrive at the equation

$$G_{\xi}'(x, \xi; \lambda) = g_{0\xi}'(x, \xi; \lambda) + B^{-1}\{Q(x) - Q(\xi)\}g_0(x, \xi; \lambda)$$
$$+ \int_{-\infty}^{\infty} G_{\xi}'(\eta, \xi; \lambda)\{Q(x) - Q(\eta)\}g_0(x, \eta; \lambda)\, d\eta,$$

which, pre-multiplied throughout by B, relying on (3.13), yields

$$BG_{\xi}'(x, \xi; \lambda) = \{\lambda I - Q(\xi)\}g_0(x, \xi; \lambda)$$
$$+ \int_{-\infty}^{\infty} BG_{\xi}'(\eta, \xi; \lambda)\{Q(x) - Q(\eta)\}g_0(x, \eta; \lambda)\, d\eta,$$

i.e.,

$$\{\lambda I - Q(\xi)\}^{-1} B G'_\xi(x, \xi; \lambda) = g_0(x, \xi; \lambda)$$
$$+ \int_{-\infty}^\infty \{\lambda I - Q(\xi)\}^{-1} B G'_\xi(\eta, \xi; \lambda) \{Q(x) - Q(\eta)\} g_0(x, \eta; \lambda) \, d\eta.$$

Comparing the last equation with (3.1), by the uniqueness of their solution, we conclude that

$$\{\lambda I - Q(\xi)\}^{-1} B G'_\xi(x, \xi; \lambda) \equiv G(x, \xi; \lambda),$$

i.e., the equation (3.5). The lemma is thus proved.   ∎

Lemma 11.3.2 *At* $\xi = x$, *the matrix* $G(x, \xi; \lambda)$ *has a discontinuity and*

$$G(x, \xi; \lambda)|_{\xi=x+0} - G(x, \xi; \lambda)|_{\xi=x-0} = -B^{-1} = B.$$

Proof. In fact, since the matrix

$$\mathcal{L}(x, \xi; \lambda) = G(x, \xi; \lambda) - g_0(x, \xi; \lambda)$$

was proved above to be differentiable with respect to all ξ, it is continuous everywhere relative to ξ. Therefore, $G(x, \xi; \lambda)$, $g_0(x, \xi; \lambda)$ have the same discontinuity with respect to ξ, and the stated lemma follows from the equality (3.10).   ∎

Lemma 11.3.3 *The matrix* $G(x, \xi; \lambda)$ *is such that*

$$G(x, y; \lambda) = G^*(y, x; \bar{\lambda}),$$

where G^* *is the conjugate matrix.*

Proof. By (3.5), we have

$$B G'_\xi(y, \xi; \bar{\lambda}) = \{\bar{\lambda} I - Q(\xi)\} G(y, \xi; \bar{\lambda}).$$

Taking the conjugate matrices and recalling that $(AB)^* = B^* A^*$, we obtain

$$G_\xi^{*'}(y, \xi; \bar{\lambda}) B^* = G^*(y, \xi; \bar{\lambda}) \{\bar{\lambda} I - Q(\xi)\}.$$

Since $B^* = -B$, the equation takes the form

$$-G_\xi^{*'}(y, \xi; \bar{\lambda}) B = G^*(y, \xi; \bar{\lambda}) \{\lambda I - Q(\xi)\},$$

which, post-multiplied throughout by $G(x, \xi; \lambda)$, and subtracted from the equation (3.5) after pre-multiplication by $G^*(y, \xi; \bar{\lambda})$, yields

$$G_\xi^{*'}(y, \xi; \bar{\lambda}) B G(x, \xi; \lambda) + G^*(y, \xi; \bar{\lambda}) B G'_\xi(x, \xi; \lambda) = 0,$$

i.e., $\dfrac{\partial}{\partial \xi}\{G^*(y,\xi,\bar{\lambda})BG(x,\xi;\lambda)\} = 0.$

Let $x < y$. Integrating the last equality with respect to ξ from $-\infty$ to ∞ and then breaking the interval of integration into three parts, $-\infty < \xi < x$, $x \le \xi \le y$, $y \le \xi < \infty$, we get

$$0 = \int_{-\infty}^{x} \frac{\partial}{\partial \xi}\{G^*(y,\xi;\bar{\lambda})BG(x,\xi;\lambda)\}\,d\xi + \int_{x}^{y} \frac{\partial}{\partial \xi}\{G^*(y,\xi;\bar{\lambda})BG(x,\xi;\lambda)\}\,d\xi$$
$$+ \int_{y}^{\infty} \frac{\partial}{\partial \xi}\{G^*(y,\xi;\bar{\lambda})BG(x,\xi;\lambda)\}\,d\xi \equiv I_1 + I_2 + I_3. \ (3.15)$$

We have already noted above that the convergence $g(x,\xi;\lambda) \to 0$ as $|\xi| \to \infty$ follows from the explicit form of the matrix. The equation (1.1) then implies the same property for the matrix $G(x,\xi;\lambda)$. Hence,

$$I_1 = G^*(y,\xi;\bar{\lambda})BG(x,\xi;\lambda)|_{\xi=x-0} = G^*(y,x;\bar{\lambda})BG(x,x-0;\lambda)$$

$$I_2 = G^*(y,y-0;\bar{\lambda})BG(x,y;\lambda) - G^*(y,x,\bar{\lambda})BG(x,x+0;\lambda)$$

$$I_3 = -G^*(y,\xi;\bar{\lambda})BG(x,\xi;\lambda)|_{\xi=y+0} = -G^*(y,y+0,\bar{\lambda})BG(x,y;\lambda).$$

Adding these equalities together and taking into account (3.15), we obtain

$$G^*(y,x;\bar{\lambda})B\{G(x,x-0;\lambda) - G(x,x+0;\lambda)\}$$
$$+\{G^*(y,y-0,\bar{\lambda}) - G^*(y,y+0;\bar{\lambda})\}BG(x,y;\lambda) = 0. \ (3.16)$$

By Lemma 11.3.2, i.e., the equality

$$G(x,x-0;\lambda) - G(x,x+0;\lambda) = +B^{-1}$$

$$G^*(y,y-0;\bar{\lambda}) - G^*(y,y+0;\bar{\lambda}) = -B^{-1}$$

we derive from (3.16) that

$$G^*(y,x;\bar{\lambda})BB^{-1} - B^{-1}BG(x,y;\lambda) = 0,$$

i.e., $G(x,y;\lambda) = G(y,x;\bar{\lambda})$ as required.

Lemmas 11.3.1, 11.3.2, 11.3.3 show that $G(x,\xi;\lambda)$ fulfils all the conditions to be Green's matrix function of the operator

$$Ly \equiv B\frac{dy}{dx} + Q(x)y, \quad -\infty < x < \infty.$$

It follows from the uniqueness of this matrix that $G(x,\xi;\lambda)$, the solution of the integral equation (3.1), is Green's matrix function of L. ∎

Theorem 11.3.2 *(Spectrum Discreteness Theorem.) If the matrix $Q(x)$ fulfils Conditions 1 to 4 of Section 1, then the spectrum of L is discrete.*

Proof. It obviously suffices to show that Green's matrix $G(x, \xi; \lambda)$ of L is a Hilbert-Schmidt kernel, i.e., that the integral

$$\int_{-\infty}^{\infty}\int_{-\infty}^{\infty} \|G(x, \xi; i\mu)\|^2 \, dx \, d\xi < \infty \qquad (3.17)$$

converges. By the estimate (1.40), we have

$$\int_{-\infty}^{\infty}\int_{-\infty}^{\infty} \|G(x, \xi; i\mu)\|^2 \, dx \, d\xi$$

$$\leq \int_{-\infty}^{\infty}\int_{-\infty}^{\infty} \left\{ e^{-\kappa(x)|x-\xi|} + \frac{C_1}{\kappa^{2-a}(x)} e^{-B_0\kappa(x)|x-\xi|} + \frac{C_2}{\kappa^r(x)} e^{-B_0|x-\xi|} \right\}^2 dx \, d\xi$$

as $\mu \to \infty$, or, removing the braces,

$$\int_{-\infty}^{\infty}\int_{-\infty}^{\infty} \|G(x, \xi; i\mu)\|^2 \, dx \, d\xi \leq \int_{-\infty}^{\infty}\int_{-\infty}^{\infty} e^{-2\kappa(x)|x-\xi|} \, dx \, d\xi$$

$$+C_1^2 \int_{-\infty}^{\infty}\int_{-\infty}^{\infty} \frac{e^{-2B_0\kappa(x)|x-\xi|}}{\kappa^{2(2-a)}(x)} \, dx \, d\xi + C_2^2 \int_{-\infty}^{\infty}\int_{-\infty}^{\infty} \frac{e^{-2B_0|x-\xi|}}{\kappa^{2r}(x)} \, dx \, d\xi$$

$$+2C_1 \int_{-\infty}^{\infty}\int_{-\infty}^{\infty} \frac{e^{-(1+B_0)\kappa(x)|x-\xi|}}{\kappa^{2-a}(x)} \, dx \, d\xi + 2C_2 \int_{-\infty}^{\infty}\int_{-\infty}^{\infty} \frac{e^{-[B_0+\kappa(x)]|x-\xi|}}{\kappa^r(x)} \, dx \, d\xi$$

$$+2C_1 C_2 \int_{-\infty}^{\infty}\int_{-\infty}^{\infty} \frac{e^{-B_0[1+\kappa(x)]|x-\xi|}}{\kappa^{2+r-a}(x)} \, dx \, d\xi \equiv \sum_{k=1}^{6} I_k. \quad (3.18)$$

To prove (3.17), it suffices to show that each integral in (3.18) converges. They are all of the same kind; therefore, we need only show that one of them converges. Consider the first. We have

$$I_1 = \int_{-\infty}^{\infty} dx \int_{-\infty}^{\infty} e^{-2\kappa(x)|x-\xi|} \, d\xi = \int_{-\infty}^{\infty} \frac{dx}{\kappa(x)} = \int_{-\infty}^{\infty} \frac{dx}{\{p^2(x) + q^2(x) + \mu^2\}^{1/2}} < +\infty,$$

where the last converges, relying on Condition 4. ∎

11.4 Derivation of the bilateral asymptotic formula

11.4.1

Here, we prove the basic theorem of the chapter, yielding the bilateral asymptotic formula for the eigenvalue distribution of the operator

$$Ly \equiv B\frac{dy}{dx} + Q(x)y, \quad -\infty < x < \infty.$$

We first prove a number of lemmas.

Lemma 11.4.1 *Let the matrix $Q(x)$ fulfil Condition 4 of Section 1, i.e., at least one of the integrals*

$$\int_{-\infty}^{\infty} |p(x)|^{-1}\, dx, \qquad \int_{-\infty}^{\infty} |q(x)|^{-1}\, dx$$

converges. Put

$$\psi(\lambda) = \frac{1}{\pi} \int_{p^2(x)+q^2(x)<\lambda^2} \{\lambda^2 - p^2(x) - q^2(x)\}^{1/2}\, dx,$$

and $\psi(-\lambda) = -\psi(\lambda)$ for $\lambda > 0$. Then the equality

$$\int_{-\infty}^{\infty} \frac{d\psi(\lambda)}{(\lambda-z)^2} = \int_{-\infty}^{\infty} \frac{p^2(x)+q^2(x)}{\{p^2(x)+q^2(x)-z^2\}^{3/2}}\, dx \qquad (4.1)$$

holds for $z = i\mu$.

Proof. Put $\sigma(\nu) = \mathrm{mes}\,\{p^2(x)+q^2(x)<\nu\}$. It is obvious that, for $\lambda > 0$,

$$\psi(\pm\lambda) = \pm\frac{1}{\pi} \int_0^{\lambda^2} (\lambda^2-\nu^2)^{1/2}\, d\sigma(\nu). \qquad (4.2)$$

Consider the left-hand side of (4.1). We have

$$\int_{-\infty}^{\infty} \frac{d\psi(\lambda)}{(\lambda-z)^2} = \int_{-\infty}^{0} \frac{d\psi(\lambda)}{(\lambda-z)^2} + \int_0^{\infty} \frac{d\psi(\lambda)}{(\lambda-z)^2}.$$

Now, calculating the value $d\psi(\lambda)$ (on the basis of (4.2)), and substituting in the integrals on the right-hand side of the last equality, we have

$$\int_{-\infty}^{\infty} \frac{d\psi(\lambda)}{(\lambda-z)^2} = -\frac{1}{\pi} \int_{-\infty}^{0} \frac{\lambda\, d\lambda}{(\lambda-z)^2} \int_0^{\lambda^2} \frac{d\sigma(\nu)}{(\lambda^2-\nu)^{1/2}} + \frac{1}{\pi} \int_0^{\infty} \frac{\lambda\, d\lambda}{(\lambda-z)^2} \int_0^{\lambda^2} \frac{d\sigma(\nu)}{(\lambda^2-\nu)^{1/2}}$$

$$= \frac{1}{\pi} \int_0^{\infty} \frac{\lambda\, d\lambda}{(\lambda+z)^2} \int_0^{\lambda^2} \frac{d\sigma(\lambda)}{(\lambda-\nu)^{1/2}} + \frac{1}{\pi} \int_0^{\infty} \frac{\lambda\, d\lambda}{(\lambda-z)^2} \int_0^{\lambda^2} \frac{d\sigma(\nu)}{(\lambda-\nu)^{1/2}}.$$

Changing the order of integration in the integrals on the right-hand side and adding them together, we obtain

$$\int_{-\infty}^{\infty} \frac{d\psi(\lambda)}{(\lambda-z)^2} = \frac{1}{\pi} \int_0^{\infty} d\sigma(\nu) \int_{\sqrt{\nu}}^{\infty} \frac{2\lambda(\lambda^2+z^2)}{(\lambda^2-z^2)^2(\lambda^2-z)^{1/2}}\, d\lambda. \qquad (4.3)$$

To calculate the inner integral, we put $z = i\mu$ and make the substitution $\lambda = (\nu+t^2)^{1/2}$. We have

$$2\int_{\sqrt{\nu}}^{\infty} \frac{\lambda(\lambda^2+z^2)}{(\lambda^2-z^2)^2(\lambda^2-z)^{1/2}}\, d\lambda = 2\int_0^{\infty} \frac{t^2+\nu-\mu^2}{(t^2+\nu+\mu^2)^2}\, dt.$$

Direct integration yields

$$2 \int_0^\infty \frac{t^2 + \nu - \mu^2}{(t^2 + \nu + \mu^2)^2}\, dt = \pi \frac{\nu}{(\nu + \mu^2)^{3/2}}.$$

Therefore, the integral (4.3) takes the form

$$\int_{-\infty}^\infty \frac{d\psi(\lambda)}{(\lambda - z)^2} = \int_0^\infty \frac{\nu\, d\sigma(\nu)}{(\nu - z^2)^{3/2}}, \qquad z = i\mu. \tag{4.4}$$

Furthermore, since $\sigma(\nu) = \mathrm{mes}\ \{p^2(x) + q^2(x) < \nu\}$,

$$\int_{-\infty}^\infty \frac{p^2(x) + q^2(x)}{\{p^2(x) + q^2(x) - z\}^{3/2}}\, dx = \int_0^\infty \frac{\nu\, d\sigma(\nu)}{(\nu - z^2)^{3/2}}. \tag{4.4$'$}$$

The last two equalities prove the lemma.   ∎

11.4.2

Let the matrix $Q(x)$ fulfil Conditions 1 to 4 of Section 1. Then, by Theorem 11.3.2, the operator

$$Ly \equiv B\frac{dy}{dx} + Q(x)y, \qquad -\infty < x < \infty,$$

has a discrete spectrum. Denote by $\lambda_0, \lambda_{\pm 1}, \lambda_{\pm 2}, \ldots, \lambda_{\pm n}, \ldots$ its eigenvalues, putting

$$N_+(\lambda) = \sum_{0 \le \lambda_n < \lambda} 1, \qquad N_-(\lambda) = \sum_{\lambda < \lambda_n < 0} 1.$$

Lemma 11.4.2 *Let the matrix $Q(x)$ fulfil Conditions 1 to 4. Then, for nonreal z, the equality*

$$\int_{-\infty}^\infty \mathrm{trace}\, \frac{\partial}{\partial z} G(x, x; z)\, dx = \int_{-\infty}^\infty \frac{dN(\lambda)}{(\lambda - z)^2}$$

holds, where

$$N(\lambda) = \begin{cases} N_+(\lambda), & \lambda > 0, \\[2mm] N_-(\lambda), & \lambda < 0. \end{cases}$$

Proof. Let $\varphi_n(x) = \begin{pmatrix} \varphi_{n1}(x) \\ \varphi_{n2}(x) \end{pmatrix}$ be the normalized vector-valued eigenfunction of L, corresponding to the eigenvalue λ_n, i.e.,

$$L\varphi_n(x) \equiv B\frac{d\varphi_n(x)}{dx} + Q(x)\varphi_n(x) = \lambda_n\varphi_n(x),$$

and

$$\int_{-\infty}^\infty \{\varphi_{n1}^2(x) + \varphi_{n2}^2(x)\}\, dx = 1. \tag{4.5}$$

It is then known (see (3.17), Chapter 8) that Green's matrix function of L admits the representation

$$G(x,y;z) = \sum_{n=-\infty}^{\infty} \frac{\varphi_n(x)\varphi_n^T(y)}{\lambda_n - z}, \tag{4.6}$$

(for nonreal z), where

$$\varphi_n(x)\varphi_n^T(y) = \begin{pmatrix} \varphi_{n1}(x)\varphi_{n1}(y) & \varphi_{n1}(x)\varphi_{n2}(y) \\ \varphi_{n2}(x)\varphi_{n1}(y) & \varphi_{n2}(x)\varphi_{n2}(y) \end{pmatrix}. \tag{4.7}$$

It follows from the equality (4.6) that

$$\frac{\partial}{\partial z}G(x,y;z) = \sum_{n=-\infty}^{\infty} \frac{\varphi_n(x)\varphi_n^T(y)}{(\lambda_n - z)^2}.$$

Therefore, because of

$$\text{trace}\,\frac{\partial}{\partial z}G(x,x;z) = \sum_{n=-\infty}^{\infty} \frac{\varphi_{n1}^2(x) + \varphi_{n2}^2(x)}{(\lambda_n - z)^2},$$

which, integrated with respect to x from $-\infty$ to ∞, relying on (4.5), yields

$$\int_{-\infty}^{\infty} \text{trace}\,\frac{\partial}{\partial z}G(x,x;z)\,dx = \sum_{n=-\infty}^{\infty} \frac{1}{(\lambda_n - z)},$$

and the statement of the lemma follows by the definition of the function $N(\lambda)$. ∎

Basic Lemma 11.2.1 and Lemmas 11.4.1, 11.4.2 imply the following.

Lemma 11.4.3 *Let the matrix $Q(x)$ fulfil the conditions of basic Lemma 11.3.1. Then, as $z = i\mu$ and $\mu \to \infty$, the asymptotic formula*

$$\int_{-\infty}^{\infty} \frac{dN(\lambda)}{(\lambda - z)^2} \sim \int_{-\infty}^{\infty} \frac{d\psi(\lambda)}{(\lambda - z)^2}$$

holds, where the functions $N(\lambda)$, $\psi(\lambda)$ are as above.

We now formulate the basic theorem of the chapter. For convenience and completeness, we specify all the conditions for $Q(x)$, and again define all the functions involved.

Theorem 11.4.1 *Consider the operator*

$$Ly \equiv B\frac{dy}{dx} + Q(x)y = \lambda y$$

$$y(x) = \begin{pmatrix} y_1(x) \\ y_2(x) \end{pmatrix}, \quad -\infty < x < \infty,$$

where $B = \begin{pmatrix} 0 & 1 \\ -1 & 0 \end{pmatrix}$, $Q(x) = \begin{pmatrix} p(x) & q(x) \\ q(x) & -p(x) \end{pmatrix}$. Let the matrix $Q(x)$ fulfil the conditions ($\|\cdot\|$ being the Hilbert-Schmidt norm) :

1. *For $|x - \xi| \leq 1$,*

$$||\{Q(x) - Q(\xi)\}Q^{-a}(x)|| \leq A_1|x - \xi|,$$

   where $A_1 > 0$, $0 < a < 2$ are two constant numbers.

2. *For $|x - \xi| \leq 1$,*

$$||Q(x)|| \leq A_2||Q(\xi)||,$$

   where $A_2 > 0$ is a constant number.

3. *For $|x - \xi| > 1$,*

$$||Q(\xi)|| \leq A_3 \exp\{c_0|x - \xi|\}||Q(x)||,$$

   where $A_3 > 0$, $0 < c_0 < 1$ are two constant numbers.

4. *For sufficiently large $|x|$,*

$$cx^\alpha < ||Q(x)||^2 < Cx^\alpha,$$

   where $c > 0$, $C > 0$ and $\alpha > 2$ are three constant numbers.

The following is then valid.

Proposition 1 *The spectrum of the operator L is discrete.*

Let the eigenvalues of L be $\lambda_0, \lambda_{\pm 1}, \lambda_{\pm 2}, \ldots, \lambda_{\pm n}, \ldots$. Put

$$N(\lambda) = \begin{cases} N_+(\lambda) = \sum_{0 \leq \lambda_n < \lambda} 1, & \lambda > 0 \\ \\ N_-(\lambda) = -\sum_{\lambda < \lambda_n < 0} 1, & \lambda < 0. \end{cases}$$

Also, let

$$\psi(\lambda) = \begin{cases} \dfrac{1}{\pi} \displaystyle\int_{p^2(x)+q^2(x)<\lambda^2} \{\lambda^2 - p^2(x) - q^2(x)\}^{1/2}\, dx, & \lambda > 0 \\ \\ -\dfrac{1}{\pi} \displaystyle\int_{p^2(x)+q^2(x)<\lambda^2} \{\lambda^2 - p^2(x) - q^2(x)\}^{1/2}\, dx & \lambda < 0. \end{cases}$$

Assume that there exist two constants a and b such that, for sufficiently large $|\lambda|$, the Tauberian conditions

$$a\psi(\lambda) < \lambda\psi'(\lambda) < b\psi(\lambda) \tag{4.8}$$

are fulfilled. The following is valid.

Proposition 2 *As $\lambda \to +\infty$ (resp. $\lambda \to -\infty$), the asymptotic formulas*

$$N_+(\lambda), N_-(\lambda) \sim \frac{1}{\pi} \int_{p^2(x)+q^2(x)<\lambda^2} \{\lambda^2 - p^2(x) - q^2(x)\}^{1/2}\, dx \tag{4.9}$$

hold.

Proof. If Conditions 1 to 4 are valid, the first statement follows from Theorem 11.3.2. On the other hand, under the conditions of the theorem, Lemma 11.4.3 holds, according to which the asymptotic formula

$$\int_{-\infty}^{\infty} \frac{dN(\lambda)}{(\lambda - z)^2} \sim \int_{-\infty}^{\infty} \frac{d\psi(\lambda)}{(\lambda - z)^2}$$

is true as $z = i\mu$ and $\mu \to \infty$. Relying on (4.8), the Tauberian theorem of Keldysh type is applicable (see A. G. Kostyuchenko and I. S. Sargsjan (1979; Chapter 10, Theorem 3.1)), according to which $N(\lambda) \sim \psi(\lambda)$ both as $\lambda \to \infty$ and $\lambda \to -\infty$; the second statement is thus proved. ∎

Remark 1 If the operator L is given by

$$Ly \equiv \begin{pmatrix} 0 & 1 \\ -1 & 0 \end{pmatrix} \frac{dy}{dx} + \begin{pmatrix} p(x) & q(x) \\ q(x) & r(x) \end{pmatrix} y, \quad -\infty < x < \infty,$$

then the formulas (4.9) assume the form

$$N_+(\lambda), N_-(\lambda) \sim \frac{1}{\pi} \int_{h(x)<\lambda^2} \{\lambda^2 - h(x)\}^{1/2} \, dx,$$

where $h(x) = \frac{1}{4}\{p(x) - r(x)\}^2 + q^2(x)$.

Remark 2 Let L be given on the half-line $[0, \infty)$ in the form

$$Ly \equiv \begin{pmatrix} 0 & 1 \\ -1 & 0 \end{pmatrix} \frac{dy}{dx} + \begin{pmatrix} p(x) & 0 \\ 0 & q(x) \end{pmatrix} y, \quad y(x) = \begin{pmatrix} y_1(x) \\ y_2(x) \end{pmatrix}$$

$$y_2(0) - hy_1(0) = 0.$$

Then the asymptotic formulas

$$N_+(\lambda), N_-(\lambda) \sim \frac{2}{\pi} \int_{\frac{1}{4}\{p(x)-q(x)\}^2<\lambda^2} \left\{\lambda^2 - \frac{1}{4}\{p(x) - q(x)\}^2\right\}^{1/2} dx \qquad (4.10)$$

hold.

Remark 3 The formulas (4.9) and the similar ones given in Remarks 1 and 2 show that, under Conditions 1 to 4, the positive and negative eigenvalues have the same asymptotic distributions.

Certainly, this is not always so. We give theorems in support of this fact below. Unfortunately, methods for their proof are beyond the scope of this book; we therefore confine ourselves to their formulation, referring the reader to the monograph of A. G. Kostyuchenko and I. S. Sargsjan (1979).

Let $N[a, b; L]$ be the number of points of the spectrum of the operator L in an interval (a, b), the operator L being of the form

$$Ly \equiv \begin{pmatrix} 0 & 1 \\ -1 & 0 \end{pmatrix} \frac{dy}{dx} + \begin{pmatrix} q_1(x) & q_3(x) \\ q_3(x) & q_2(x) \end{pmatrix} y \equiv B\frac{dy}{dx} + Q(x)y, \quad -\infty < x < \infty.$$

Theorem 11.4.2 *Let $\lambda_1(x) \leq \lambda_2(x)$ be the eigenvalues of the matrix $Q(x)$.*

$$M(\lambda) = \frac{1}{\pi} \int_{\lambda_2(x) \leq \lambda} \{[\lambda - \lambda_1(x)][\lambda - \lambda_2(x)]\}^{1/2} dx,$$

and let the following conditions be fulfilled :

1.

$$\sup_{-\infty < x < \infty} \lambda_1(x) \leq c < \infty, \qquad \lim_{|x| \to \infty} \lambda_2(x) = +\infty,$$

2.

$$\sup_{|x-t| \leq 1} \left| \frac{q_i(x) - q_i(t)}{f(|x-t|)} \right| < c\{1 + \min[|\lambda_1(x)|, |\lambda_2(x)|]\},$$

   where $f(x)$ is a continuous function in $(-\infty, \infty)$, $f(0) = 0$ and c is a constant.

3.

$$\lim_{\lambda \to +\infty} M^{-1}(\lambda) \int_{\lambda_2(x) < \lambda} dx = 0,$$

4.

$$\lim_{\delta \to 0} \limsup_{\lambda \to +\infty} \frac{M[\lambda(1+\delta)]}{M(\lambda)} = 1.$$

Then the continuous spectrum of L is bounded from above, and the asymptotic formula

$$N_+(\lambda) \equiv N[b+1, \lambda; L] = M(\lambda)[1 + o(1)]$$

holds as $\lambda \to +\infty$ (b being the supremum of the spectrum), i.e.,

$$N_+(\lambda) \sim \frac{1}{\pi} \int_{\lambda_2 \leq \lambda} \{[\lambda - \lambda_1(x)][\lambda - \lambda_2(x)]\}^{1/2} dx$$

as $\lambda \to \infty$.

It can be seen from this theorem statement that a continuous spectrum is admitted.

Theorem 11.4.3 *With the same values of $\lambda_1(x)$, $\lambda_2(x)$, suppose that*

$$\tilde{M}(\lambda) = \frac{1}{\pi} \int_{\lambda_1(x) \geq \lambda} \{[\lambda - \lambda_1(x)][\lambda - \lambda_2(x)]\}^{1/2} dx$$

and

1.

$$\inf_{-\infty < \lambda < \infty} \lambda_2(x) > -c > -\infty, \qquad \lim_{|x| \to \infty} \lambda_1(x) = -\infty,$$

2.

$$\sup_{|x-t|\leq 1} \left| \frac{q_i(x) - q_i(t)}{f(|x-t|)} \right| \leq c\{1 + \min[|\lambda_1(x)|, |\lambda_2(x)|]\},$$

where $f(x)$ is a continuous function in $(-\infty, \infty)$, $f(0) = 0$ and c is a constant.

3.

$$\lim_{\lambda \to -\infty} M^{-1}(\lambda) \int_{\lambda_1(x) \geq \lambda} dx = 0,$$

4.

$$\lim_{\delta \to 0} \limsup_{\lambda \to -\infty} \frac{\tilde{M}[\lambda(1+\delta)]}{\tilde{M}(\lambda)} = 1.$$

Then the continuous spectrum of the operator L is bounded from below, and the asymptotic formula

$$N_-(\lambda) \equiv N[\lambda, b_1; L] = \tilde{M}(\lambda)\{1 + o(1)\}$$

holds as $\lambda \to -\infty$ (b_1 being a certain number), i.e.,

$$N_-(\lambda) \sim \frac{1}{\pi} \int_{\lambda_1(x) \geq \lambda} \{[\lambda - \lambda_1(x)][\lambda - \lambda_2(x)]\}^{1/2} dx.$$

Notes

The results of this chapter are due to I. S. Sargsjan (1972), the formulas (4.10) to G. V. Dikhamindzhiya (1976), and Theorems 11.4.2, 11.4.3 to M. Otelbayev (1973). The complete proofs of the formulas (4.10) and of Theorems 11.4.2, 11.4.3 are given in the monograph of A. G. Kostyuchenko and I. S. Sargsjan (1979).

Chapter 12

THE INVERSE PROBLEM ON THE HALF-LINE, FROM THE SPECTRAL FUNCTION

12.1 Stating the problem. Auxiliary propositions

12.1.1

Consider the boundary-value problem

$$B\frac{dy}{dx} + Q(x)y = \lambda y, \quad 0 < x < \infty, \quad y = \begin{pmatrix} y_1 \\ y_2 \end{pmatrix} \tag{1.1}$$

$$y_1(0) = 0, \tag{1.2}$$

where

$$B = \begin{pmatrix} 0 & 1 \\ -1 & 0 \end{pmatrix}, \qquad Q(x) = \begin{pmatrix} p(x) & q(x) \\ q(x) & -p(x) \end{pmatrix}.$$

We assume that $p(x)$ and $q(x)$ are two real-valued functions, continuous in each finite interval on the half-line $[0, \infty)$.

Let $\varphi(x, \lambda) = \begin{pmatrix} \varphi_1(x, y) \\ \varphi_2(x, y) \end{pmatrix}$ be the solution of the equation (1.1), satisfying the initial conditions

$$\varphi_1(0, \lambda) = 0, \qquad \varphi_2(0, \lambda) = -1. \tag{1.3}$$

Let $f(x)$ be an arbitrary vector-valued function of class $\mathcal{L}^2(0, \infty)$. If we put

$$F_n(\lambda) = \int_0^n f^T(x)\varphi(x, \lambda)\, dx,$$

then, by Theorem 8.1.1, there exists a monotonically increasing function $\rho(\lambda)$, $-\infty < \lambda < \infty$, not depending on $f(x)$ and called the *spectral function* of the problem (1.1), (1.2), so that the *Parseval equation*

$$\int_0^\infty f^T(x)f(x)\, dx = \int_{-\infty}^\infty F^2(\lambda)\, d\rho(\lambda) \tag{1.4}$$

holds, where $\lim_{n\to\infty} \int_{-\infty}^\infty \{F(\lambda) - F_n(\lambda)\}^2\, d\rho(\lambda) = 0$.

In this chapter, we consider the problem of reconstructing (1.1), (1.2) from its spectral function $\rho(\lambda)$.

12.1.2

In the sequel, we shall need the following.

Lemma 12.1.1 *If $\varphi(x,\lambda)$ is the solution of (1.1) with the initial conditions (1.3), then a matrix function $K(x,t)$ exists, so that*

$$\varphi(x,\lambda) = \begin{pmatrix} \sin \lambda x \\ -\cos \lambda x \end{pmatrix} + \int_0^x K(x,t) \begin{pmatrix} \sin \lambda t \\ -\cos \lambda t \end{pmatrix} dt, \qquad (1.5)$$

where $K(x,t)$ is the solution of the problem

$$BK_x'(x,t) + K_t'(x,t)B = -Q(x)K(x,t) \qquad (1.6)$$

$$BK(x,x) - K(x,x)B = -Q(x) \qquad (1.7)$$

$$K_{11}(x,0) = K_{12}(x,0) = 0. \qquad (1.8)$$

Conversely, if a matrix function $K(x,t)$ is a solution of the problem (1.6–8), then the vector-valued function $\varphi(x,\lambda)$ determined by (1.5) is the solution of (1.1) with the initial conditions (1.3).

Proof. We apply the transformation operator constructed in Section 3, Chapter 10. Let

$$A_1 = B\frac{d}{dx} + Q(x), \qquad A_2 = B\frac{d}{dx}$$

be two matrix operators acting from E to E (where E is the space of vector-valued functions $F(X)$ which are continuous and continuously differentiable), and let $E_1 = E_2$ be the subspace of E, containing all vector-valued functions $f(x)$ fulfilling the boundary condition $f_1(0) = 0$.

We showed in Item 2, Section 3, Chapter 10, that the transformation operator X (for a pair of operators A_1, A_2) mapping E_1 onto E_1 can be realized as

$$Xf(x) = f(x) + \int_0^x K(x,t)f(t)\, dt, \qquad (1.9)$$

where the kernel $K(x,t)$ is the solution of the problem (1.6–8), or, equivalently, of the problem (3.24–26), Chapter 10.

Let $\varphi_\lambda \in E_1$ be the eigenvector of the operator A_2, corresponding to the eigenvalue λ, i.e., $A_2\varphi_\lambda = \lambda\varphi_\lambda$. Then $\psi_\lambda = X\varphi_\lambda$ is the eigenvector of the operator A_1, corresponding to the same eigenvalue λ.

In fact, since $A_1X = XA_2$, by the definition of a transformation operator,

$$A_1\psi_\lambda = A_1X\varphi_\lambda = XA_2\varphi_\lambda = \lambda X\varphi_\lambda = \lambda\psi_\lambda. \qquad (1.10)$$

The eigenvector of the operator $A_2 = B\frac{d}{dx}$, belonging to E_1, is the vector-valued function

$$s(x,\lambda) = \begin{pmatrix} \sin \lambda x \\ -\cos \lambda x \end{pmatrix}. \qquad (1.11)$$

Therefore, according to (1.10), $\varphi(x,\lambda) = Xs(x,\lambda)$ is the eigenvector of $A_1 = B\frac{d}{dx} + Q(x)$, i.e., by (1.9),

$$\varphi(x,\lambda) = s(x,\lambda) + \int_0^x K(x,t)s(t,\lambda)\,dt, \tag{1.12}$$

and the formula (1.5) is thus obtained; the proof is complete. ∎

Remark If A_1 and A_2 are interchanged in the previous argument, we obtain the formula

$$s(x,\lambda) = \varphi(x,\lambda) + \int_0^x L(x,t)\varphi(t,\lambda)\,dt \tag{1.13}$$

instead of (1.12), where the kernel $L(x,t)$ is the solution of the problem

$$BL_x'(x,t) + L_t'(x,t)B = L(x,t)Q(t) \tag{1.14}$$

$$BL(x,x) - L(x,x)B = Q(x) \tag{1.15}$$

$$L_{11}(x,0) = L_{21}(x,0) = 0. \tag{1.16}$$

12.1.3

We have already noted in Item 1 that the reconstruction of the boundary-value problem (1.1), (1.2) from its spectral function $\rho(\lambda)$ is considered in the present chapter.

We now clarify the conditions for $\rho(\lambda)$. The following is valid.

Lemma 12.1.2 *Let $g(x)$ be an arbitrary vector-valued function of class $\mathcal{L}^2(0,\infty)$ with bounded support. Put*

$$G(\lambda) = \int_0^\infty g^T(x)s(x,\lambda)\,dx. \tag{1.17}$$

If

$$\int_{-\infty}^\infty G^2(\lambda)\,d\rho(\lambda) = 0, \tag{1.18}$$

then $g(x) \equiv 0$.

Proof. It follows from the formulas (1.13), (1.17) that

$$G(\lambda) = \int_0^\infty g^T(x)s(x,\lambda)\,dx = \int_0^\infty g^T(x)\left\{\varphi(x,\lambda) + \int_0^x L(x,t)\varphi(t,\lambda)\,dt\right\}dx$$

$$= \int_0^\infty \left\{g^T(x) + \int_x^\infty g^T(t)L(t,x)\,dt\right\}\varphi(x,\lambda)\,dx.$$

Therefore, the function $G(\lambda)$ is the φ-transform of the vector-valued function

$$h^T(x) = g^T(x) + \int_x^\infty g^T(t)L(t,x)\,dt$$

whose components have bounded support and are in the class $\mathcal{L}^2(0, \infty)$. Hence, by the Parseval equation (1.4),

$$\int_{-\infty}^{\infty} G^2(\lambda) \, d\rho(\lambda) = \int_0^{\infty} h^T(x) h(x) \, dx = \int_0^{\infty} \{ h_1^2(x) + h_2^2(x) \} \, dx;$$

if the condition (1.18) is fulfilled, then $h_1(x) = h_2(x) \equiv 0$, i.e.,

$$g^T(x) + \int_x^{\infty} g^T(t) L(t, x) \, dt = 0.$$

Since $g(x)$ is a vector-valued function with bounded support, the integral is actually with finite limits; therefore, the above is a Volterra integral equation and $g(x) \equiv 0$. The proof is now complete. ∎

12.2 Derivation of the basic integral equation

Let $\rho(\lambda)$ be the spectral function of the problem (1.1), (1.2). It can be seen by straightforward calculation that the function $\rho_0(\lambda) = \frac{1}{\pi} \lambda$ is spectral for (1.1), (1.2) with $Q(x) \equiv 0$. Put

$$\sigma(\lambda) = \rho(\lambda) - \frac{1}{\pi} \lambda \tag{2.1}$$

$$c(x, \lambda) = \int_0^x s(x, \lambda) \, dx = \int_0^x \left(\begin{array}{c} \sin \lambda x \\ -\cos \lambda x \end{array} \right) dx = \left(\begin{array}{c} [1 - \cos \lambda x]/\lambda \\ -\sin \lambda x/\lambda \end{array} \right). \tag{2.2}$$

Lemma 12.2.1 *The integral*

$$\int_{-\infty}^{\infty} c(x, \lambda) c^T(y, \lambda) \, d\sigma(\lambda) = f(x, y) \tag{2.3}$$

exists, and the matrix function $f(x, y)$ has a continuous second derivative $f''_{xy}(x, y)$, where

$$f''_{xy}(x, y) \equiv F(x, y) = L(x, y) + \int_0^y L(x, t) L^T(y, t) \, dt \tag{2.4}$$

$$F(x, y) = \frac{1}{2} \{ F(x + y, 0) + F(|x - y|, 0) + B[F(x + y, 0) - F(|x - y|, 0)] B \}. \tag{2.5}$$

Proof. Integrating both sides of (1.12) with respect to x from 0 to x (taking into account (2.2)), and changing the order of integration, we have

$$c(x, \lambda) = \int_0^x \Phi(x, t) \varphi(x, t) \, dt, \tag{2.6}$$

where

$$\Phi(x, t) = \begin{cases} I + \int_t^x L(u, t) \, du, & t \le x \\ \\ 0, & t > x \end{cases}$$

and I is the 2×2 identity matrix.

We derive from (2.6) on the basis of the Parseval equation that, for $y \leq x$,

$$\int_{-\infty}^{\infty} c(x,\lambda)c^T(x,\lambda)\, d\rho(\lambda) = \int_0^{\infty} \Phi(x,t)\Phi^T(y,t)\, dt = \int_0^y dt + \int_0^y dt \int_t^y L^T(u,t)\, du$$
$$+ \int_0^y dt \int_t^x L(u,t)\, du + \int_0^y dt \int_t^x L(u,t)\, du \int_t^y L^T(v,t)\, dv. \quad (2.7)$$

Furthermore, if $Q(x) \equiv 0$, (2.7) takes the form

$$\int_{-\infty}^{\infty} c(x,\lambda)c^T(y,\lambda)\, d\left(\frac{1}{\pi}\lambda\right) = \int_0^y dt, \quad (2.8)$$

since $\rho(\lambda) = \rho_0(\lambda) = \frac{1}{\pi}\lambda$, $L(x,t) \equiv 0$.

Now, subtracting the equality (2.8) from (2.7) and taking into account the notation (2.1), we obtain

$$\int_{-\infty}^{\infty} c(x,\lambda)c^T(y,\lambda)\, d\sigma(\lambda) = \int_0^y dt \int_t^y L^T(u,t)\, du$$
$$+ \int_0^y dt \int_t^x L(u,t)\, du + \int_0^y dt \int_t^x L(u,t)\, du \int_t^y L^T(v,t)\, dv.$$

It follows that the right-hand side, and, therefore, the matrix function $f(x,y)$, have continuous second derivative, making the formula (2.4) valid.

With the use of (1.4), we can easily see that the matrix function $F(x,y)$ satisfies the equation

$$BF'_x(x,y) + F'_y(x,y)B = 0. \quad (2.9)$$

Furthermore, it follows from (2.4) that

$$F(x,0) = L(x,0). \quad (2.10)$$

∎

We now derive a linear integral equation satisfied by the kernel $K(x,y)$ of the transformation operator (1.9) for each fixed x.

Theorem 12.2.1 *The matrix function $K(x,y)$ for $y \leq x$ satisfies the linear integral equation*

$$F(x,y) + K(x,y) + \int_0^x K(x,s)F(s,y)\, ds = 0, \quad (2.11)$$

where $F(x,y)$ is determined by the formulas (2.3), (2.4), or

$$F(x,y) = \frac{\partial^2}{\partial x \partial y} \int_{-\infty}^{\infty} \begin{pmatrix} (1 - \cos \lambda x)/\lambda \\ -\sin \lambda x/\lambda \end{pmatrix} ((1 - \cos \lambda y)/\lambda, \ -\sin \lambda y/\lambda)\, d\sigma(\lambda),$$

with $\sigma(\lambda) = \rho(\lambda) - \lambda/\pi$.

Proof. We first show that, for $b < y < a < x$, the vector-valued functions

$$\psi(\lambda) = \int_0^x \varphi(t, \lambda)\, dt, \qquad \omega(\lambda) = \int_0^y s^T(t, \lambda)\, dt$$

are orthogonal with respect to the measure $d\rho(\lambda)$, i.e.,

$$\int_{-\infty}^{\infty} \psi(\lambda)\omega(\lambda)\, d\rho(\lambda) = 0. \tag{2.12}$$

Indeed, applying the formula (1.13), we have

$$\omega(\lambda) = \int_b^y s^T(t, \lambda)\, dt = \int_b^y \varphi^T(t, \lambda)\, dt + \int_b^y dt \int_0^t \varphi^T(s, \lambda) L^T(t, s)\, ds$$

$$= \int_b^y \varphi^T(t, \lambda)\, dt + \int_0^b \varphi^T(s, \lambda)\, ds \int_b^y L^T(t, s)\, dt + \int_b^y \varphi^T(s, \lambda)\, ds \int_s^y L^T(t, s)\, dt,$$

or the vector-valued function $\omega(\lambda)$ is the φ^T-Fourier transform of the matrix function vanishing outside the interval (b, y). On the other hand, $\psi(\lambda)$ is the Fourier φ-transform of the function vanishing outside the interval (a, x). Since (b, y), (a, x) are disjoint, (2.12) immediately follows from the Parseval equation.

Furthermore, relying on (1.5), (1.11), we have

$$\psi(\lambda) = \int_a^x \varphi(t, \lambda)\, dt$$

$$= \int_a^x s(t, \lambda)\, dt + \int_0^a \left\{ \int_a^x K(t, \tau)\, dt \right\} s(\tau, \lambda)\, d\tau + \int_a^x \left\{ \int_\tau^x K(t, \tau)\, dt \right\} s(\tau, \lambda)\, d\tau.$$

Therefore, the equality (2.12) is

$$\int_{-\infty}^{\infty} \left\{ \int_a^x s(t, \lambda)\, dt \right\} \left\{ \int_b^y s^T(t, \lambda)\, dt \right\} d\rho(\lambda)$$

$$+ \int_{-\infty}^{\infty} \left\{ \int_0^{\infty} H(x, t)s(t, \lambda)\, dt \right\} \left\{ \int_b^y s^T(t, \lambda)\, dt \right\} d\rho(\lambda) = 0, \tag{2.13}$$

where

$$H(x, t) = \begin{cases} \int_a^x K(\tau, t)\, d\tau, & 0 \le t \le a \\[2mm] \int_t^x K(\tau, t)\, d\tau, & a \le t \le x \\[2mm] 0, & t > x \end{cases} \tag{2.14}$$

For $Q(x) \equiv 0$, (2.7) assumes the form (in which case $\rho_0(\lambda) = \lambda/\pi$ and $K(x, t) \equiv 0$; therefore, also $H(x, t) \equiv 0$)

$$\frac{1}{\pi} \int_{-\infty}^{\infty} \left\{ \int_0^x s(t, \lambda)\, dt \right\} \left\{ \int_b^y s^T(t, \lambda)\, dt \right\} d\lambda = 0.$$

Now, subtracting the equality from (2.13) and taking into account (2.1), we obtain

$$\int_{-\infty}^{\infty} \left\{ \int_a^x s(t,\lambda)\, dt \right\} \left\{ \int_b^y s^T(t,\lambda)\, dt \right\} d\sigma(\lambda)$$

$$+ \int_{-\infty}^{\infty} \left\{ \int_0^{\infty} H(x,t)s(t,\lambda)\, dt \right\} \left\{ \int_b^y s^T(t,\lambda)\, dt \right\} d\sigma(\lambda)$$

$$+ \frac{1}{\pi} \int_{-\infty}^{\infty} \left\{ \int_0^{\infty} H(x,t)s(t,\lambda)\, dt \right\} \left\{ \int_b^y s^T(t,\lambda)\, dt \right\} d\lambda = 0. \quad (2.15)$$

With the purpose of transforming each addend, by the definition of the matrix function $f(x,y)$, i.e., (2.3), taking into account (2.2), we get

$$\int_{-\infty}^{\infty} \left\{ \int_a^x s(t,s)\, dt \right\} \left\{ \int_b^y s^T(t,s)\, dt \right\} d\sigma(\lambda) = f(x,y) - f(x,b) - f(a,y) + f(a,b). \quad (2.16)$$

Furthermore, applying the Parseval equation and taking into account (2.14), the third addend of (2.15) is reduced to the form

$$\frac{1}{\pi} \int_{-\infty}^{\infty} \left\{ \int_0^{\infty} H(x,t)s(t,\lambda)\, dt \right\} \left\{ \int_b^y s^T(t,\lambda)\, dt \right\} d\lambda$$

$$= \int_b^y H(x,t)\, dt = \int_b^y dt \int_a^x K(\tau,t)\, d\tau. \quad (2.17)$$

Finally, we consider the second addend in (2.15). Since $H(x,x) = 0$ and $H_t'(x,t)$ is bounded, integrating by parts and then changing the order of integration, on the basis of (2.3), we obtain

$$\int_{-\infty}^{\infty} \left\{ \int_0^{\infty} H(x,t)s(t,\lambda)\, dt \right\} \left\{ \int_b^y s^T(t,\lambda)\, dt \right\} d\sigma(\lambda)$$

$$= - \int_0^x H_t'(x,t)\{f(t,y) - f(t,b)\}\, dt.$$

Hence, integrating by parts on the right-hand side and then changing the order of integration, we obtain

$$- \int_0^x H_t'(x,t)\{f(t,y) - f(t,b)\}\, dt = \int_a^x ds \int_0^s K(s,t)[f_t'(t,y) - f_t'(t,b)]\, dt. \quad (2.18)$$

By (2.16), (2.17) and (2.18), the equality (2.15) takes the form

$$f(x,y) - f(x,b) - f(a,y) + f(a,b) + \int_a^x ds \int_0^s K(s,t)[f_t'(t,y) - f_t'(t,b)]\, dt$$

$$+ \int_b^y dt \int_a^x K(s,t)\, ds = 0,$$

which, differentiated once with respect to x and then once with respect to y, yields

$$\frac{\partial^2 f(x,y)}{\partial x\, \partial y} + \int_0^x K(x,t)\frac{\partial^2 f(t,y)}{\partial t\, \partial y}\, dt + K(x,y) = 0$$

and

$$F(x,y) + K(x,y) + \int_0^x K(x,t)F(t,y)\, dt = 0$$

because of the notation (2.4), thus deriving (2.11). The theorem is now proved. ∎

12.3 Solvability of the basic integral equation

The kernel (absolute term) of the integral equation

$$F(x,y) + \int_0^x K(x,t)F(t,y)\,dt + K(x,y) = 0 \tag{3.1}$$

can directly be expressed in terms of the spectral function $\rho(\lambda)$ of the boundary-value problem under consideration, as

$$F(x,y) = \frac{\partial^2}{\partial x \partial y} \int_{-\infty}^{\infty} c(x,\lambda)c^T(y,\lambda)\,d\left\{\rho(\lambda) - \frac{\lambda}{\pi}\right\}, \tag{3.2}$$

where $c(x,\lambda) = \begin{pmatrix} [1 - \cos \lambda x]/\lambda \\ -\sin \lambda x/\lambda \end{pmatrix}$. Having solved (3.1) (we show below that, for any fixed x, it has a unique solution), we reconstruct the transformation operator kernel $K(x,t)$, and, simultaneously, by the equality (1.7) (or by $Q(x) = K(x,x)B - BK(x,x)$), the potential matrix $Q(x)$; therefore, we can reconstruct the boundary-value problem (1.1), (1.2) from its spectral function $\rho(\lambda)$. The uniqueness of the boundary-value problem reconstructed from the given spectral function just follows from the uniqueness of the solution of (3.1).

Theorem 12.3.1 *Let the function $\rho(\lambda)$ satisfy the conditions :*

1. *If $g(x)$ is an arbitrary vector-valued function in the class $\mathcal{L}^2(0,\infty)$ and if*
$$\int_{-\infty}^{\infty} G^2(\lambda)\,d\rho(\lambda) = 0, \text{ where}$$

$$G(\lambda) = \int_0^{\infty} g^T(x)s(x,\lambda)\,dx, \quad s(x,\lambda) = \begin{pmatrix} \sin \lambda x \\ -\cos \lambda x \end{pmatrix},$$

   then $g(x) \equiv 0$.

2. *The matrix function*

$$f(x,y) = \int_{-\infty}^{\infty} c(x,\lambda)c^T(y,\lambda)\,d\sigma(\lambda),$$

   where

$$\sigma(\lambda) = \rho(\lambda) - \frac{1}{\pi}\lambda, \quad c(x,\lambda) = \begin{pmatrix} [1 - \cos \lambda x]/\lambda \\ -\sin \lambda x/\lambda \end{pmatrix},$$

   has continuous second derivative $f''_{xy}(x,y) \equiv F(x,y)$.

Then, for each fixed $x \geq 0$, (3.1) has a unique solution $K(x,y)$ which is continuous with respect to both variables.

Proof. By the Fredholm alternative, the solvability of (3.1) is proved if we show that the corresponding homogeneous equation

$$g(y) + \int_0^x g(t)F(t,y)\,dt = 0, \quad 0 \le y \le x, \tag{3.3}$$

has only the zero solution.

First, we assume that the function $g(y)$, $0 \le y \le x$, satisfies the conditions :

(a) $g(x) = 0$,

(b) $g'(y)$ is continuous in the interval $0 \le y \le x$.

Now, multiplying the equation throughout scalarly by $g(y)$ and integrating from 0 to x, we have

$$\int_0^x (g(y), g(y))\,dy + \int_0^x \int_0^x (g(t)F(t,y), g(y))\,dt\,dy = 0.$$

Taking into account (3.2) and integrating the second integral by parts, relying on the condition (a) and (2.2), we obtain

$$\int_0^x (g(y), g(y))\,dy + \int_{-\infty}^\infty G^2(\lambda)\,d\rho(\lambda) - \frac{1}{\pi} \int_{-\infty}^\infty G^2(\lambda)\,d\lambda = 0, \tag{3.4}$$

where

$$G(\lambda) = \int_0^x g'(y)c(y,\lambda)\,dy = \int_0^x g(y)s(y,\lambda)\,dy.$$

However, by the Parseval equation,

$$\frac{1}{\pi} \int_{-\infty}^\infty G^2(\lambda)\,d\lambda = \int_0^x (g(y), g(y))\,dy;$$

therefore, we infer from (3.4) that

$$\int_{-\infty}^\infty G^2(\lambda)\,d\rho(\lambda) = 0.$$

It then follows that $g(y) \equiv 0$, relying on Condition 1.

Now, let $g(y)$ be a function which is continuous in the interval $[0, x]$ but does not satisfy the conditions (a), (b). Then there exists a sequence of $g_n(y)$ fulfilling (a), (b), which converges to $g(y)$ in the sense of the norm on $\mathcal{L}^2(0,x)$. Therefore,

$$\lim_{n\to\infty} \int_{-\infty}^\infty G_n^2(\lambda)\,d\rho(\lambda) = 0. \tag{3.5}$$

Furthermore, since $g_n(y) \to g(y)$ in the sense of the norm on $\mathcal{L}^2(0,x)$, $G_n(\lambda) \to G(\lambda)$ uniformly in any interval $[-a, a]$. We also derive from (3.5) that

$$0 = \lim_{n\to\infty} \int_{-a}^a G_n^2(\lambda)\,d\rho(\lambda) = \int_{-a}^a G^2(\lambda)\,d\rho(\lambda),$$

and, since a is arbitrary,

$$\int_{-\infty}^{\infty} G^2(\lambda)\, d\rho(\lambda) = 0.$$

It follows that $g(y) = \equiv 0$, again by Condition 1. ∎

Theorem 12.3.2 *For a monotonically increasing function $\rho(\lambda)$ to be the spectral function of a boundary-value problem of the form (1.1), (1.2) with a continuous matrix function $Q(x)$, it is necessary and sufficient that the following conditions should hold :*

1. *If $g(x)$ is an arbitrary vector-valued function of the class $\mathcal{L}^2(0, \infty)$ with bounded support, and if*

$$\int_{-\infty}^{\infty} G^2(\lambda)\, d\rho(\lambda) = 0,$$

where $G(\lambda) = \int_0^{\infty} g^T(x)s(x, \lambda)\, dx$, then $g(x) \equiv 0$.

2. *The matrix function*

$$f(x, y) = \int_{-\infty}^{\infty} c(x, \lambda)c^T(y, \lambda)d\left\{\rho(\lambda) - \frac{\lambda}{\pi}\right\}$$

has continuous second derivative $f''_{xy}(x, y) \equiv F(x, y)$, with

$$F_{11}(x, 0) = F_{21}(x, 0) = 0.$$

Proof. The necessity was proved by Lemmas 12.1.2 and 12.2.1. To prove the sufficiency, we construct (3.1) with the matrix function $F(x, y)$ determined by the formula (3.2). By Theorem 12.3.1, (3.1) for any $x \geq 0$ has a unique solution $K(x, y)$, continuous with respect to both variables, so that the matrix function

$$Q(x) = K(x, x)B - BK(x, x) \tag{3.6}$$

exists and is continuous. The proof is thus complete. ∎

12.4 Derivation of the differential equation

The matrix function $K(x, y)$ was determined in the preceding section from the given monotone function satisfying Conditions 1 and 2 in Theorem 12.3.2. Put

$$\varphi(x, \lambda) = s(x, \lambda) + \int_0^x K(x, t)s(t, \lambda)\, dt \tag{4.1}$$

(analogous to the formula (1.12)).

Theorem 12.4.1 *The vector-valued function $\varphi(x, \lambda)$ determined by the formula (4.1), satisfies the differential equation*

$$B\frac{d\varphi}{dx} + Q(x)\varphi = \lambda\varphi, \quad (0 \le x < \infty), \tag{4.2}$$

where the matrix function $Q(x)$ is determined by the formula (3.6).

Proof. We first assume that $F(x, y)$ has continuous derivatives F'_x and F'_y. Then it is not hard to show that $K(x, y)$ also has continuous derivatives K'_x and K'_y. It therefore follows from the identity

$$I(x, y) \equiv F(x, y) + K(x, y) + \int_0^x K(x, t)F(t, y)\,dt \equiv 0 \tag{4.3}$$

that

$$BI'_x(x,y) = BF'_x(x,y) + BK'_x(x,y) + BK(x,x)F(x,y) + \int_0^x BK'_x(x,t)F(t,y)\,dt \equiv 0 \tag{4.4}$$

$$I'_y(x,y)B = F'_y(x,y)B + K'_y(x,y)B + \int_0^x K(x,t)F'_y(t,y)B\,dt \equiv 0. \tag{4.5}$$

Since $K(x, 0) = -F(x, 0)$, we derive from (2.10), relying on (1.16), that $K_{11}(x, 0) = K_{21}(x, 0) = 0$. On the other hand, it follows from the explicit form of $F(x, y)$ that $F(x, y) = F^T(y, x)$. Hence, $F_{11}^T(0, y) = F_{12}^T(0, y) = 0$, and

$$K(x, 0)BF(0, y) = 0. \tag{4.6}$$

Now, replacing $F'_y(t, y)B$ by $-BF'_t(t, y)$ in (4.5) (relying on the equation (2.9)), integrating by parts and using the equality (4.6), we rewrite (4.5) in the form

$$I'_y(x,y)B = F'_y(x,y)B + K'_y(x,y)B - K(x,x)BF(x,y) + \int_0^x K'_t(x,t)BF(t,y)\,dt \equiv 0. \tag{4.7}$$

Pre-multiplying the identity (4.3) by $Q(x)$ and adding the product to (4.4), (4.7), we obtain

$$BI'_x(x, y) + I'_y(x, y) + Q(x)I(x, y) \equiv 0;$$

hence, because of (4.3), (4.4), (4.7) and (2.9),

$$BK'_x(x,y) + K'_y(x,y)B + Q(x)K(x,y)$$
$$+ \int_0^x \{BK'_x(x,t) + K'_t(x,t)B + Q(x)K(x,t)\}F(t,y)\,dt = 0,$$

which is a homogeneous integral equation with respect to the matrix function

$$BK'_x(x, y) + K'_y(x, y)B + Q(x)K(x, y),$$

and coincides with the equation (3.3); therefore,

$$BK'_x(x,y) + K'_x(x,y)B + Q(x)K(x,y) \equiv 0. \tag{4.8}$$

On the other hand, $K(x,y)$ satisfies the conditions

$$K(x,x)B - BK(x,x) = Q(x) \tag{4.9}$$

$$K_{11}(x,0) = K_{21}(x,0) = 0 \tag{4.10}$$

(see (3.6)). Therefore, by Lemma 12.1.1, the vector-valued function $\varphi(x,\lambda)$ determined by the formula (4.1) satisfies the differential equation (4.2).

Thus, if $F(x,y)$ has continuous derivatives $F'_x(x,y)$, $F'_y(x,y)$, then the theorem is proved. ∎

We now assume that $F(x,0)$ is no more than continuous. The matrix function

$$F_\delta(x,0) = \frac{1}{\delta} \int_x^{x+\delta} F(t,0)\,dt$$

is then continuously differentable. It is obvious that, as $\delta \to 0$, $F_\delta(x,0) \to F(x,0)$ uniformly in any finite interval. Therefore, the equations

$$F_\delta(x,y) + K_\delta(x,y) + \int_0^x K_\delta(x,t)F_\delta(t,y)\,dt = 0, \quad 0 \le y \le x,$$

where

$$F_\delta(x,y) = \frac{1}{2}\{F_\delta(x+y,0) + F_\delta(|x-y|,0) + B[F_\delta(x+y,0) - F_\delta(|x-y|,0)]B\},$$

have unique solutions for sufficiently small δ, while

$$\lim_{\delta \to 0} K_\delta(x,y) = K(x,y)$$

uniformly in any bounded domain of the variables x, y. Furthermore, it follows from the above that the vector-valued functions

$$\varphi_\delta(x,\lambda) = s(x,\lambda) + \int_0^x K_\delta(x,y)s(t,\lambda)\,dt$$

satisfy the equations

$$B\frac{dy}{dx} + Q_\delta(x)y = \lambda y,$$

where $Q_\delta(x) = K_\delta(x,x)B - BK_\delta(x,x)$, and also the initial conditions $\varphi_\delta(0,\lambda) = \begin{pmatrix} 0 \\ -1 \end{pmatrix}$.

Passing to the limit as $\delta \to 0$, we prove the theorem.

12.5   Derivation of the Parseval equation

12.5.1

We have constructed in sections 3 and 4 a boundary-value problem of the form
(1.1), (1.2) from the given monotone function $\rho(\lambda)$ satisfying Conditions 1 and 2 in
Theorem 12.3.2. To solve the inverse problem completely, it remains to prove that
$\rho(\lambda)$ is, in fact, spectral for the boundary-value problem. It then suffices to show
that the Parseval equation holds, or, for an arbitrary vector-valued function $g(x)$
with bounded support, in the class $\mathcal{L}^2(0,\infty)$, the equality

$$\int_0^\infty g^T(x)g(x)\,dx = \int_{-\infty}^\infty G^2(\lambda)\,d\rho(\lambda) \tag{5.1}$$

holds, where

$$G(\lambda) = \int_0^\infty g^T(x)\varphi(x,\lambda)\,dx. \tag{5.2}$$

12.5.2

Let a matrix function $F(x,y)$ be determined by the formula (3.2), let $K(x,y)$ be the
solution of (3.1), and let the vector-valued function $\varphi(x,\lambda)$ be determined by the
formula (4.1), i.e.,

$$\varphi(x,\lambda) = s(x,\lambda) + \int_0^x K(x,t)s(t,\lambda)\,dt. \tag{5.3}$$

If $s(x,\lambda) \equiv \begin{pmatrix} \sin\lambda x \\ -\cos\lambda x \end{pmatrix}$ is expressed in terms of $\varphi(x,\lambda)$, then

$$s(x,\lambda) = \varphi(x,\lambda) + \int_0^x L(x,y)\varphi(y,\lambda)\,dy. \tag{5.4}$$

Lemma 12.5.1 *For the kernel $L(x,y)$, the identity*

$$L^T(x,y) = F(x,y) + \int_0^y K(y,s)F(s,x)\,ds \tag{5.5}$$

holds.

Proof. Assume that the matrix function $F(x,y)$ has continuous derivatives $F'_x(x,y)$,
$F'_y(x,y)$ (in the general case, the lemma is proved by passing to the limit as in
Theorem 12.4.1). Then $K(x,y)$ satisfies the equation (4.8) and the conditions (4.9),
(4.10) whereas $L^T(x,y)$ is the solution of the problem

$$B\frac{\partial}{\partial y}L^T(x,y) + \frac{\partial}{\partial x}L^T(x,y)B - Q(y)L^T(x,y) = 0 \tag{5.6}$$

$$BL^T(x,x) - L^T(x,x)B = Q(x) \tag{5.7}$$

$$L_{11}^T(x,0) = L_{12}^T(x,0) = 0. \tag{5.8}$$

We introduce the notation

$$H(x,y) = F(y,x) + \int_0^y K(y,s)F(s,x)\,ds. \tag{5.9}$$

Putting $y = 0$, we obtain $H(x,0) = F(0,x)$. On the other hand, $F_{11}(0,x) = F_{12}(0,x) = 0$; therefore, $H(x,y)$ satisfies the condition (5.8). Furthermore, we derive from (5.9), (4.3) that

$$H(x,x) = F(x,x) + \int_0^x K(x,s)F(s,x)\,ds = -K(x,x),$$

i.e., $H(x,y)$ also fulfils the condition (5.7).

It remains to verify that $H(x,y)$ satisfies the equation (5.6). By (5.9), we have

$$BH_y'(x,y) = BF_y'(y,x) + BK(y,y)F(y,x) + \int_0^y BK_y'(y,s)F(s,x)\,ds$$

$$H_x'(x,y)B = F_x'(y,x)B - K(y,y)BF(y,x) + \int_0^y K_s'(y,s)BF(s,x)\,dx.$$

In calculating $H_x'(x,y)B$, $F_x'(s,x)B$ was replaced in the integrand (relying on the equation (2.9)), by the expression $-BF_s'(s,x)$, then integration by parts was carried out, and (4.6) was taken into account.

By the formulas (2.9), (3.6) and (4.8), it follows from the last two equalities that

$$BH_y'(x,y) + H_x'(x,y)B - Q(y)H(x,y)$$
$$= \int_0^y \{BK_y'(y,s) + K_s'(y,s)B - Q(y)K(y,s)\}F(s,x)\,ds = 0,$$

i.e., $H(x,y)$ satisfies the equation (5.6); the lemma is thus proved. ∎

12.5.3

We now obtain the Parseval equation (5.1).

By the formula (4.1), we derive from (5.2) that

$$G(\lambda) = \int_0^\infty g^T(x)\varphi(x,\lambda)\,ds = \int_0^\infty g^T(x)\left\{s(x,\lambda) + \int_0^x K(x,t)s(t,\lambda)\,dt\right\}\,ds$$

$$= \int_0^\infty \left\{g^T(x) + \int_t^\infty g^T(\tau)K(\tau,t)\,d\tau\right\}s(t,\lambda)\,dt = \int_0^\infty h^T(t)s(t,\lambda)\,dt, \tag{5.10}$$

where

$$h^T(t) = g^T(t) + \int_t^\infty g^T(\tau)K(\tau,t)\,d\tau. \tag{5.11}$$

Using the Parseval equation with respect to the measure $\rho_0(\lambda) = \pi/\lambda$,

$$\int_{-\infty}^{\infty} G^2(\lambda) d\left(\frac{\lambda}{\pi}\right) = \int_0^{\infty} h^T(x)h(x)\, dx,$$

we infer from (5.10), also taking (3.2) into account, that

$$\int_{-\infty}^{\infty} G^2(\lambda)\, d\rho(\lambda) = \int_{-\infty}^{\infty} G^2(\lambda) d\left(\frac{\lambda}{\pi}\right) + \int_{-\infty}^{\infty} G^2(\lambda) d\left[\rho(\lambda) - \frac{\lambda}{\pi}\right]$$

$$= \int_0^{\infty} h^T(x)h(x)\, dx + \int_{-\infty}^{\infty} \left[\int_0^{\infty} h^T(x)s(x,\lambda)\, dx\right] \left[\int_0^{\infty} h^T(t)s(t,\lambda)\, dt\right] d\left\{\rho(\lambda) - \frac{\pi}{\lambda}\right\}$$

$$= \int_0^{\infty} h^T(x)h(x)\, dx + \int_0^{\infty} h^T(x) \left\{\int_0^{\infty} \left[\int_{-\infty}^{\infty} s(x,\lambda)s^T(t,\lambda)\, d\left(\rho(\lambda) - \frac{\pi}{\lambda}\right)\right] h(t,\lambda)\, dt\right\} dx$$

$$= \int_0^{\infty} h^T(x)h(x)\, dx + \int_0^{\infty} \int_0^{\infty} h^T(x)F(x,t)h(t)\, dx\, dt.$$

Consider the integral

$$\int_0^{\infty} h^T(x)F(x,t)\, dx = \int_0^{\infty} \left\{g^T(x) + \int_x^{\infty} K(\tau,x)\, d\tau\right\} F(x,t)\, dx$$

$$= \int_0^{\infty} g^T(x)F(x,t)\, dx + \int_0^{\infty} g^T(x) \left\{\int_0^x K(x,s)F(s,t)\, ds\right\} dx$$

$$= \int_0^{\infty} g^T(x) \left\{F(x,t) + \int_0^x K(x,s)F(s,t)\, ds\right\} dx$$

$$= \int_0^t g^T(x) \left\{F(x,t) + \int_0^x K(x,s)F(s,t)\, ds\right\} dx$$

$$\qquad + \int_t^{\infty} g^T(x) \left\{F(x,t) + \int_0^x K(x,s)F(s,t)\, ds\right\} dx.$$

Therefore,

$$\int_0^{\infty} h^T(x)F(x,t)\, dx = \int_0^t g^T(x)L^T(x,t)\, dt - \int_t^{\infty} g^T(x)K(x,t)\, dx$$

(as it follows from the equalities (5.5), (4.3)), and

$$\int_{-\infty}^{\infty} G^2(\lambda)\, d\rho(\lambda) = \int_0^{\infty} h^T(x)h(x)\, dx + \int_0^{\infty} \left\{\int_0^x g^T(y)L^T(y,x)\, dy\right\} h(x)\, dx$$

$$- \int_0^{\infty} \left\{\int_x^{\infty} g^T(y)K(y,x)\, dy\right\} h(x)\, dx = \int_0^{\infty} h^T(x)h(x)\, dx$$

$$+ \int_0^{\infty} g^T(x) \left\{\int_x^{\infty} L^T(x,y)h(y)\, dy\right\} dx - \int_0^{\infty} \left\{\int_x^{\infty} g^T(y)K(y,x)\, dy\right\} h(x)\, dx.$$

We derive from (5.11) that

$$\int_x^{\infty} g^T(y)K(y,x)\, dy = h^T(x) - g^T(x)$$

$$\int_x^\infty L^T(y,x)h(y)\,dy = g(x) - h(x).$$

Therefore, we obtain

$$\int_{-\infty}^\infty G^2(\lambda)\,d\rho(\lambda) = \int_0^\infty h^T(x)h(x)\,dx + \int_0^\infty g^T(x)\{g(x) - h(x)\}\,dx$$

$$- \int_0^\infty \{h^T(x) - g^T(x)\}h(x)\,dx = \int_0^\infty g^T(x)g(x)\,dx,$$

i.e, the Parseval equation (5.1).

Notes

The results of this chapter are due to M. G. Gasymov and B. M. Levitan (1966), and are given here in detail for the very first time. At the authors' request, this chapter was written by M. G. Gasymov.

References

Abdukadyrov, E. (1967). Computation of the regularized trace for a Dirac system. *Vestnik Moskov. Univ. Ser. Mat. Mekh.* **22**, (4) 17–24 *(Russian)*.

Akhiezer, N. I. and Glazman, I. M. (1961). *Theory of Linear Operators in Hilbert Space.* Ungar, New-York.

Ambarzumian, V. (1929). Über eine Frage der Eigenwerttheorie. *Z. Physik* **53** 690–695.

Borg, G. (1946). Eine Umkehrung der Sturm-Liouvilleschen Eigenwertaufgabe. *Acta Math* **78**(1) 1–96.

Coddington, E. and Levinson, N. (1955). *Theory of Ordinary Differential Equations.* McGraw-Hill, New-York.

Conte, S. and Sangren, W. (1953). An asymptotic solution for a pair of first-order equations. *Proc. Amer. Math. Soc.* **4** 696–702.

Conte, S. and Sangren, W. (1954). An expansion theorem for a pair of singular first-order equations. *Canadian J. Math.* **6** 554–560.

Delsarte, J. (1938a). Sur une extension de la formule de Taylor. *J. Math. Pure Appl.* **17** 213–231.

Delsarte, J. (1938b). Sur certaines transformations fonctionelles relatives aux équations linéaires aux dérivées partielles du second ordre. *C. R. Hebd. Acad. Sci.* **206** 178–182.

Delsarte, J. (1938c). Une extension nouvelle de la théorie des fonctions presque périodiques de Bohr. *Acta Math.* **59** 259–317.

Delsarte, J. and Lions, J. (1957). Transmutations d'opérateurs différentiels dans le domaine complexe. *Comm. Math. Helv.* **32**(2) 113–128.

Dikii, L. A. (1955). The zeta function of an ordinary differential equation on a finite interval. *Izv. Akad. Nauk SSSR. Ser. Mat.* **19** 187–200 *(Russian)*.

Dikhamindzhiya, G. V. (1976). Two-sided asymptotic behaviour of the number of eigenvalues of differential operators. *Sakharth. SSR Mekh. Akad. Moambe* **81**(1) 25–28 *(Russian)*.

Dikhamindzhiya, G. V. (1978). Some problems of the spectral theory of the one-dimensional Dirac operator. Thesis, Tbilisi Univ. *(Russian)*.

Dubrovin, B. A. (1975). Periodic problems for the Korteweg-de Vries equation in the class of finite-band potentials. *Funct. Anal. Appl.* 9 275.

Gasymov, M. G. (1967). The inverse scattering problem for a system of Dirac equations of order 2n. *Soviet Physics Dokl.* 11 676–678.

Gasymov, M. G. and Levitan, B. M. (1966). The inverse problem for the Dirac system. *Dokl. Akad. Nauk SSSR* 167 967–970 *(Russian)*.

Gelfand, I. M. (1967). On identities for eigenvalues of a differential operator of the second order. *Uspekhi Mat. Nauk (N.S.)* 11(1) 191–198 *(Russian)*.

Gelfand, I. M. and Levitan, B. M. (1951). On the determination of a differential equation from its spectral function. *Amer. Math. Soc. Trans.* 1 253–304.

Gelfand, I. M. and Levitan, B. M. (1953). On a simple identity for the eigenvalues of a differential operator of second order. *Dokl. Akad. Nauk SSSR* 88 593–596 *(Russian)*.

Glazman, I. M. (1966). Direct methods of qualitative spectral analysis of singular differential operators. Translated from the Russian by IPST staff. Saniel Davey & Co., Inc., New-York.

Hamel, G. (1913). Lineare Diff.-Gleichgn. 2. Ordng. mit period. Koeff. *Mathem. Ann.* 73 371–412.

Haupt, O. (1919). Lineare homog. Diff.-Gleichgn. 2. Ordng. mit period. Koeff. *Mathem. Ann.* 79 278–285.

Hilbert, D. (1904–1910). Grundzüge einer allgem. Theorie d. linearen Integralgleichgn. Göttingen, *Ges. d. Wiss. Nachr. Phys. Kl.* I 49-91, II 213–259, III 307–338, IV 157–227, V 439–480, VI 355–417.

Hochstadt, H. (1965). On the determination of a Hill's equation from its spectrum. *Arch. Rat. Mech. Anal.* 19 353–362.

Hoheisel, G. (1951). *Gewöhnliche Differentialgleichungen.* W. de Gruyter, Berlin.

Hurwitz, W. (1921). An expansion theorem for a system of linear differential equations of the first order. *Trans. Amer. Math. Soc.* 22 526–543.

Keldysh, M. V. (1951). On a Tauberian theorem. *Trudy Mat. Inst. Steklova* 38 77–86 *(Russian)*.

Kostyuchenko, A. G. and Sargsjan, I. S. (1979). *Distribution of Eigenvalues.* Nauka, Moscow *(Russian)*.

Krein, M. G. (1951a). Solution of the inverse Sturm-Liouville problem. *Dokl. Akad. Nauk SSSR* 76(1) 21–24 *(Russian)*.

Krein, M. G. (1951b). Determination of the density of a nonhomogeneous symmetric cord from its frequency spectrum. *Dokl. Akad. Nauk SSSR* 76(3) 345–348 *(Russian)*.

Krein, M. G. (1952). On inverse problems for a nonhomogeneous cord. *Dokl. Akad. Nauk SSSR* 82(5) 669-672 *(Russian)*.

Krein, M. G. (1953a). On the transfer function of a one-dimensional boundary problem of second order. *Dokl. Akad. Nauk SSSR* 88(3) 405–408 *(Russian)*.

343

Krein, M. G. (1953b). On some cases of effective determination of the density of a nonhomogeneous cord from its spectral function. *Dokl. Akad. Nauk SSSR* 93(4) 617–620 *(Russian)*.

Krein, M. G. (1954). On a method of efficient solution of an inverse boundary-value problem. *Dokl. Akad. Nauk SSSR* 105(3) 433–436 *(Russian)*.

Krein, M. G. (1955). On the determination of the potential of a particle from its S-function. *Dokl. Akad. Nauk SSSR* 105(3) 433–436 *(Russian)*.

Levin, B. Y. (1959). Interpolation by entire functions of exponential type. *Trudy Fiz.-Tekhnich. Inst. Nizkikh Akad. Nauk SSSR* 1 83–92 *(Russian)*.

Levin, B. Y. (1980). Distribution of zeros of entire functions (revised). *Am. Math. Soc. Trans.* 5.

Levinson, N. (1949a). The inverse Sturm-Liouville problem. *Math. Tidsskr.* B 25 25–30.

Levinson, N. (1949b). Criteria for the limit-point case for second-order linear differential operators. *Časop. pěstov. mat. fys.* 74 17–20.

Levinson, N. (1951). A simplified proof of the expansion theorem for singular second-order linear differential equations. *Duke Math. J.* 18 57–71.

Levinson, N. (1954). The expansion theorem for singular self-adjoint differential operators. *Ann. of Math.* 59(2) 300–315.

Levitan, B. M. (1950a). *Eigenfunction Expansion*. Gosthekhizdat, Moscow *(Russian)*.

Levitan, B. M. (1950b). Proof of the theorem on the expansion in eigenfunctions of self-adjoint differential equations. *Dokl. Akad. Nauk SSSR (N.S.)* 73 651–654 *(Russian)*.

Levitan, B. M. (1964a). On the determination of the Sturm-Liouville differential equation from two spectra. *Izv. Akad. Nauk SSSR, Ser. Mat.* 28(1) 63–78 *(Russian)*.

Levitan, B. M. (1964b). Regularized trace calculation for the Sturm-Liouville operator. *Uspekhi Mat. Nauk* 19(115) 161–165 *(Russian)*.

Levitan, B. M. (1968). Investigation of the Green's function of a Sturm-Liouville equation with an operator coefficient. *Mat. Sb. (N.S.)* 76(118) 239–270.

Levitan, B. M. (1973). *Theory of Generalized Shift Operators*. 2nd ed. Nauka, Moscow *(Russian)*.

Levitan, B. M. (1981). Regularized traces and smooth periodicity conditions for the potential of the Sturm-Liouville equation. *Sib. Mat. Zhurn.* 22(2) 137–148 *(Russian)*.

Levitan, B. M. (1987). *Inverse Sturm-Liouville Problems*. Translated from the Russian by O. Efimov. VNU Science Press BV, Utrecht.

Levitan, B. M. and Gasymov, M. G. (1964). The determination of a differential equation from two spectra. *Uspekhi Mat. Nauk* 19(2) 3–63 *(Russian)*.

Levitan, B. M. and Sargsjan, I. S. (1960). Some problems in the theory of the Sturm-Liouville equation. *Russian Math. Surveys* 151 95.

Levitan, B. M. and Sargsjan, I. S. (1965). On the continuation of solutions of the one-dimensional Dirac system. *Soviet Math. Dokl.* 6 1588-1591.

Levitan, B. M. and Sargsjan, I. S. (1975). Introduction to spectral theory : self-adjoint differential operators. *Am. Math. Soc. Trans.* 39.

Magnus, W. and Winkler, W. (1966). *Hill's Equation.* Interscience Publishers, New-York.

McKean, H. and Van Moerbeke, P. (1975). The spectrum of Hill's equation. *Invent. Math.* 30(3) 217-274.

Macleod, R. (1966). Some problems in the theory of eigenfunctions. Thesis, Oxford.

Marchenko, V. A. (1950). Some problems in the theory of the second-order differential operator. *Dokl. Akad. nauk SSSR* 72(3) 457-460 *(Russian)*.

Marchenko, V. A. (1952). Some problems in the theory of one-dimensional linear differential operators of second order, I. *Trudy Mosk. Mat. Obshch.* 1 327-420 *(Russian)*.

Marchenko, V. A. (1953). Some problems in the theory of one-dimensional linear differential operators of second order, II. *Trudy Mosk. Mat. Obshch.* 2 3-83 *(Russian)*.

Marchenko, V. A. (1972). *Spectral Theory of the Sturm-Liouville Operator.* Naukova Dumka, Kiev *(Russian)*.

Marchenko, V. A. (1977). Sturm-Liouville Operators and Their Applications. Naukova Dumka, Kiev *(Russian)*.

Marchenko, V. A. and Ostrovsky, I. V. (1975). A characterization of the spectrum of the Hill operator. *Math. USSR-Sb.* 26 493-554.

Markushevich, A. I. (1950). *Theory of Analytic Functions.* Gosthekhizdat, Moscow-Leningrad *(Russian)*.

Martynov, V. V. (1965). Conditions of discreteness and continuity of the spectrum in the case of a self-adjoint first-order system of differential equations. *Dokl. Akad. Nauk SSSR* 165 996-999 *(Russian)*.

Molchanov, A. M. (1953). On conditions of discreteness for the spectrum of self-adjoint differential equations of second order. *Trudy Mosk. Mat. Obshch.* 2 169-199 *(Russian)*.

Naimark, M. A. (1967). *Linear differential operators.* Translated by E. R Dawson. Frederick Ungar Publishing Co., New-York.

Otelbayev, M. O. (1973). Distribution of the eigenvalues of the Dirac operator. *Mat. Zametki* 14 843-852 *(Russian)*.

Petrovsky, I. G. (1954). *Lectures on partial differential equations.* Translated by A. Shenitzer. Interscience Publishers, New-York.

Povzner, A. Y. (1948). On differential equations of Sturm-Liouville type on the semi-axis. *Mat. Sbornik* 23(1) 3–52 *(Russian)*.

Prats, F. and Toll, J. (1959). Construction of the Dirac equation central potential from phase shifts and bound states. *Phys. Rev.* 113(1) 363–370.

Roos, B. and Sangren, W. (1961). Spectra for a pair of singular first-order differential equations. *Proc. Amer. Math. Soc.* 12 468–476.

Sargsjan, I. S. (1966a). A theorem of the completeness of the eigenfunctions of the generalized Dirac system. *Dokl. Akad. Nauk Arm. SSR*, 42(2) 77–82 *(Russian)*.

Sargsjan, I. S. (1966b). The asymptotic behaviour of the spectral matrix of a one-dimensional Dirac system. *Dokl. Akad. Nauk SSSR* 166(5) 1058–1061 *(Russian)*.

Sargsjan, I. S. (1966c). Decomposition into a series of eigenfunctions of a one-dimensional Dirac system. *Dokl. Akad. Nauk SSSR* 166(6) 1292–1295 *(Russian)*.

Sargsjan, I. S. (1966d). Solution of the Cauchy problem for a one-dimensional Dirac system. *Izv. Akad. Nauk. Arm. SSSR Ser. Mat.* 1(6) 392–436 *(Russian)*.

Sargsjan, I. S. (1972). Two-sided asymptotic analysis of the number of eigenvalues of a non-semibounded Dirac operator. *Izv. Akad. Nauk. SSSR Ser. Mat.* 36(6) 1402–1436 *(Russian)*.

Sears, D. and Titchmarsh, E. (1950). Some eigenfunction formulas. *Quart. J. Math. Oxford Ser.* 2 1 165–175.

Stankevich, I. V. (1970). A certain inverse spectral analysis problem of Hill's equation. *Sov. Math. Dokl.* 11 582–586.

Steklov, V. A. (1896). Problème du refroidissement d'une tige solide hétérogène. *Kharkov math. Soc. Commun.* 5 136–181.

Titchmarsh, E. (1949). On the uniqueness of the Green's function associated with a second-order differential equation. *Canad. J. Math* 1 191–198.

Titchmarsh, E. (1950). Eigenfunction problems with periodic potentials. *Proc. Roy. Soc. Ser. A* 203 501–514.

Titchmarsh, E. (1961). Some eigenfunction expansion formulas. *Proc. London Math. Soc.* 11 159–168.

Titchmarsh, E. (1962). *Eigenfunction expansions associated with Second-Order Differential Equations*. 2nd ed. Clarendon Press, Oxford.

Wet, J. de and Mandl, F. (1950). On the asymptotic distribution of eigenvalues. *Proc. Roy. Soc. Ser. A200* 572–580.

Weyl, H. (1909–1910a). Gewöhnl. lineare Diff. Gleichgn m. Singulären Stellen & ihre Eigenfunkt. *Göttingen, Ges. d. Wiss. Nachr.* 37–64, 442–467.

Weyl, H. (1910b). Gewöhnl. Diff. Gleichgn m. Singularitäten d. zugehör. Entwick. willkür. Funktn. *Math. Ann.* 68 220-269.

Yosida, K. (1950). On Titchmarsh-Kodaira formula concerning Weyl-Stone's eigenfunction expansion. *Nagoya Math. J.* 1 49–58.

Name Index

Subject Index